Sources, Biocatalytic Characteristics and Bioprocesses of Marine Enzymes

Sources, Biocatalytic Characteristics and Bioprocesses of Marine Enzymes

Edited by Darius Pearson

SYRAWOOD
PUBLISHING HOUSE
New York

Published by Syrawood Publishing House,
750 Third Avenue, 9th Floor,
New York, NY 10017, USA
www.syrawoodpublishinghouse.com

Sources, Biocatalytic Characteristics and Bioprocesses of Marine Enzymes
Edited by Darius Pearson

International Standard Book Number: 978-1-64740-394-2 (Hardback)

Trademark Notice: Registered trademark of products or corporate names are used only for explanation and identification without intent to infringe.

Cataloging-in-publication Data

Sources, biocatalytic characteristics and bioprocesses of marine enzymes / edited by Darius Pearson.
    p. cm.
Includes bibliographical references and index.
ISBN 978-1-64740-394-2
1. Enzymes. 2. Enzymes--Biotechnology. 3. Marine biology. 4. Biosynthesis. 5. Biocatalysis. I. Pearson, Darius.
TP248.65.E59 S68 2023
660.634--dc23

TABLE OF CONTENTS

PREFACE

This book aims to highlight the current researches and provides a platform to further the scope of innovations in this area. This book is a product of the combined efforts of many researchers and scientists, after going through thorough studies and analysis from different parts of the world. The objective of this book is to provide the readers with the latest information of the field.

Certain enzymes that are secreted by microbes such as fungi and bacteria present in the marine environment are called marine enzymes. These microbes secrete different enzymes based on their ecological functions and habitat. Marine microorganisms and their enzymes can function in environments with high pressure and light, high salt concentrations, and in a wide variety of temperatures. Marine enzymes are a significant source of biocatalysis. Fermentation process can be used to produce and extract marine enzymes from animals, microorganisms, and plants. Enzymes generated by marine microorganisms are used in a variety of sectors, including textiles, chemical, pharmaceuticals, biomedicine, agriculture, and food processing. Marine enzymes exhibit a wide range of pharmacological uses, and can be used as pesticides, mycotoxins, pigments, antiparasitics, antibiotics, herbicides, antitumor agents, toxins, and growth promoters for both plants and animals. This book is a compilation of chapters that discuss the most vital concepts and emerging trends with respect to the sources, biocatalytic characteristics, and bioprocesses of marine enzymes. It will help the readers in keeping pace with the rapid changes in this area of study.

I would like to express my sincere thanks to the authors for their dedicated efforts in the completion of this book. I acknowledge the efforts of the publisher for providing constant support. Lastly, I would like to thank my family for their support in all academic endeavors.

Editor

Exploring Marine Environments for the Identification of Extremophiles and their Enzymes for Sustainable and Green Bioprocesses

Paola Di Donato [1,2,*], **Andrea Buono** [3]🆔, **Annarita Poli** [1], **Ilaria Finore** [1]🆔,
Gennaro Roberto Abbamondi [1], **Barbara Nicolaus** [1] **and Licia Lama** [1,*]🆔

[1]   Institute of Biomolecular Chemistry, National Research Council of Italy, Via Campi Flegrei 34,
     80078 Pozzuoli, Naples, Italy; apoli@icb.cnr.it (A.P.); ilaria.finore@icb.cnr.it (I.F.);
     roberto.abbamondi@icb.cnr.it (G.R.A.); bnicolaus@icb.cnr.it (B.N.)

[2]   Department of Science and Technology, University of Naples "Parthenope", Centro Direzionale Isola C4,
     80143 Naples, Italy

[3]   Department of Engineering, University of Naples "Parthenope", Centro Direzionale Isola C4, 80143 Naples,
     Italy; andrea.buono@uniparthenope.it

*   Correspondence: paola.didonato@uniparthenope.it (P.D.D.); licia.lama@icb.cnr.it (L.L.)

Abstract: Sea environments harbor a wide variety of life forms that have adapted to live in hard and sometimes extreme conditions. Among the marine living organisms, extremophiles represent a group of microorganisms that attract increasing interest in relation to their ability to produce an array of molecules that enable them to thrive in almost every marine environment. Extremophiles can be found in virtually every extreme environment on Earth, since they can tolerate very harsh environmental conditions in terms of temperature, pH, pressure, radiation, etc. Marine extremophiles are the focus of growing interest in relation to their ability to produce biotechnologically useful enzymes, the so-called extremozymes. Thanks to their resistance to temperature, pH, salt, and pollutants, marine extremozymes are promising biocatalysts for new and sustainable industrial processes, thus representing an opportunity for several biotechnological applications. Since the marine microbioma, i.e., the complex of microorganisms living in sea environments, is still largely unexplored finding new species is a central issue for green biotechnology. Here we described the main marine environments where extremophiles can be found, some existing or potential biotechnological applications of marine extremozymes for biofuels production and bioremediation, and some possible approaches for the search of new biotechnologically useful species from marine environments.

Keywords: extremophiles; extremozymes; biofuels; bioremediation; microwave; satellite remote sensing

1. Introduction

Seas and oceans cover more than 70 % of the Earth's surface and harbor a wide variety of life forms that have adapted to live in hard and sometimes extreme conditions. The 'marine microbiome', i.e., the microbial species living in marine environments, play an important ecological role and possess an enormous potential for several biotechnological applications. The global bacterial marine biomass accounts for about 5.4×1029 cells and it is distributed in all the marine ecosystem from the open waters (ocean and seas), to the tidal regions, the seafloor and the sub-seafloor, the polar sea ice masses, and brines [1]. The complex of microorganisms hosted in marine environments belong to all of the three domains of life, i.e., Eukarya, Bacteria, and Archaea (Figure 1), although the great majority of them belong to the latter two domains. The sequencing of 16S rRNA of marine microbial species has evidenced a high taxonomical diversity of marine Bacteria and Archaea; indeed, it has been possible to identify as prevailing phyla the following ones: Alphaproteobacteria, Actinobacteria, Acidobacteria Cyanobacteria, Deltaproteobacteria, Gammaproteobacteria, and Flavobacteria [1] A significant share of marine Bacteria and Archaea belong to the group of extremophiles, i.e., those microorganisms that are able to live and thrive in extreme chemical and physical conditions [2]. The extremophiles are classified according to the physical or chemical parameters that characterize the environmental conditions in which they survive and optimally grow. Hence, on the basis of temperature values we can identify the thermophiles, i.e., microorganisms living at temperatures ranging from 60 °C to 80 °C; the hyperthermophiles, that live at T > 80 °C; the psychrophiles, i.e., living at temperatures below 15 °C. On the basis of pH values, extremophiles are classified as acidophiles (living at pH < 3) or alkaliphiles (living at pH > 9). The bacterial species able to survive in the presence of high NaCl concentrations are the so-called halophiles; those living at low oxygen tension or in its total absence, are the microanaerobes and the anaerobes, respectively. The microorganisms able to survive to low water activity are defined as x erophiles. Finally, those microorganisms living under high pressure or in the presence of high radiations levels are defined as barophiles and radioresistant, respectively. Many extremophiles can also be defined as poly-extremophiles since some of them are actually able to simultaneously resist different extreme conditions, for example the thermoacidophilic bacteria that thrives at high temperature and low pH; some species halophilic species that tolerate both high salt concentration and alkaline pH; or finally some radioresistant species that resist also other extreme conditions like very low temperature, absence of water, and vacuum. Different kinds of extremophiles can be found in several marine ecosystems that are characterized by more than one extreme condition such as hypersaline habitats, high pressures, and extreme temperature. Some examples include the shallow vents, the submarine hydrothermal vents and the black smokers; the cold seas in both Arctic or Antarctic regions, the ocean depths and some hypersaline lakes of oceanic origin. Over the last few decades, the extremophiles have attracted a great deal of interest since they produce a wide array of biotechnologically useful molecules like the enzymes (also called extremozymes). Competition for space and nutrients in the marine environment constitutes a selective force leading to evolution and generating multiple enzyme systems to adapt to the different environments. Many marine extremophiles are capable of overcoming such extreme conditions and are a source of enzymes with special characteristics. Therefore, these microorganisms are of great interest for industrial processes, mainly in biocatalysis [3,4]. Metagenomic studies have revealed that extremophile prokaryotes from marine habitats are a source of novel genes and consequently a source of new bioproducts, including enzymes and other active metabolites. Therefore, it is important to study and understand these microorganisms in order to be able to use the biochemical, ecological, evolutionary, and industrial potential of these marine microbes [5,6].

This vast variation in marine habitats has led to the development of new hydrolases—such as proteases, lipases, glycoside hydrolases, etc.—with novel specificities and properties including tolerance to extreme conditions used in industrial processes [7,8] Thanks to their peculiar features, the enzymes from marine extremophiles can be exploited for several industrial processes in the agricultural, chemical, food, textile, pharmaceutical, bioenergy, and cosmetic fields. Indeed, extremozymes are active at the harsh pH, temperature, and pressure conditions typical of many industrial environments. For these reasons, their application enables the implementation of new biotechnologies that are the key approach to a more sustainable industrial system. Therefore, the variety of marine ecosystems in which extremophiles can be found represent an interesting source of industrially useful for manifold applications. With the advent of biotechnology, enzymatic engineering, and the introduction of other innovative technologies, it is possible to achieve efficient management of our rich marine microbial biodiversity towards the creation of new enzymes that could be recovered from marine microorganisms and efficiently exploited, not only as a cost effective biocatalyst but also as an ecofriendly reagent in the coming years. Considering the enormous microbial diversity native to the vast marine environments of this planet earth, efforts channeled into the discovery of new enzymes from marine microbes are inadequate and justify the launch of intensive screening programs by scientists globally. Such a mammoth attempt alone can return a large number of new enzymes for various human purposes and services, for the simple reason that marine environments are rich in new enzymes that probably could also avoid the need for enzymatic engineering or molecular cloning to design new enzymes for specific needs. In the long term, probably, processes based on marine microbial enzymes will replace many of the current chemical processes [4].

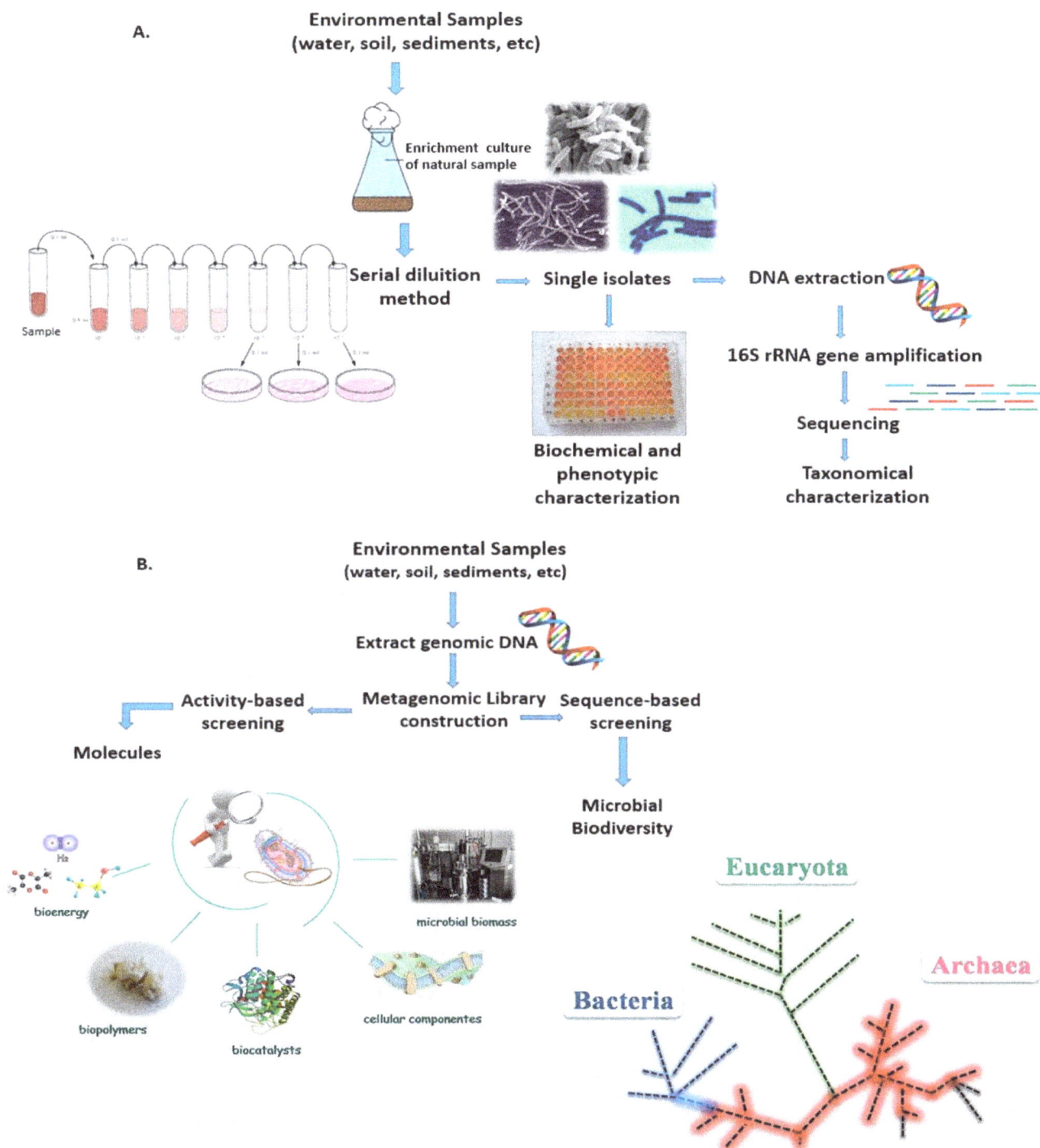

Figure 1. Microbial diversity studies. **(A)** Cultivation-dependent approach: scheme of a traditional isolation of strains through serial dilution method and their genetic, phenotypic and biochemical studies for microbial characterization; **(B)** Metagenomic approach: scheme of a metagenomic library construction with identification of molecules (activity-screening) and microbial communities (sequence-based screening).

In the following section, we will give an overview of the main marine extremophiles and of the environments in which they are found, of the actual or possible industrial applications of marine extremozymes and finally of the strategies to identify new interesting species based on identification of new sampling sites.

2. Extreme Ecosystems

The biodiversity of ecosystems is a subject of intensive study; consequently, an affluence of information has been gathered on the distribution of microorganisms in the world. In addition, there is a growing interest about the role of marine microorganisms in biogeochemical processes, biotechnology, pollution, and pharmaceutical fields. In recent years, many authors have focused on the potential of marine microorganisms as prolific producers of bioactive substances and exploiting the vast marine microbial treasure for their utilization as novel drug delivery systems [9]. Extremophilic microorganisms are in several extreme marine environments, such as hydrothermal vents, hot springs, salty lakes, and deep-sea floors. The ability of these microorganisms to support extremes of temperature, salinity, and pressure demonstrates their great potential for biotechnological processes. Several different extreme environments, characterized by geochemical and physical extremes, are found in the ocean and in seas and many of them appeared to be hot spots for microbial abundance and diversity, thanks to the overwhelming presence of substrates and energy sources that support microbial metabolism. The most studied extreme oceanic environments are the vent ecosystems, such as the hot deep-sea hydrothermal vents (DSHVs) or cold seeps and mud volcanoes, and the hypersaline ecosystems such as the deep anoxic hypersaline lakes, brine lakes on mud volcanoes, and brines contained within sea ice. However, new fascinating extreme habitats for microbial life in the ocean are being discovered continuously such as water droplets entrapped in oil deposits. These environments comprise a large variety of extreme physicochemical conditions, which contribute importantly to the composition and shaping of the residing microbial communities and select for extremophile populations of microorganisms. These polyextremophiles are the key players of the element cycles in these environments, often responsible for primary productivity and endemic [10]. The development of more automated and affordable techniques for isolating and characterizing marine microbial bioactive metabolites, make marine products more accessible. Actually, the marine habitat represents the most studied environment for the richness related to the diversity of microorganisms and for the potential source of molecules possessing biological activities [9]. In particular, extreme marine environments constitute peculiar ecological niches in which extremophiles thrive developing unique biochemical strategies and features of industrial interest. The first studies on extremophiles from extreme habitats based on strains isolation using classic culture-dependent approaches (Figure 1A). With this method, only microorganisms whose metabolic and physiological requirements can be duplicated in the laboratory could be isolated. To overcome this limitation related to culture-dependent method, recently metagenomic approaches have been developed in order to explore and to access the uncultured microbial community (Figure 1B).

2.1. Cold Environments

Psychrophiles are extremophilic bacteria or archaea which are cold-loving, having an optimal temperature for growth at about 15 °C or lower, a maximal temperature for growth at about 20 °C and a minimal temperature for growth at 0 °C or lower. Psychrotrophs, also termed psychrotolerant, are cold-tolerant bacteria or archaea that have the ability to grow at low temperatures, but have optimal and maximal growth temperatures above 15 °C and 20 °C, respectively. Most of the Earth's biosphere is cold. Approximately 14% of the Earth's surface is in the polar region, whereas 71% is marine. By volume, more than 90% of the ocean is 5 °C or colder. Below the thermocline, the ocean maintains a constant temperature, a maximum of 4–5 °C, regardless of latitude. Therefore, all pressure-loving microorganisms (i.e., barophiles) are either psychrophilic or psychrotrophic and this is to be expected because the water below the thermocline of the ocean is under hydrostatic pressure. Moyer and

Morita [11] underlined the importance to take environmental samples where the in-situ temperature never exceeds the psychrophilic range and to ensure that the growth medium, pipettes, inoculating loops, etc. are kept cold before use. This could explain why early microbiologists failed to isolate psychrophiles. Even if the term psychrophiles was first reported in 1884, most of the early literature actually dealt with psychrotrophic bacteria and not with true psychrophiles. Since investigators were not working with extreme cold-loving bacteria, there was much debate and, as a result, many terms were coined to designate psychrophiles such as cryophile, psychrorobe, facultative psychrophile, psychrocartericus, psychrotrophic, and psychrotolerant [11]. The Antarctic marine environments for example are characterized by an average temperature of about $-1\ °C$. This low temperature produces two main physicochemical effects: the rate of chemical reactions decreases exponentially according to the Arrhenius law and strong effect on the viscosity of the medium, thereby contributing to further slow-down reaction rates. The Antarctic continent, considered as an uncontaminated site, is unfortunately experiencing increasing contaminant influxes that are likely to become more severe in the future. Heavy metals have been mainly detected in the 2% of ice-free lands of the continent, where most of the human activities occur in different areas, including Terra Nova Bay in addition to the concentrations accumulated in the biota [12]. Currently, the Arctic sea ice bacterium *Psychromonas ingrahamii* has demonstrated the lowest growth temperature $(-12\ °C)$ of any organism authenticated by a growth curve [13]. The first and only truly psychrophilic archaeon to be isolated is *Methanogenium frigidum*, a methanogen from Ace Lake, Antarctica [14]. Taxonomically, cold-loving microbes are found both in Archaea and Eucarya domains; they have been found in free or associated form with sponges for example, and distributed in numerous genera such as *Arthrobacter, Colwellia, Exiguobacterium, Gelidibacter, Glaciecola, Halobacillus, Halomonas, Hyphomonas, Listeria, Marinobacter, Methanococcoides, Methanogenium, Moritella, Planococcus, Pseudoalteromonas, Pseudomonas, Psychrobacter, Psychroflexus, Psychromonas, Psychroserpens, Shewanella* and *Sphingomonas* [9,15,16]. In order to understand the adaptation of psychrophiles to these extreme environments, their enzymes should be isolated, cloned, and characterized to gain further insight into their biotechnological potential. The first genomes studied were that of *Colwellia psychrerythraea* from Arctic marine sediments [17] and *Pseudoalteromonas haloplanktis* from Antarctic seawaters [18]. Nowadays, many genomes are available for a wide range of psychrophilic bacteria and archaea. As of early 2017, approximately 130 cold-adapted species have genome sequences [19]. Trait surveys, limited to the perspective of gene gain, reveal prevalence of genes demonstratively providing better growth at low temperature including compatible solute uptake and synthesis, antifreeze proteins and polyunsaturated fatty acids. This includes the presence of anti-freeze DUF3494-type proteins that occur in all domains of life but is limited to cold-adapted taxa and is absent in higher-temperature adapted life [19]. Among these adaptations, the factors responsible for the adjustment of membrane fluidity are of prime importance, whereas the large diversity of factors contributing to limit the toxicity of highly concentrated dissolved oxygen has been already investigated. These studies indicate a high content of enzymes involved in oxygen consumption such as desaturases, superoxide dismutases, and catalases whereas an unusual feature observed in the Antarctic *Pseudoalteromonas haloplanktis* is the elimination of the ubiquitous molybdopterin-dependent metabolism which is usually responsible for the production of reactive oxygen species. Nevertheless, further genome sequences are needed to detect whether there are some general trends in cold-adaptation or, if on the contrary, each microorganism has its own specific strategy. Proteomic analyses are also progressing and are required to establish the relationships that should exist between the expression of regulatory proteins and the environmental temperature [19].

2.2. Hydrothermal Vent Habitats

Hydrothermal vent fields occur mainly along the boundaries of tectonic plates, in regions known as mid-ocean ridge ranges, 90 % of which are under the ocean, such as the East Pacific Rise and the Mid-Atlantic Ridge. Mid-ocean ridges are the sites of oceanic spreading centers, where magma rises from the mantle forming new crust as it cools and spreads away from the ridge. For the first time, by the submersible Alvin, at a depth of 2500 m on the Galapagos Rift of the Pacific Ocean, the chimneys were seen, from which black water at a temperature of about 300 °C and saturated with minerals shot out (Martin et al., 2008). These hydrothermal vent, called black smoker, located on the basaltic rock bottom and originating from fresh lava flows, were chemically reactive environments able to support suitable conditions for sustained prebiotic synthesis. The hot fluid is acidic, anoxic, rich in Fe, Mn, Cu, Zn, Ba etc., and poor in magnesium, nitrates, and phosphates. Volatile compounds from magma (H_2S, CO_2, CH_4, H_2) may be added, further modifying the fluid composition [20]. The conditions around the vent systems can change quite rapidly with spreading rates changing the size of the edifices and ecology. Shallow water submarine hydrothermal vents represent easily accessible natural systems. Venting is well known off volcanic islands and provinces and are commonly detected by the presence of streams of gas bubbles. Deep-sea vents (>200 m) and shallow-water vents (<200 m) differ in community structure (in deep-sea dominance of symbiotrophic forms and in shallow-water a higher ratio of vent obligate taxa), composition, and environmental parameters. The temperature of fluids in shallow water vents is between 10 and 119 °C whereas sediments can reach up to 95.8 °C, both light and hydrothermal energy support a complex microbial community [21]. At shallow depths hydrothermal vents lack the typically sulphide structures with some exceptions; fluid with lower concentrations of CH_4 and H_2 with respect to deep-sea vents, are enriched in N, P, and Si. Thermophilic and hyperthermophilic Archaea and Bacteria are a common feature in these sites and more than 35 species have been found at west Pacific and Mediterranean vents. Coastal zones of the Southern Tyrrhenian Sea (Flegrean area, Cape Palinuro, Eolian Islands) represent easily accessible vents by diving. Analysis using Fluorescent in Situ Hybridization molecular and DGGE technique found that bacterial richness and biodiversity at two Volcano vents are greater than archeabacteria [22]. Pyrococcus, Thermotoga, Thermococcus, Archeoaglobus, Methanococcus, Pyrodictium, Aquifex and Igneococcus thrive in both shallow-water and in deep-sea hydrothermal systems. Members of Thermococcus (*T. celer* and *T. litoralis*) have been isolated from coastal hydrothermal systems [9]. In 2000, during a National Science Foundation expedition in the Mid Atlantic Ocean Zone 30° N, a new type of vent system, named The Lost City hydrothermal field, was discovered. This vent system is one of the best examples of serpentinization processes in a marine environment [23]. The fluids venting in the Lost City chimneys range in temperature from 40 °C to 90 °C and are highly alkaline (pH 9–11), with high concentration of dissolved H_2, CH_4, low molecular weight hydrocarbons but almost no dissolved CO_2. These characteristics and extreme conditions are the result of chemo- and biosynthetic reactions.

The production of fluids enriched in CH_4 and H_2 during serpentinization suggest that geochemical and geological processes should be support by moderately thermophilic as well as sulfate-reducing bacteria [24]. Deep-sea hydrothermal vents are unique environments that provide partial or complete energy/nutrient fluxes necessary to support diverse microbial communities that are distributed along the temperature range and reduced compound gradients more or less correlated with the transition from anoxic to oxic conditions. The large population of animals that surround the volcanically driven warm vents (mainly tubeworm communities and an array of crabs, shrimp, giant clam, and gastropods) is supported by the growth of chemoautotrophic microorganisms. These bacteria are the bases of the hydrothermal trophic food chain and they can exist: free-living associated with the dismissed vent fluids and probably growing and reproducing within the sub-seabed system, free-living microbial carpets growing on the surface exposed to flowing vent waters, through endo- and exosymbiotic associations with invertebrates, and within the deep sea hydrothermal vent plumes [25]. The most-studied physical and chemical parameter that limits microbial life is the temperature, indeed Bacteria and Archaea from hydrothermal vents have been extensively investigated in smoker fluids, black smoker sulfides and sediments that have higher temperatures [26] in order to hypothesize and speculate on the origin of life. The main bacterial species isolated and that grow at strict anaerobic and extreme temperature conditions belong to the Archaea kingdom: euryarchaeota and crenarchaeota. Euryarchaeota include methanogens (Methanococcus, Methanopyrus), and sulfate and iron-reducers (Archaeoglobus), whereas crenarchaeota include thermophilic and hyperthermophilic heterotrophs (Hyperthermus, Thermococcus, Staphylothermus, Pyrococcus, Desulforococcus). Most of the heterotrophic species exhibit maximal growth temperatures less than 105 °C, while Pyrodictium and the methanogen Methanopyrus species grow at 110 °C, *Pyrolobus fumarii* at 113 °C and strain 121 (member of the Desulfurococcales) at 121 °C [9]. In The Lost City hydrothermal areas, the porous walls of the structures host hyperthermophilic, thermophilic and mesophilic bacteria. A methane-metabolizing Archaea related to the Methanosarcinales, growing at 80 °C, is the predominant group that thrive in these edifices, forming biofilms of about 10 cm thickness, adjoining to hydrothermal flow. Within the bacterial domain, cultured and identified thermophilic microorganisms make up a short list with respect to the domain Archaea. The thermophilic Eubacteria most frequently isolated in a deep-sea chimney belong to the orders Thermales, Aquificales and Bacillales; the main genera, Thermotoga and Desulforobacterium, sulphur-reducing, thermophilic, anaerobic and strictly autotrophic, grow at the temperature range of the 60–80 °C. Thermus and Bacillus thermophilic are heterotrophic aerobes and grow at the range of 60–75 °C. Representative species assigned to the order Thermotogales are *Thermotoga maritima, Thermotoga neapolitana, Thermoanaerobacter ethanolics, Thermosipho melansiesis*.

2.3. Hypersaline Environments

Hypersaline environments are extreme habitats where the salinity is much higher than that of seawater and depending on whether they are originated from seawater or not, can be divided into two main types of environments, thalassolohaline and athalassohaline, respectively. Examples of thalassohaline environments, concentrated salt solutions (brines), are some lakes, such as the Great Salt Lake in Utah, marine ponds and salt marshes subject to evaporation for high temperature. Salt marshes can be found in inland areas and coastal (marine) marshes occur in sheltered sites (frequently estuaries) where wave action is slight and deposition of silt allows higher plant to root. The estuaries exposed to intensive evaporation can also become extremely saline [27]. Human activity also creates highly saline habitats such as solar salterns (used for the production of salt by evaporation of seawater), which may have a NaCl concentration at saturation in some ponds [28]. Chemically, thalassohaline environments are characterized by a clear predominance of Cl^- and Na^+ (responsible for 49% and 42% of the total molarity, respectively). Other important ions are Mg^{2+}, SO_4^{2-}, K^+, Br^-, HCO_3^-, and F^-. The average salinity of seawater is 3.5%; when it concentrates (as in a solar saltern) its composition changes due to the serial precipitations. The class of extremophilic microorganisms specialized for living in extreme hypersaline environments, are designated as halophiles. Different authors use different definitions for what constitutes a halophile; the most popular definition of halophiles identifies microorganisms which grow optimally at Na^+ concentrations greater than 0.2 M. According to the optimal salt concentration for growth, they are classified in three categories: extreme halophiles, that grow in an environment with 3.4–5.1 M (20% to 30%, *w/v*) NaCl; moderate halophiles, that grow in an environment with 0.85–3.4 M (3% to 25%, *w/v*) NaCl; slightly halophiles that grow in an environment with 0.2–0.85 M (1% to 5%, *w/v*) NaCl. Halotolerant microorganisms do not show an absolute requirement for salt but grow well in high salt concentrations [29]. Marine salterns are habitats for a large variety of halophilic or halotolerant bacteria that develop throughout the entire gradient of salt concentration [30]. In the first ponds most bacteria are slightly halophilic, whereas in the intermediary ponds, where the seawater is concentrated to a salinity of about 10 to 20% NaCl, most of the bacteria are moderately halophilic. This intermediate environment contains the greatest numbers of organisms. The last ponds are inhabited by extremely halophilic organisms including aerobic members of the Archaea belong to the genera Halobacterium, Natronobacterium, Haloferax and Haloarcula in addition to several species pertaining to the Bacteria and Eucarya. Only one methanogenic species of the Archaea was reported to grow optimally at NaCl concentrations over 20% [28]. Halophiles have developed different adaptive strategies to support the osmotic pressure induced by the high NaCl concentrations. Some extremely halophilic bacteria accumulate inorganic ions (K^+, Na^+, Cl^-) in the cytoplasm, which is a type of 'salt-in' strategy to balance the osmotic pressure of the environment. Moreover, they have also developed specific proteins that are stable and active in the presence of salts [31–35]. The halophilic microorganisms contain enzymes that maintain their activity at high salt concentrations, alkaline pH and high temperatures. The stability of the enzymes depends on the negative charge on the surface of the protein due to acidic amino acids, the hydrophobic groups in the presence of high salt concentrations and the hydration of the protein surface due to carboxylic groups present in aspartic and glutamic acids. In addition, negative surface charges are thought to be important for the solvation of halophilic proteins, to prevent denaturation, aggregation, and precipitation [36].

3. Marine Extremozymes: Current and Potential Applications for Biofuels Production and Bioremediation Processes

The development of the modern biotechnology has generated an increasing request for enzymes with novel properties for multiple industrial processes. Extremozymes are produced by bacteria that have adapted to harsh environmental conditions, like extreme temperature and pH variations, high salt concentration and pressure. Therefore, extremozymes are quite attractive because they are more resistant when compared to the terrestrial mesophilic homologs. Indeed, their peculiar features make them suitable for different biotechnological applications [36]: in this section, attention is paid to the exploitation of extremozymes for the production of renewable energy and the bioremediation of polluted sites.

The green biotechnology for the sustainable production of bioenergy can take advantages from marine extremozymes that indeed can be useful for the production of biofuels like bioethanol, biodiesel, and biohydrogen [37]: some remarkable examples are reported in Table 1.

Table 1. Marine enzymes for biofuels production.

Source	Enzyme	Application	Ref.
Thermococcus sp.	Amylase (Fuelzyme®)	1G-Bioethanol production	[36]
Bacillus carboniphilus CAS 3 *B. subtilis* subsp. *subtilis* A-53 *Bacillus licheniformis* AU01 *G. thermoleovorans* IT-08	Cellulase Carboxymethylcellulase Cellulase Cellulase	2G-Bioethanol production	[38] [39] [40] [41]
Bacillus sp. H1666 *Cobetia* sp. NAP1 *Exiguobacterium* sp. Alg-S5 *Microbulbifer thermotolerans* JAMB-A94	Cellulase Alginate lyase Alginate lyase and cellulase Agarase	3G-Bioethanol production	[42] [43] [44] [45]
Aeromonas sp. EBB-1 *Candida antarctica* *Photobacterium lipolyticum*	Lipase	Biodiesel production	[46] [47] [48]
Catenovulum sp. X3	Amylase	Biohydrogen production	[49]

Bioethanol production from starchy materials, the so-called first generation (1G) bioethanol, is a consolidated industrial process in those countries using it as a transportation fuel, like USA and Brazil. The conversion of starchy materials in glucose is catalyzed by amylases, particularly by thermostable amylases that allow to accelerate the conversion rates with low risks of contamination. Different commercial amylases are available, like the alpha-amylase produced by a *Thermococcus* sp. isolated from a deep-sea hydrothermal vents. The latter is commercialized as Fuelzyme®—Verenium Corporation (San Diego, CA, USA) and it is usually exploited for the mash liquefaction, the first step of starchy biomass conversion [36].

The concerns about bioethanol production due to its impact on food chain, in recent years, has driven the search for a more sustainable production of bioethanol based on non-food biomass like the lignocellulosic or the algal biomass. The enzymes from marine extreme bacteria are among the most promising biocatalysts, since they are resistant to temperature, salt concentration and contaminants. The production of bioethanol from lignocellulosic biomass, the so-called second generation bioethanol (2G-bioethanol), relies on the exploitation of cellulase and xylanase enzymes. The bacterium *Bacillus carboniphilus* CAS 3, a species isolated from marine sediments collected from Parangipettai coast in

India, has shown to be a cellulase activity producer. Its cellulase activity has been shown to carry out an extensive saccharification of pretreated rice straw, yielding about 15.56 g/L of reducing sugar after 96 h [38]. The species *Bacillus subtilis* subsp. *subtilis* A-53, isolated from seawater of the seashore in the Kyungsang (Korea), produces a carboxymethylcellulase that is a salt tolerant enzyme able to hydrolyze cellulosic materials in the typical severe conditions of the industrial conversion of biomass to fermentable sugars [39]. *Bacillus licheniformis* AU01, a species isolated from marine sediments in India, is able to produce a cellulose enzyme by using cellulosic wastes as carbon sources. The cellulase purified from this bacterium is thermostable and resists high pHs and several types of detergents, thus it could be useful for hydrolysis of lignocellulosic materials for ethanol production [40]. The species xylanase activity from *Geobacillus thermoleovorans* IT-08, isolated from a hot spring in Indonesia, has been investigated for its application in lignocellulose degradation. After expression in *Escherichia coli* DH5, the xylanase has been used for the hydrolysis of corncob and oat spelt xylan: this enzyme is able to degrade insoluble lignocellulosic materials to produce xylooligosaccharides and monomer sugars (xylose and arabinose) to be fermented to ethanol [41].

The production of third generation bioethanol (3G-bioethanol)—i.e., bioethanol from algal biomass—is receiving growing interest in relation to the search for more sustainable processes of renewable energy production. Algae store carbohydrates mainly as agarose, but also in the form of starch or cellulose, whose conversion to fermentable sugars is of interest in relation to the production of biofuels. The biotransformation of algal biomass to fermentable sugars is carried out by means of cellulose, agarose, and alginate lyase enzymes. Some recent examples of marine enzymes for 3G processes include the cellulase activity from *Bacillus* sp. H1666, whose applicability was tested on dried green seaweed (*Ulva lactuca*). The saccharification of the algal biomass by means of this enzyme was carried out in a single step processes, affording an increase of 450 mg/g in glucose yield [42]. The bacterium *Cobetia* sp. NAP1, isolated from the brown algae *Padina arborescens* Holmes, produces an alginate lyase (AlgCPL7) whose optimal temperature and pH are 45 °C and 8, respectively. This enzyme is thermostable and salt tolerant, is promising for the production of biofuels since it afforded high yields of alginate's degradation [43]. The species *Exiguobacterium* sp. Alg-S5 is the first example of bacterium able to co-produce alginate lyase and cellulase enzymes. These enzymes have been showed to be of potential application for the bioconversion of alginate and cellulose containing wastes into value-added products and biofuels [44]. The deep-sea bacterium *Microbulbifer thermotolerans* JAMB-A94 produces an endo-β-agarase that has been modified by fusion with a carbohydrate-binding module (CBM) derived from the species *Catenovulum agarivorans*. The fusion resulted in a significant agar depolymerisation (45.3%, with a reducing sugar yield of 14.2%) that showed the potential application of tis enzyme for a new enzymatic prehydrolysis process for the saccharification of agar, a major component of red algal biomass, one of the main feedstocks for 3G bioethanol production [45].

Marine extremozymes find applications also for the production of other alternative fuels like biodiesel and biohydrogen. Biodiesel is produced via alcohol transesterification of oils of vegetable or animal sources, catalyzed by lipase enzymes. A thermophilic lipase is produced by *Aeromonas* sp. EBB-1, isolated from marine sludge in Thailand: this enzyme is stable at pH 6.0–8.0 and T 30–80 °C, and is very active in the hydrolysis of long chain esters, thus being a promising and robust biocatalyst for industrial applications [46]. An industrially relevant lipase of marine bacterial origin is represented by lipase B from *Candida antarctica* (Novozym435). In a recent study, the lipase B, immobilized by means of functionalized nanoparticles was used to generate biodiesel from waste cooking oil. The immobilized preparation of the enzyme enhanced the reusability of the enzyme, that indeed kept 100% of its starting activity after six cycles of the reaction [47]. Lipases can undergo loss of activity in the presence of high concentrations of methanol, usually required for the efficient production of biodiesel. A methanol-tolerant lipase is produced by *Photobacterium lipolyticum*: this enzyme was tested with waste oil and olive oil, and it proved to be very effective also in the presence of either water or high alcohol concentrations, achieving a biodiesel yield of about the 70% of the possible maximum yield [48]. Amylases from marine bacteria have been tested for the production of biohydrogen, a biofuel produced by fermentation of carbohydrate-rich biomass like starch. The marine bacterial strain *Catenovulum* sp. X3, isolated from seawater in China, produces an interesting amylase enzyme that is active at alkaline pHs and in the presence of organic solvents. This enzyme was used to produce fermentable sugars for *Clostridium* species, and it allowed to gain a 3.73-fold higher yields of biohydrogen production [49].

The extremozymes of marine origin have proved to be useful also for bioremediation applications. Different kinds of wastes and contaminants are produced from the industrial activities, the mining activities for oils extraction or the accidental oil spills. All these activities release in the marine environments several pollutants like hydrocarbons, polycyclic aromatic hydrocarbons, chlorinated hydrocarbons, pesticides, heavy metals, etc. [50]. The removal and detoxification of these contaminants and wastes can be achieved by means of extremozymes: some examples are listed in Table 2.

Table 2. Marine enzymes for bioremediation processes.

Source	Enzyme	Application	Ref.
Bacillus safensis (CFA-06)	Oxidoreductase	Biodegradation of aromatic compounds	[51]
Marine metagenome	Laccase	Degradation of industrial dyes	[52]
Alcanivorax borkumensis SK2T *Alcanivorax dieselolei* B-5T *Alcanivorax venustensis* ISO4T *Bacillus licheniformis* ATCC 14580T *Bacillus litoralis* DSM 16303T *Bacillus oshimensis* JCM *Halomonas ventosae* Al12T *Idiomarina baltica* DSM 15154T	Alkane hydroxylases/Cytochrome P450	Degradation of alkanes	[53]
Nocardioides sp. strain KP7	Dioxygenase	Degradation of PAH	[54]
Pseudomonas stutzeri DEH130 *Paracoccus* sp. DEH99 *Psychromonas ingrahamii* *Alcanivorax dieselolei* strain B-5	Haloalkane dehalogenases	Degradation of halogenated pollutants	[55,56] [57] [55,58] [59]

Legend. PAH, polycyclic aromatic hydrocarbons.

The oxidoreductases are a group of enzymes that display a potential role for bioremediation of dyes, the main contaminants released by the textile industry. The bacterial species *Bacillus safensis* (CFA-06), isolated from petroleum in Campos Basin (Brazil) produces two oxidoreductases, a catalase and a new oxidoreductase. The two enzymes showed to be very active in the degradation of aromatic hydrocarbons thus suggesting promising application for petroleum removal [51]. Another interesting group of marine bacterial oxidoreductases is represented by the laccases—i.e., those enzymes able to catalyze the oxidation of phenolic and non-phenolic aromatic compounds. Such enzymes find manifold applications in bioremediation of textile dyes or as biosensors. Recently, a bacterial laccase gene (lac21) was screened by a metagenomic approach from a marine library: this enzyme showed unusual properties like high stability at 40 °C, for pHs ranging from 5.5 to 9.0; high activity in the presence of chloride, high decolorization capability toward azo dyes [52].

Natural oil seepage and oil spills are the main causes of hydrocarbons' contamination of marine environments. The biodegradation of these polluting agents can be carried out by marine microorganisms producing enzymes able to catalyze the oxidization of medium-length alkanes, like the non-haem diiron alkane hydroxylases (AlkA, AlkB) and the cytochrome P450 belonging to the CYP153 family. Several bacterial species belonging to many genera, have been isolated from marine environments as producers of alkane degrading enzymes: some remarkable examples of these enzymes and of their producers are listed in Table 2 [53] The degradation of aromatic compounds is another key issue in bioremediation of oil contaminated sites: the species *Nocardioides* sp. strain KP7, isolated from a Kuwait beach, produces a dioxygenase that is able to degrade phenanthrene and that has been identified thanks to its detoxification action after an oil spill accident [54]. Numerous marine species have been identified as producers of enzymes catalyzing the degradation of halogenated compounds that have a significantly negative impact on the health and the environment [55]. Some interesting haloacid dehalogenases—i.e., enzymes able to catalyze the de-halogenation of 2-alkanoic acids—have been isolated from the marine bacterium *Pseudomonas stutzeri* DEH130 [56] and from *Paracoccus* sp. DEH99 [57]. The bacterium *Psychromonas ingrahamii*, isolated from the sea ice interface, has been described as a producer of a haloacid dehalogenase active against chlorinated and brominated short chain (<C3) haloacids [55,58]. *Alcanivorax dieselolei* strain B-5, isolated for the first time from surface water of the Bohai Sea, produces different alkane hydroxylase systems that enables it to degrade either chlorinated or brominated alkanes with different chain lengths, thus displaying an interesting potential for biodegradation and other industrial applications [59]

4. Satellite Microwave Remote Sensing to Support the Identification of Potential Sampling Sites

Remote sensing tools represent a key resource to identify potential sites harboring extremophiles and to support the planning of in-situ sampling campaign. Satellite remote sensing provides, in a cost-effective way, updated and synoptic maps of the oceans ensuring global coverage and sufficient revisit time. In most cases, in fact, the harsh conditions where extremophiles are present make very difficult to organize in-situ campaigns, while satellite remote sensing allows monitoring and exploring even unsafe and almost inaccessible areas—e.g., polar regions or polluted areas. Among the different remote sensing instruments actually operating from space, microwave sensors can provide information on the sea surface day- and night-time in almost any weather conditions and, therefore, represent the most suitable tool to identify potential sites for extremophiles' sampling. In this framework, both active and passive microwave remote sensing sensors can be exploited [60–62]. The formers are equipped with their own source of radiation, i.e., laser imaging detection and ranging (LiDAR) and synthetic aperture radar (SAR), while the latter only receive the radiation emitted by the Earth's surface in the microwave range of the electromagnetic spectrum, i.e., radiometer.

The always growing interest in the operational exploitation of satellite remote sensing data to monitor the oceans is witnessed by the European Space Agency (ESA) Sentinel-3 mission started in 2016 with the launch of the Sentinel-3A satellite and continued in 2018 with the launch of the Sentinel-3B one. This space mission represents a cutting-edge technology to monitor the oceans from space since it is equipped with a payload composed by four main sensors, namely the sea and land surface temperature radiometer (SLSTR), the ocean land and color instrument (OLCI), the SAR altimeter (SRAL), and the microwave radiometer (MRW) [63]. The Sentinel-3 mission, operated by EUMETSAT, represents, up to now, the most advanced ocean mission that aims at providing a comprehensive understanding of ocean processes taking full benefits by a synergistic approach that combines both active and passive remote sensing sensors covering a broad spectral range (including optical, near infrared, thermal infrared, and microwaves). In this framework, several operational high-quality products are delivered by the Sentinel-3 mission that include, but are not limited to, sea surface temperature (SST) and wind speed, significant wave height, sea ice concentration, and algal pigment concentration [64]. Nevertheless, in this study, only spaceborne microwave tools are addressed as a source of information for supporting the identification of potential extremophiles sites.

Nowadays, satellite microwave remote sensing sensors are currently exploited to deliver added-value products operationally. Spaceborne SAR measurements are used to monitor, on a local scale, critical infrastructures belonging to oil and gas companies, including rigs, drilling platforms, and pipelines, as well as polluted sea areas [65,66]. Satellite radiometers are used to monitor, on a global scale, sea surface parameters as temperature and salinity, as well as ice-covered sea areas [67,68] In this study, a brief overview on the exploitation of satellite microwave remote sensing sensors to identify potential sites of extremophiles living in harsh environments is proposed. In this study, three different showcases are presented and discussed:

- Identification of oil-polluted sea surface areas by means of satellite polarimetric SARs.
- Identification of sea surface areas characterized by extreme temperature/salinity conditions by means of multi-frequency polarimetric radiometers.
- Identification of extreme ice-infested sea surface areas by means of single- and multi-frequency polarimetric radiometers.

SARs measure, for each resolution cell, the associated normalized radar cross sections (NRCS). The latter is a measure of how much electric field is scattered off the elements within the resolution cell towards the SAR antenna. An example of NRCS measurements, collected on 5 February 2016 over the Ross Sea (Antarctica) by the C-band (5.4 GHz) polarimetric SAR on-board of the ESA Sentinel-1 mission is shown Figure 2.

Figure 2. HH-polarized NRCS excerpt of a Sentinel-1 SAR image, shown in radar coordinates, collected on 5 February 2016 over the Ross Sea (Antarctica).

The SAR-based detection of polluted sea surface areas is possible since the presence of the oil layers over the ocean dampen the short gravity and capillary waves which, being resonant with the wavelength of the incident electromagnetic wave, are responsible of the backscattered NRCS measured by the SAR antenna. As a result, due to the oil damping properties that reduce the roughness of clean sea surface, the NRCS related to the oil slick is lower than the one coming from the surrounding sea, i.e., oil-affected sea areas appear as patches darker than the sea background in grey-tones intensity image. Nevertheless, there are several marine features that may call for similar dark patches over the sea surface, including low-wind areas, shadow regions, biogenic films, algal blooms, and other weak-damping surfactants which are all termed as oil look-alikes. However, it was widely demonstrated that polarimetric information allows significantly reducing false alarms due to oil look-alikes by exploiting the different scattering properties that characterize slick-free and oil slick-covered sea surface [69,70]. Among the several polarimetric features used to emphasize the presence of an oil slick with respect to the clean sea background in order to support oil-polluted sea area detection, the standard deviation (std) of the co-polarized phase difference (CPD) was shown to be very effective [71]. The latter provides a reliable estimation of the correlation between HH and VV backscattering channels (horizontally- and vertically-polarized transmitting/receiving electromagnetic wave). In fact, the large degree of correlation between co-polarized channels that characterizes clean sea surface scattering results in CPD std values lower than the corresponding ones that characterize the oil-polluted area calling for weak HH-VV correlation. As an example, Figure 3 shows the oil polluted sea areas identified by the X-band (9.6 GHz) polarimetric SAR on-board of the German Aerospace Center (DLR) TerraSAR-X (TSX) mission, which provides NRCS measurements with 0.5 decibel (dB) radiometric sensitivity, in dual-polarimetric HH-VV mode over an area of about 50×30 km, with approximately 3 m spatial resolution and an average revisit time of 4.5 days.

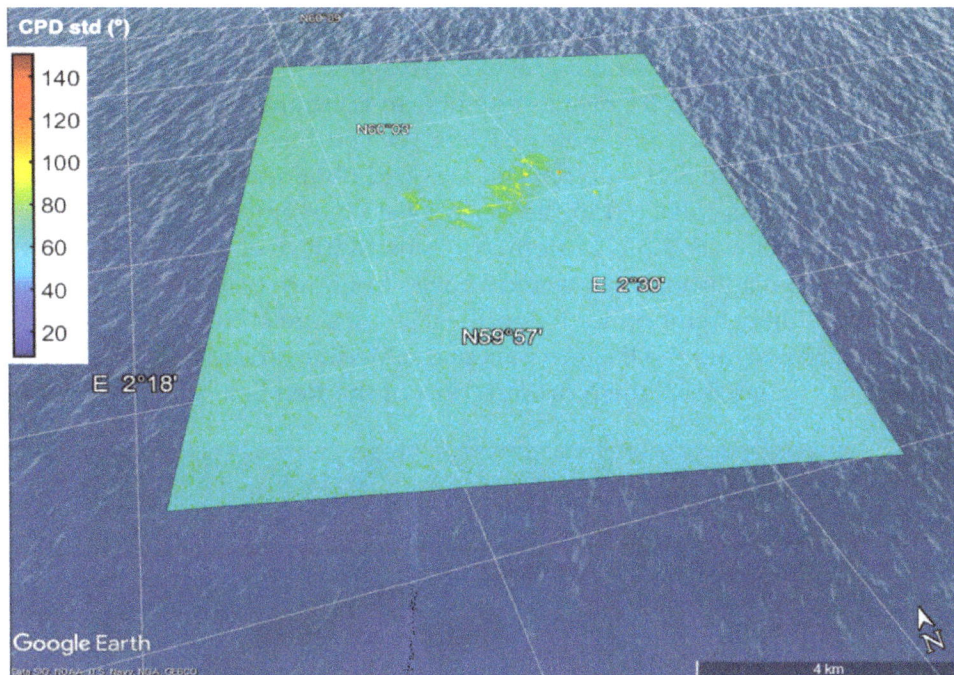

Figure 3. CPD std map obtained from the a TSX SAR scene collected on 8 June 2011 over an oil-polluted area off the southwestern coast of Norway and superimposed on a reference Google Earth optical image.

Two of the most important sea surface geophysical parameters useful for the identification of extremophiles are SST and sea surface salinity (SSS). Microwave radiometers equipped on-board of satellite platforms allow providing global information to identify sea surface areas calling for extreme SST and SSS values where thermophiles/hyperthermophiles and psychrophiles (SST larger than $60°$ and lower than 15 °C, respectively), and halophiles (SSS > 30 psu) can be potentially found. They can be both retrieved from brightness temperatures (BT) measured from satellite microwave radiometers, that is a measurement of the radiance associated to the microwave radiation traveling upward from the top of the atmosphere to the radiometer antenna. An example of BT measurements, collected at 89.000 GHz over the north Atlantic area by the advanced microwave scanning radiometer for EOS (AMSR-E) on-board of the National Aeronautics and Space Administration (NASA)/Japanese Aerospace eXploration Agency (JAXA) Aqua space mission on 2 July 2011 is shown Figure 4.

The derivation of SST standard products from microwave radiometers is possible since the intensity of the radiations naturally emitted by sea surface in the microwave range of the electromagnetic spectrum changes with temperature [67,68]. In fact, the V-polarized BT measured by microwave radiometers over ocean areas calls for significant sensitivity to SST, surface roughness, and atmospheric temperature. Nonetheless, the effects of the latter are removed exploiting the polarimetric and multi-frequency capabilities of the most up-to-date spaceborne radiometers. As an example, Figure 5 shows a global SST map on a 25 km-grid obtained from AMSR-E measurements collected in the period 2–4 October 2011. AMSR-E provides multi-frequency (6.925 GHz, 10.650 GHz, 18.700 GHz, 23.800 GHz, 36.500 GHz, and 89.000 GHz) BT measurements, with 0.3–1.1 K radiometric sensitivity, a swath width of approximately 1445 km, with about 6 km $\times$ 4 km (at 89.000 GHz) to 75 km $\times$ 43 km (at 6.925 GHz) spatial resolution on a daily basis. In Figure 5, it can be observed that the most promising sites for sampling extremophiles thermophiles and hyperthermophiles, i.e., sea areas characterized by extreme temperature values, are the polar regions, the Pacific coast of Mexico, the Gulf of Mexico and the Caribbean Sea, the Red Sea and the Gulf of Aden, the coasts of India, and the Arabian Sea.

The derivation of SSS standard products from microwave radiometers is possible since the microwave radiation emitted by sea surface is partly dependent on the electric permittivity that, in turn, is related to the degree of salinity [72,73]. Nevertheless, a unique SSS value can be estimated from the BT only when SST is known. Generally, given the operating frequency and viewing angle of the satellite microware radiometer, larger salinity levels correspond to lower BT values—given SST. As for the derivation of SST products from BT measurements, an accurate SSS retrieval needs the estimation of different factors influencing the radiometer BT (e.g., sea surface roughness) to be removed. As an example, Figure 6 shows a global SSS map on a 40 km-grid obtained from NASA soil moisture active passive (SMAP) measurements collected in the period 4–11 November 2018. SMAP provides L-band (1.41 GHz) BT measurements, with 1.5 K radiometric sensitivity, over an area of about 1000×100 km, with approximately 40 km resolution on a weekly basis. In Figure 6, it can be observed that the most promising sites for sampling extremophiles halophiles, i.e., sea areas characterized by extreme salinity values (e.g., larger than 38 psu) are the Eastern part of the Mediterranean Sea, the Mediterranean coasts of Spain and France, the Italian Seas, the Northern part of the Red Sea, and the Persian Gulf.

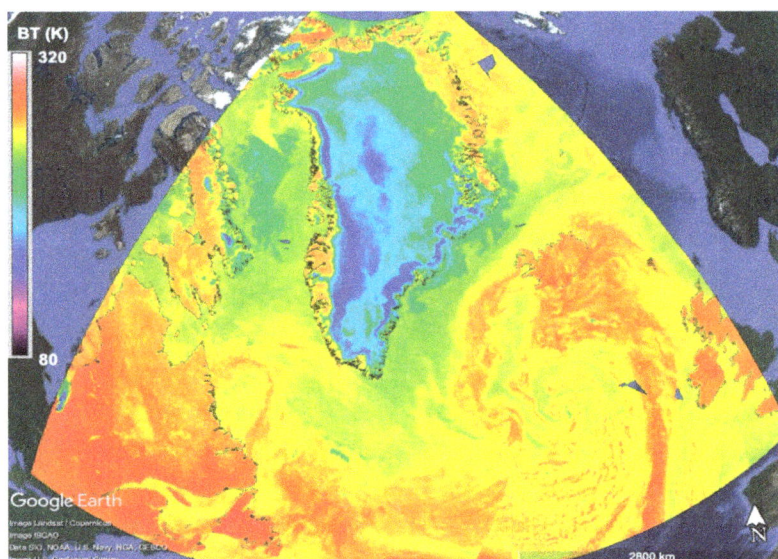

Figure 4. AMSR-E BT measurements, collected at 89.000 GHz over the north Atlantic area on 2 July 2011.

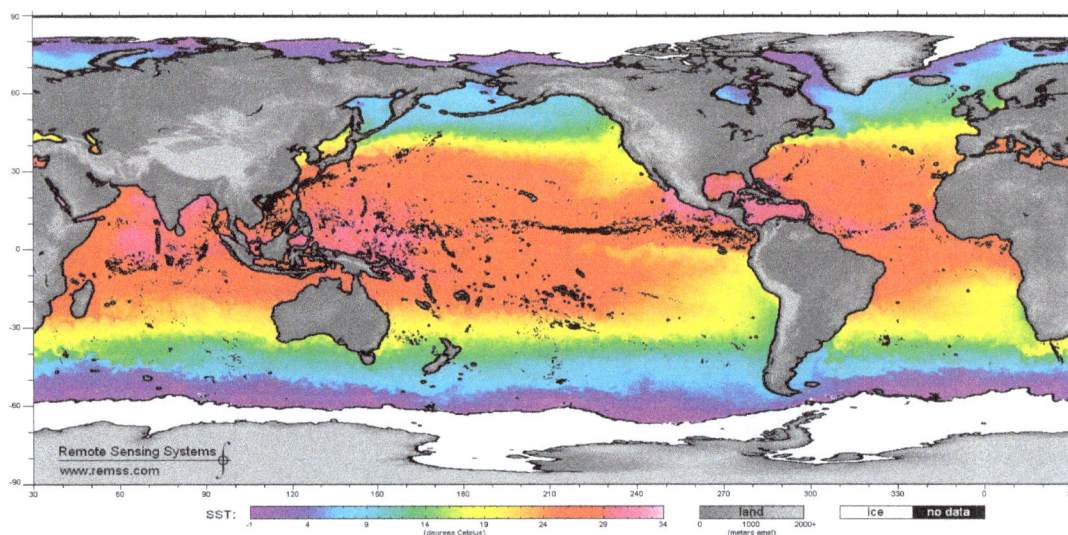

Figure 5. SST global map on 25 km resolution cell obtained by three-day averaging AMSRE-E measurements collected in the period 2–4 October 2011.

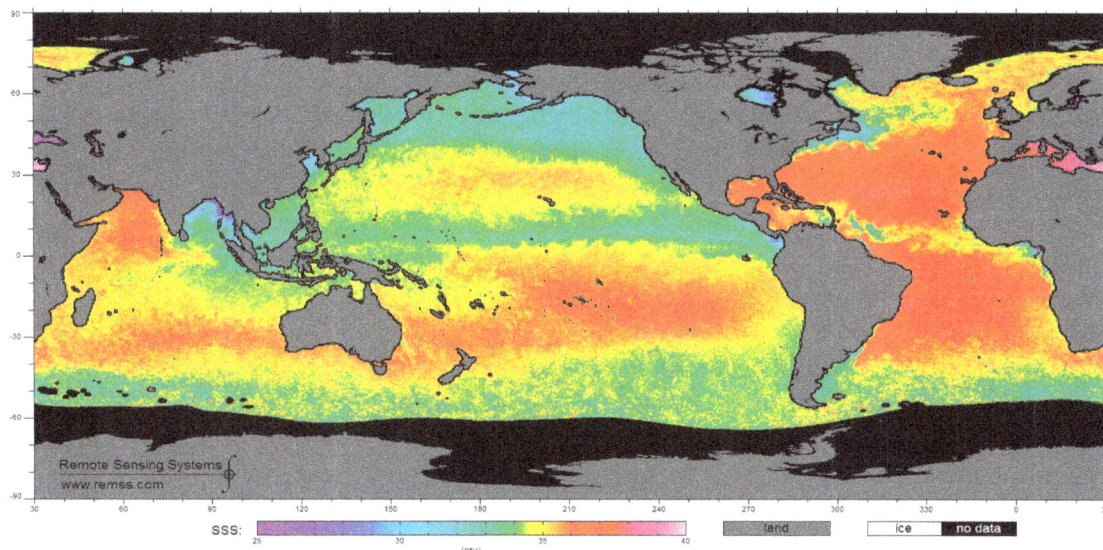

Figure 6. SSS global map on 40 km resolution cell obtained by eight-day averaging SMAP measurements collected in the period 4–11 November 2018.

The derivation of SIC standard products from microwave radiometers is possible since the microwave emissions from ice over sea surface, due to its crystalline structure, are usually much larger than the surrounding sea [74,75]. Even when deriving SIC products, BT measurements need corrections to reduce the influence of several factors as atmospheric water vapor and sea surface roughness. As an example, Figure 7 shows a SIC map on a 10 km-grid obtained from AMSR-2 measurements collected over Antarctica region on 22 November 2018. AMSR-2 provides multi-frequency (6.930 GHz, 7.300 GHz, 10.650 GHz, 18.700 GHz, 23.800 GHz, 36.500 GHz, and 89.000 GHz) BT measurements, with 0.3–1.1 K radiometric sensitivity, a swath width of approximately 1450 km, with about 5×3 km (at 89.000 GHz) to 62×35 km (at 6.930 GHz) spatial resolution on a daily basis. In Figure 7 it can be noted that the highest SIC values are experienced by the Larsen ice shelf and Wilkes land areas, which can be, therefore, potential sites of extremophiles psychrophiles, while very low SIC values characterize the Ross ice shelf, the Pine island bay and the eastern part of the Ronne ice shelf.

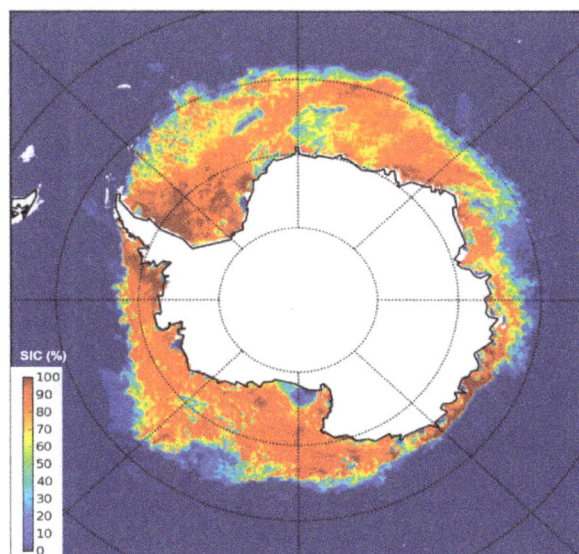

Figure 7. SIC map of Antarctica on 10 km resolution cell obtained by daily averaging AMSR-2 measurements collected on 22 November 2018.

5. Conclusions

The extremozymes, i.e., the enzymes produced by microorganisms living in extreme environments, are the focus of great interest for their manifold applications in several fields ranging from the food and pharmaceutical industries, to the conversion of biomass to energy, to the bioremediation of polluted areas. The marine extremozymes are produced by species living in extreme habitats like cold or hypersaline environments, hydrothermal vents or polluted sites. Therefore, marine extremozymes have adapted to harsh conditions and are active in the conditions typical of many industrial processes like extreme temperatures and pH values, high saline concentrations, and presence of metals and of organic solvents. Indeed, many examples of application of marine extremozymes can be found including the green conversion of different kinds of biomass to biofuels (ethanol, diesel, hydrogen) and the bioremediation of polluted sites in consequence of accidental oil spills or release of contaminants by industrial activities. The enormous biodiversity of marine extremophiles, and therefore the great variety of useful enzymes, is still underexplored. Hence, the identification of new enzyme producing species is a key issue for modern biotechnology that take advantages from exploitation of extremozymes. The identification of new species could be strongly pushed by the combination of different search approaches like metagenomic and remote sensing. Indeed, satellite microwave remote sensing, including active polarimetric SAR sensors and passive radiometers, can effectively support, on a regional/global scale and on a regular basis, a continuous and updated identification of sea sites affected by oil pollution, sea areas characterized by extreme SST and SSS values, and sea ice-infested areas. Such environments are promising sources of extremozymes producers that, thanks to the metagenomic tools, could be easily selected for their potential biotechnology applications.

Author Contributions: For research articles with several authors, a short paragraph specifying their individual contributions must be provided. The following statements should be used "Conceptualization, P.D.D., L.L., A.B.; A.P., A.B.; Investigation, P.D.D., I.F., G.R.A.; Writing-Original Draft Preparation, P.D.D., A.B., L.L., A.P.; Writing-Review & Editing, P.D.D., A.B., B.N.; Supervision, P.D.D., B.N., A.B.

Acknowledgments: The authors thank the European Space Agency who provided Sentinel-1 SAR data through the Copernicus scientific hub and the German Space Agency who provided TerraSAR-X SAR data under the AO OCE3201. AMSR-E data and SST/SSS products are produced by Remote Sensing Systems and sponsored by the NASA Earth Science MEaSUREs DISCOVER Project and the AMSR-E Science Team (freely available at www.remss.com). Sea ice concentration products are produced by the Satellite Application Facility on Ocean and Sea Ice freely available at www.osi-saf.org). The authors also acknowledge the Italian Antarctic Research Program and the project PNRA16_00274.

References

1.   Bolhuis, H.; Cretoiu, M.S. What is so Special about Marine Microorganisms? Introduction to the Marine Microbiome—From Diversity to Biotechnological Potential. In *The Marine Microbiome an Untapped Source of Biodiversity and Biotechnological*; Stal, L.J., Cretoiu, M.S., Eds.; Springer: Cham, Switzerland, 2016; Chapter 1; pp. 3–20.

2.   MacElroy, R.D. Some comments on evolution of extremophiles. *Biosystems* **1974**, *6*, 74–75. [CrossRef]

3.   Rothschild, L.J.; Mancinelli, R.L. Life in extreme environments. *Nature* **2001**, *409*, 1092–1101. [CrossRef] [PubMed]

4.   Chandrasekaran, M.; Kumar, S.R. Marine microbial enzymes. In *Biotechnology*; Werner, H., Roken, S., Eds.; EOLSS: Paris, France, 2010; Volume 9, pp. 47–49.

5.   Russo, R.; Giordano, D.; Riccio, A.; di Prisco, G.; Verde, C. Cold-adapted bacteria and the globin case study in the Antarctic bacterium *Pseudoalteromonas haloplanktis* TAC125. *Mar. Genom.* **2010**, *3*, 125–131. [CrossRef]

6.   Trincone, A. Potential biocatalysts originating from sea environments. *J. Mol. Catal. B-Enzym.* **2010**, *66*, 241–256. [CrossRef]

7. Fulzele, R.; Desa, E.; Yadav, A.; Shouche, Y.; Bhadekar, R. Characterization of novel extracellular protease produced by marine bacterial isolate from the Indian Ocean. *Braz. J. Microbiol.* **2011**, *42*, 1364–1373. [CrossRef] [PubMed]

8. Samuel, P.; Raja, A.; Prabakaran, P. Investigation and application of marine derived microbial enzymes: Status and prospects. *Int. J. Oceanogr. Mar. Ecol. Syst.* **2012**, *1*, 1–10. [CrossRef]

9. Poli, A.; Finore, I.; Romano, I.; Gioiello, A.; Lama, L.; Nicolaus, B. Microbial Diversity in Extreme Marine Habitats and Their Biomolecules. *Microorganisms* **2017**, *5*, 25. [CrossRef]

10. Mapelli, F.; Crotti, E.; Molinari, F.; Daffonchio, D.; Borin, S. Extreme Marine Environments (Brines, Seeps, and Smokers). In *the Marine Microbiome*; Stal, L.J., Cretoiu, M.S., Eds.; Springer: Cham, Switzerland, 2016; Chapter 9; pp. 251–282. ISBN 978-3-319-32998-7, 978-3-319-33000-6.

11. Moyer, C.L.; Morita, R.Y. Psychrophiles and Psychrotrophs. In *Encyclopedia of Life Sciences*; John Wiley & Sons, Ltd.: Hoboken, NJ, USA, 2007.

12. Caruso, C.; Rizzo, C.; Mangano, S.; Rappazzo, A.C.; Poli, A.; Di Donato, P.; Nicolaus, B.; Di Marco, G.; Michaud, L.; Lo Giudice, A. Extracellular polymeric substances with metal adsorption capacity produced by *Pseudoalteromonas* sp. MER144 from *Antarctic seawater*. *Environ. Sci. Pollut. Res. Int.* **2018**, *25*, 4667–4677. [CrossRef]

13. Breezee, J.; Cady, N.; Staley, J.T. Subfreezing growth of the sea ice bacterium 'Psychromonas ingrahamii'. *Microb. Ecol.* **2004**, *47*, 300–304. [CrossRef]

14. Franzmann, P.D.; Liu, Y.; Balkwill, D.L.; Aldrich, H.C.; De Macario, E.C.; Boone, D.R. *Methanogenium frigidum* sp. nov.; a psychrophilic, H2-using methanogen from Ace Lake, Antarctica. *Int. J. Syst. Bacteriol.* **1997**, *47*, 1068–1072. [CrossRef]

15. Caruso, C.; Rizzo, C.; Mangano, S.; Poli, A.; Di Donato, P.; Finore, I.; Nicolaus, B.; Di Marco, G.; Michaud, L.; Lo Giudice, A. Production and biotechnological potentialities of extracellular polymeric substances from sponge-associated Antarctic bacteria. *Appl. Environ. Microbiol.* **2018**, *84*, e01624-17.

16. Caruso, C.; Rizzo, C.; Mangano, S.; Poli, A.; Di Donato, P.; Nicolaus, B.; Finore, I.; Di Marco, G.; Michaud, L.; Lo Giudice, A. First evidence of extracellular polymeric substance production by a cold-adapted *Marinobacter* isolate from Antarctic seawater. *Antarct. Sci.* **2018**, in press.

17. Methé, B.A.; Nelson, K.E.; Deming, J.W.; Momen, B.; Melamud, E.; Zhang, X.J.; Moult, J.; Madupu, R.; Nelson, W.C.; Dodson, R.J.; et al. The psychrophilic lifestyle as revealed by the genome sequence of *Colwellia psychrerythraea* 34H through genomic and proteomic analyses. *Proc. Natl. Acad. Sci. USA* **2005**, *102*, 10913–10918. [CrossRef]

18. Médigue, C.; Krin, E.; Pascal, G.; Barbe, V.; Bernsel, A.; Bertin, P.N.; Cheung, F.; Cruveiller, S.; D'Amico, S.; Duilio, A.; et al. Coping with cold: The genome of the versatile marine *Antarctica bacterium Pseudoalteromonas haloplanktis* TAC125. *Genome Res.* **2005**, *15*, 1325–1335. [CrossRef]

19. Bowman, J.P. Genomics of Psychrophilic Bacteria and Archaea. In *Psychrophiles: From Biodiversity to Biotechnology*; Margesin, R., Ed.; Springer: Cham, Switzerland, 2017.

20. Martin, W.; Baross, J.; Kelley, D.; Russell, M.J. Hydrothermal vents and the origin of life. *Nat. Rev. Microb.* **2008**, *6*, 805–814. [CrossRef]

21. Tarasov, V.G.; Propp, M.V.; Propp, L.N. Functions of the coastal marine ecosystems in relation of hydrothermal venting. *Ecol. Chim.* **2002**, *11*, 1–15.

22. Maugeri, T.L.; Bianconi, G.; Canganella, F.; Danovaro, R.; Gugliandolo, C.; Italiano, F.; Lentini, V.; Manini, E.; Nicolaus, B. Shallow hydrothermal vents in the southern Tyrrhenian Sea. *Chem. Ecol.* **2010**, *26*, 285–298. [CrossRef]

23. Hekinian, R.; Renard, V.; Cheminee, J.L. Hydrothermal deposits on the East Pacific Rise near 13°N: Geological setting and distribution of active sulfide chimneys. In *Hydrothermal Processes at Seafloor Spreading Centers*; Rona, P.A., Bostrom, K., Laubier, L., Smith, K.L., Jr., Eds.; Plenum Press: New York, NY, USA, 1984; pp. 571–594.

24. Tarasov, V.G.; Gebruk, A.V.; Mironov, A.N.; Moskalev, L.I. Deep-sea and shallow-water hydrothermal vent communities: Two different phenomena? *Chem. Geol.* **2005**, *224*, 5–39. [CrossRef]

25. Karl, D.M. *The Microbiology of Deep-Sea Hydrothermal Vents*; Karl, D.M., Ed.; CRC Press: Boca Raton, FL, USA, 1995.

26. Takai, K.; Komatsu, T.; Inagaki, F.; Horikoshi, K. Distribution of Archaea in a black smoker chimney structure. *Appl. Environ. Microbiol.* **2001**, *67*, 618–629. [CrossRef]

27. Oren, A. Industrial and environmental applications of halophilic microorganisms. *Environ. Tech.* **2010**, *31*, 825–834. [CrossRef]

28. Oren, A. Life at high salt conditions. In *the Prokaryotes. A Handbook on the Biology of Bacteria: Ecophysiology and Biochemistry*; Dworkin, M., Falkow, S., Rosenberg, E., Schleifer, K.H., Stackebrandt, E., Eds.; Springer: New York, NY, USA, 2006; Volume 2, pp. 263–282.

29. DasSarma, S.; DasSarma, P. Halophiles. In *Encyclopedia of Life Sciences*; John Wiley & Sons, Ltd.: Chichester, UK, 2012.

30. Poli, A.; Kazak, H.; Gürleyendag, B.; Tommonaro, G.; Pieretti, G.; Toksoy Öner, E.; Nicolaus, B. High level synthesis of levan by a novel *Halomonas* species growing on defined media. *Carbohydr. Polym.* **2009**, *78*, 651–657. [CrossRef]

31. Hasan, U.Ö.; Berna, S.A.; Burak, A.; Poli, A.; Denizci, A.A.; Utkan, G.; Nicolaus, B.; Kazan, D. Moderately Halophilic Bacterium *Halomonas* sp. AAD12: A promising candidate as a Hydroxyectoine Producer. *J. Microbial. Biochem. Technol.* **2015**, *7*, 262–268.

32. Romano, I.; Poli, A.; Finore, I.; Huertas, F.J.; Gambacorta, A.; Pelliccione, S.; Nicolaus, G.; Lama, L.; Nicolaus, B. *Haloterrigena hispanica* sp. nov.; an extremely halophilic archaeon from Fuente de Piedra, Southern Spain. *Int. J. Syst. Evol. Microbiol.* **2007**, *57*, 1499–1503. [CrossRef] [PubMed]

33. Romano, I.; Finore, I.; Nicolaus, G.; Huertas, F.J.; Lama, L.; Nicolaus, B.; Poli, A. *Halobacillus alkaliphilus* sp. nov.; a halophilic bacterium isolated from a salt lake in Fuente de Piedra, Southern Spain. *Int. J. Syst. Evol. Microbiol.* **2008**, *58*, 886–890. [CrossRef] [PubMed]

34. Romano, I.; Orlando, P.; Gambacorta, A.; Nicolaus, B.; Dipasquale, L.; Pascual, J.; Giordano, A.; Lama, L. *Salinivibrio sharmensis* sp. nov, a novel haloalkaliphilic bacterium from a saline lake in Ras Mohammed Park (Egypt). *Extremophiles* **2011**, *15*, 213–220. [CrossRef] [PubMed]

35. Moreno, M.L.; Pérez, D.; García, M.T.; Mellado, E. Halophilic bacteria as a source of novel hydrolytic enzymes. *Life* **2013**, *3*, 38–51. [CrossRef] [PubMed]

36. Dalmaso, G.Z.L.; Ferreira, D.; Vermelho, A.B. Marine extremophiles: A source of hydrolases for biotechnological applications. *Mar. Drugs* **2015**, *13*, 1925–1965. [CrossRef]

37. Trincone, A. Enzymatic Processes in Marine Biotechnology. *Mar. Drugs* **2017**, *15*, 93. [CrossRef]

38. Annamalai, N.; Rajeswari, M.; Balasubramanian, T. Enzymatic saccharification of pretreated rice straw by cellulase produced from *Bacillus carboniphilus* CAS 3 utilizing lignocellulosic wastes through statistical optimization. *Biomass Bioenergy* **2014**, *68*, 151–160. [CrossRef]

39. Kim, B.; Lee, B.; Lee, Y.; Jin, I.; Chung, C.; Lee, J. Purification and characterization of carboxymethylcellulase isolated from a marine bacterium, *Bacillus subtilis* subsp. *subtilis A-53*. *Enzym. Microb. Technol.* **2009**, *44*, 411–416. [CrossRef]

40. Annamalai, N.; Rajeswari, M.V.; Elayaraja, S.; Thavasi, R.; Vijayalakshmi, S.; Balasubramanian, T. Purification and Characterization of Thermostable Alkaline Cellulase from Marine Bacterium *Bacillus licheniformis* AU01 by Utilizing Cellulosic Wastes Waste Biomass. *Valorization* **2012**, *3*, 305–310. [CrossRef]

41. Purwani, N.N.; Darmokoesoemo, H.; Tri Puspaningsih, N.N. Hydrolysis of Corncob Xylan using -xylosidase GbtXyl43B from *Geobacillus thermoleovorans* IT-08 Containing Carbohydrate Binding Module (CBM). *Procedia Chem.* **2016**, *18*, 75–81. [CrossRef]

42. Harshvardhan, K.; Mishra, A.; Jha, B. Purification and characterization of cellulase from a marine *Bacillus* sp. H1666: A potential agent for single step saccharification of seaweed biomass. *J. Mol. Catal. B Enzym.* **2013**, *93*, 51–56. [CrossRef]

43. Yagi, H.; Fujise, A.; Itabashi, N.; Ohshiro, T. Purification and characterization of a novel alginate lyase from the marine bacterium *Cobetia* sp. NAP1 isolated from brown algae. *Biosci. Biotechnol. Biochem.* **2016**, *80*, 2338–2346. [CrossRef]

44. Mohapatra, B.R. Kinetic and thermodynamic properties of alginate lyase and cellulase co-produced by *Exiguobacterium* species Alg-S5. *Int. J. Biol. Macromol.* **2017**, *98*, 103–110. [CrossRef]

45. Alkotaini, B.; Han, N.S.; Kim, B.S. Enhanced catalytic efficiency of endo-β-agarase I by fusion of carbohydrate-binding modules for agar prehydrolysis. *Enzym. Microb. Technol.* **2016**, *93–94*, 142–149. [CrossRef]

46. Charoenpanich, J.; Suktanarag, S.; Toobbucha, N. Production of a thermostable lipase by *Aeromonas* sp. EBB-1 isolated from marine sludge in Angsilla. *Thail. Sci. Asia* **2011**, *37*, 105–114. [CrossRef]

47. Mehrasbi, M.; Mohammadi, J.; Peyda, M.; Mohammadi, M. Covalent immobilization of *Candida antarctica* lipase on core-shell magnetic nanoparticles for production of biodiesel from waste cooking oil. *Renew. Energy* **2017**, *101*, 593–602. [CrossRef]

48. Yang, K.; Sohn, J.; Kim, H. Catalytic properties of a lipase from *Photobacterium lipolyticum* for biodiesel production containing a high methanol concentration. *J. Biosci. Bioeng.* **2009**, *107*, 599–604. [CrossRef]

49. Wu, Y.; Mao, A.; Sun, C.; Shanmugam, S.; Li, J.; Zhong, M. Catalytic hydrolysis of starch for biohydrogen production by using a newly identified amylase from a marine bacterium *Catenovulum* sp. X3. *Int. J. Biol. Macromol.* **2017**, *104 Pt A*, 716–723. [CrossRef]

50. Sivaperumal, P.; Kamala, K.; Rajaram, R. Bioremediation of Industrial Waste through Enzyme Producing Marine Microorganisms. *Adv. Food Nutr. Res.* **2017**, *80*, 165–179. [PubMed]

51. da Fonseca, F.; Angolini, C.; Zezzi Arruda, M.; Junior, C.; Santos, C.; Saraiva, A.; Pilau, E.; Souza, A.; Laborda, P.; de Oliveira, P.; et al. Identification of oxidoreductases from the petroleum *Bacillus safensis* strain. *Biotechnol. Rep.* **2015**, *8*, 152–159. [CrossRef] [PubMed]

52. Fang, Z.; Li, T.; Chang, F.; Zhou, P.; Fang, W.; Hong, Y.; Zhang, X.; Peng, H.; Xiao, Y. A new marine bacterial laccase with chloride-enhancing, alkaline-dependent activity and dye decolorization ability. *Bioresour. Technol.* **2012**, *111*, 36–41. [CrossRef] [PubMed]

53. Wang, L.; Wang, W.; Lai, Q.; Shao, Z. Gene diversity of CYP153A and AlkB alkane hydroxylases in oil-degrading bacteria isolated from the Atlantic Ocean. *Environ. Microbiol.* **2010**, *12*, 1230–1242. [CrossRef] [PubMed]

54. Saito, A.; Iwabuchi, T.; Harayama, S. A novel phenanthrene dioxygenase from *Nocardioides* sp. Strain KP7: expression in *Escherichia coli*. *J. Bacteriol.* **2000**, *182*, 2134–2141. [CrossRef] [PubMed]

55. Nikolaivits, E.; Dimarogona, M.; Fokialakis, N.; Topakas, E. Marine-Derived Biocatalysts: Importance, Accessing, and Application in Aromatic Pollutant Bioremediation. *Front. Microbiol.* **2017**, *8*, 265. [CrossRef] [PubMed]

56. Zhang, J.; Cao, X.; Xin, Y.; Xue, S.; Zhang, W. Purification and characterization of a dehalogenase from *Pseudomonas stutzeri* DEH130 isolated from the marine sponge *Hymeniacidon perlevis*. *World J. Microbiol. Biotechnol.* **2013**, *29*, 1791–1799. [CrossRef]

57. Zhang, J.; Xin, Y.; Cao, X.; Xue, S.; Zhang, W. Purification and characterization of 2-haloacid dehalogenase from marine bacterium *Paracoccus* sp. DEH99, isolated from marine sponge *Hymeniacidon perlevis*. *J. Ocean Univ. China* **2014**, *13*, 91–96. [CrossRef]

58. Novak, H.R.; Sayer, C.; Panning, J.; Littlechild, J.A. Characterisation of an l-haloacid dehalogenase from the marine psychrophile Psychromonas ingrahamii with potential industrial application. *Mar. Biotechnol.* **2013**, *15*, 695–705. [CrossRef]

59. Li, A.; Shao, Z. Biochemical characterization of a haloalkane dehalogenase DadB from *Alcanivorax dieselolei* B-5. *PLoS ONE* **2014**, *9*, e89144. [CrossRef]

60. Ulaby, F.T.; Moore, R.K.; Fung, A.K. *Microwave Remote Sensing: Active and Passive*; Artech House: Norwood, MA, USA, 1981.

61. Fung, K.; Chen, K.S. *Microwave Scattering and Emission Models for Users*; Artech House: Norwood, MA, USA, 1996.

62. Kanevsky, M.B. *Radar Imaging of Ocean Waves*; Elsevier: Jordan Hill, UK, 2009.

63. Donlon, C.; Berruti, B.; Buongiorno, A.; Ferreira, M.-H.; Femenias, P.; Frerick, J.; Goryl, P.; Klein, U.; Laur, H.; Mavrocordatos, C.; et al. The global monitoring for environment and security (GMES) Sentinel-3 mission. *Remote Sens. Environ.* **2012**, *120*, 37–57. [CrossRef]

64. Bonekamp, H.; Montagner, F.; Santacesaria, V.; Noddo, C.N.; Wannop, S.; Tomazic, I.; O'Carroll, A.; Kwiatkowska, E.; Scharroo, R.; Wilson, H. Core operational Sentinel-3 marine data product services as part of the Copernicus space component. *Ocean Sci.* **2016**, *12*, 787–795. [CrossRef]

65. Lee, J.-S.; Pottier, C. *Polarimetric Radar Imaging: From Basics to Applications*; CRC Press: Boca Raton, FL, USA, 2009.

66. Van Zyl, J.; Kim, Y. *Synthetic Aperture Radar Polarimetry*; JPL Space Science and Technology Series; NASA: Washington, DC, USA, 2010.

67. Wentz, F.J.; Gentemann, C.L.; Smith, D.K.; Chelton, D. Satellite measurements of sea surface temperature through clouds. *Science* **2000**, *288*, 847–850. [CrossRef]

68. Hosoda, K. Review of satellite-based microwave observations of sea surface temperatures. *J. Oceanogr.* **2010**, *66*, 439–473. [CrossRef]

69. Migliaccio, M.; Nunziata, F.; Buono, A. SAR polari0metry for sea oil slick observation. *Int. J. Remote Sens.* **2015**, *36*, 3243–3273. [CrossRef]

70. Nunziata, F.; Buono, A.; Migliaccio, M. COSMO-SkyMed Synthetic Aperture Radar data to observe the Deepwater Horizon oil spill. *Sustainability* **2018**, *10*, 3959. [CrossRef]

71. Nunziata, F.; de Macedo, C.R.; Buono, A.; Velotto, D.; Migliaccio, M. On the analysis of a time series of X-band TerraSAR-X SAR imagery over oil seepages. *Int. J. Remote Sens.* 2018, in press. [CrossRef]

72. Meissner, T.; Wentz, F.J.; Le Vine, D.M. The salinity retrieval algorithms for the NASA Aquarius version 5 and SMAP version 3 releases. *Remote Sens.* **2018**, *10*, 1121. [CrossRef]

73. Lagerloef, G.S.E. *Satellite Remote Sensing: Salinity Measurements, Encyclopedia of Ocean Sciences*; Academic Press: Cambridge, MA, USA, 2009.

74. Tikhonov, V.V.; Raev, M.D.; Sharkov, E.A.; Boyarskii, D.A.; Repina, I.A.; Komarova, N. Satellite microwave radiometry of sea ice of polar regions: A review. *Atmos. Ocean. Phys.* **2016**, *52*, 1012–1030. [CrossRef]

75. Ivanova, N.; Pedersen, L.T.; Tonboe, R.T.; Kern, S.; Heygster, G.; Lavergne, T.; Sørensen, A.; Saldo, R.; Dybkjær, G.; Brucker, L.; et al. Inter-comparison and evaluation of sea ice algorithms: Towards further identification of challenges and optimal approach using passive microwave observations. *Cryosphere* **2015**, *9*, 1797–1817. [CrossRef]

Marine Organisms as Potential Sources of Bioactive Peptides that Inhibit the Activity of Angiotensin I-Converting Enzyme

Dwi Yuli Pujiastuti [1],*, Muhamad Nur Ghoyatul Amin [1], Mochammad Amin Alamsjah [1],* and Jue-Liang Hsu [2,3]

[1] Department of Marine, Faculty of Fisheries and Marine, Universitas Airlangga, Surabaya 60115, Indonesia
[2] Department of Biological Science and Technology, National Pingtung University of Science and Technology, Pingtung 91201, Taiwan
[3] Research Center for Austronesian Medicine and Agriculture,
National Pingtung University of Science and Technology, Pingtung 91201, Taiwan
* Correspondence: dwiyp@fpk.unair.ac.id (D.Y.P.); alamsjah@fpk.unair.ac.id (M.A.A.)

Abstract: Angiotensin I-converting enzyme (ACE) is a paramount therapeutic target to treat hypertension. ACE inhibitory peptides derived from food protein sources are regarded as safer alternatives to synthetic antihypertensive drugs for treating hypertension. Recently, marine organisms have started being pursued as sources of potential ACE inhibitory peptides. Marine organisms such as fish, shellfish, seaweed, microalgae, molluscs, crustaceans, and cephalopods are rich sources of bioactive compounds because of their high-value metabolites with specific activities and promising health benefits. This review aims to summarize the studies on peptides from different marine organisms and focus on the potential ability of these peptides to inhibit ACE activity.

Keywords: ACE inhibitory peptide; antihypertensive; bioactive peptides; hypertension; marine resources

1. Introduction

Hypertension or high blood pressure is generally caused by behavioral risk factors, ageing, and population growth. It emerged in upper-middle income countries among adults aged >25 years. Hypertension causes 9.4 million deaths each year worldwide [1]. Currently, hypertension is one of the leading causes of morbidity and mortality globally, followed by metabolic disorder [2]. It is a key risk factor for cardiovascular disease, heart attack, stroke, and arteriosclerosis. The common examination used to diagnose hypertension is the measurement of blood pressure; a systolic blood pressure (SBP) and diastolic blood pressure (DBP) higher than 140 mm Hg and 90 mm Hg, respectively, indicates hypertension. To mitigate the aberrations in blood pressure and restore normal physiological function, functional molecules derived from food have been widely pursued.

The renin angiotensin aldosterone system (RAAS) plays a significant role in the maintenance of arterial blood pressure and fluid balance and is regarded as the major target to combat hypertension [3]. In RAAS, angiotensinogen is cleaved by renin, producing angiotensin I. Angiotensin I is then converted to angiotensin II, a strong vasoconstrictor, by angiotensin I-converting enzyme (ACE). In addition, ACE inactivates the vasodilator bradykinin, which acts as a mediator of inflammation, a natriuretic peptide, and a potent stimulator of vasodilator prostaglandins, and is involved in nitric oxide synthesis [4]. Because the production of angiotensin II increases blood pressure [5,6], the inhibition of ACE is a reliable strategy to control hypertension [7]. ACE inhibitors decrease ACE activity and indirectly reduce the angiotensin II level, thereby exerting a vasorelaxation effect on blood vessels [8]. Captopril, enalapril, lisinopril, and benazepril are commonly used as effective synthetic ACE inhibitors and have been developed for treating hypertension. However, synthetic drugs usually cause undesirable side effects [9,10]. To reduce these side effects, food-derived ACE inhibitory peptides are preferred over synthetic drugs to combat hypertension. ACE inhibitory peptides are considered as potent antihypertensive drugs, and they do not have any undesirable side effects. ACE inhibitors are more effective than other hypertensive drugs in retarding the progression of renal damage and reducing proteinuria. Two health organization, namely the international society of hypertension-world health organization (ISHWHO) and the Canadian society of hypertension recommend ACE inhibitors as the first line of treatment for hypertension [11].

Proteins are an important macronutrient as they provide the necessary energy and amino acids essential for growth and the maintenance of normal bodily functions. Many physiological and functional properties of proteins are attributed to bioactive peptides [8]. Bioactive peptides derived from food protein have been growing attractive because of awareness of their health-boosting properties. Bioactive peptides from several natural and processed foods have now been isolated and characterized. They function as potential physiological modulators in the process of metabolism during intestinal digestion and are liberated depending on their structure, composition, and amino acid sequence. Some bioactive peptides have been identified to possess nutraceutical potential and promote overall human health [12], with the potential of being used as candidates for treating conditions, such as hypertension [13].

Bioactive peptides are usually isolated from milk and cheese. They are also isolated from other animal sources, such as meat, gelatin, eggs, and various fish species (salmon, sardine, tuna, and herring), and plant sources, such as mushroom, wheat, pumpkin, and sorghum [14]. For example, ACE inhibitory peptides derived from fish have been shown to have a favorable effect on blood pressure [7,15,16]. Unlike many synthetic ACE inhibitors, which cause dry cough and angioedema, natural peptide-inhibitors have no side effects and are considered to be safer and healthier [17]. In recent years, ACE inhibitors have been derived from food proteins, such as milk [18,19], corn [20,21], ovalbumin [22], legume [23,24], Chinese soft-shelled turtle eggs [25,26], bitter melon seeds [27], cheese [28,29], chicken eggs [30–33], casein [34–36], fish [37–39], and algae [40,41].

Oceans cover >70% of the earth's surface and are a rich resource for humans. There is increasing interest in marine organisms as new sources of natural products. Several compounds with unique biological activities have been isolated from marine organisms. The marine environment is rich in biological as well as chemical diversity; compounds isolated from marine organisms have been used as pharmaceuticals, nutraceuticals, cosmeceuticals, molecular probes, fine chemicals, and agrochemicals. Macro-and microorganisms in marine habitats possess a wide array of secondary metabolites, including terpenes, steroids, polyketides, peptides, alkaloids, polysaccharides, proteins, and porphyrins. Because the environment surrounding marine organisms is extreme, aggressive, and competitive, these organisms produce several secondary metabolites with a promising potential for use as drugs, nutritional supplements, and therapeutic agents [42–44]. Marine organisms, such as fish, shellfish, seaweed, microalgae, molluscs, crustaceans, and cephalopods, are rich sources of several functional compounds, such as bioactive peptides, enzymes, polyunsaturated fatty acids, vitamins, minerals, phenolic phlorotannins, and polysaccharides. Moreover, as some marine organisms, especially fish, are particularly rich sources of protein, they are ideal for generating protein-derived bioactive peptides [45,46]. Marine bioactive peptides have gained significant attention for their health promoting effects, such as antihypertensive, antioxidant, anticoagulant, antimicrobial, antithrombotic, and hypocholesterolemic properties [47]. Furthermore, compounds isolated from marine organisms have been commercially distributed in health markets [48]. In this review, we discuss the ACE inhibitory peptides derived from marine resources and provide information on their production, characterization, and potential health benefits. We also review the future prospects of ACE inhibitory peptides derived from marine organisms as therapeutic drugs to combat hypertension.

2. ACE Inhibitory Peptides Derived from Marine Organisms

Zinc ion (Zn^{2+})-dependent dipeptidyl carboxypeptidase, also known as ACE (EC 3.4.15.1), plays a pivotal role in the regulation of blood pressure because of its action in RAAS [49]. ACE is present in biological fluids, such as plasma and semen, and in many tissues, such as testis, intestinal epithelial cells, proximal renal tubular cells, brain, lungs, stimulated macrophages, vascular endothelium, and the medial and adventitial layers of blood vessel walls [4]. In humans, ACE exists in two isoforms: somatic ACE (sACE) and germinal ACE (gACE). sACE is distributed in many types of endothelial and epithelial cells, whereas gACE occurs in germinal cells in the testis, and is therefore also known as testicular ACE [6]. In RAAS, ACE cleaves the decapeptide angiotensin I (Asp-Arg-Val-Tyr-Ile-His-Pro-Phe-His-Leu) into the octapeptide angiotensin II (Asp-Arg-Val-Tyr-Ile-His-Pro-Phe) by removing the C-terminal dipeptide His-Leu. Angiotensin II stimulates the release of aldosterone and antidiuretic hormone or vasopressin, consequently increasing the retention of sodium and water; it also acts as a potent vasoconstrictor (Figure 1). These phenomena act in concert to directly increase the blood pressure [6]. Substrates of ACE include not only angiotensin I in RAAS and bradykinin in the kinin–kallikrein system, but also the haemoregulatory peptide N-acetyl-Ser-Asp-Lys-Pro, which is a putative bone marrow suppressor. It contributes to haemopoietic cell differentiation, regulating tissue and blood levels of the vasoactive hormones angiotensin II and bradykinin [50]. In addition, ACE shows endopeptidase activity against a wide range of substrates, such as cholecystokinin, substance P, and luliberin. The inhibition of ACE enzymatic activity on angiotensin I is one of the major challenges to combat hypertension-related disorders [51].

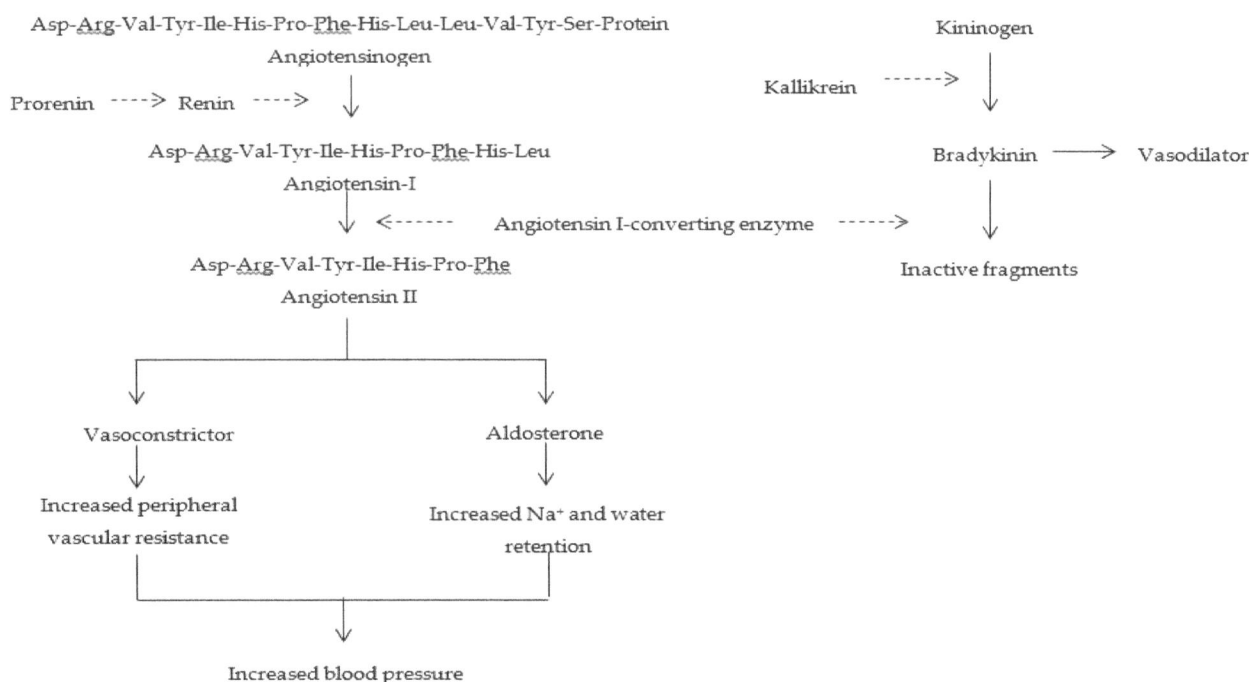

Figure 1. Role of angiotensin I-converting enzyme in the renin angiotensin aldosterone system and the kinin–kallikrein system [15].

Recently, natural marine products have been investigated as alternative synthetic drugs; they have been the topic of interest for many researchers due to their numerous beneficial effects, and some novel ACE-inhibitory compounds have been isolated from algae [52,53]. Marine proteins, such as Heshiko, a fermented mackerel product [38], sardine muscle [9], shark meat [54], Alaska pollock skin [55], marine shrimp [56], and chum salmon [57], exhibit ACE inhibitory activity. ACE inhibitory peptides usually contains 2–12 amino acid residues [10,58,59]. However, some studies have identified up to 27 amino acid residues in ACE inhibitory peptides [60,61]. Proteases, such as pepsin, chymotrypsin, alcalase, and trypsin, are frequently used in hydrolysis for generating ACE inhibitory peptides [9,10,55]. List of identified peptides derived from marine resources; origin, sequence peptides, and IC_{50} value, can be seen in Table 1.

Table 1. List of identified peptides derived from marine resources; origin, sequence peptides, and IC_{50} value.

Origin	Enzyme	Sequence Peptide	IC_{50} (μM)	Reference
Fish				
Sea bream	Alkaline Protease	GY VY GF VIY	265 16 708 7.5	[62]
Lizard fish	Neutral Protease	MKCAF RVCLP	45.7 175	[63] [64]
Alaska pollock (*Theragra chalcogramma*)	Alcalase, Pronase E and Collagenase	GPL GPM	2.6 17.3	[55]
Grass carp	Alcalase	VAP	19.9	[10]
Atlantic salmon (*Salmo salar* L.)	Alcalase and Papain	AP VR	356.9 1301.1	[65]
Skipjack (*Katsuwonus pelamis*)	Alcalase	DLDLRKDLYAN MCYPAST MLVFAV	67.4 58.7 3.07	[66]
Yellowfin sole (*Limanda aspera*)	Chymotrypsin	MIFPGAGGPEL	268.3	[67]
Pacific cod	Pepsin	GASSGMPG LAYA	6.9 14.5	[68]
Paralichthys alivaceus	Pepsin	MEVFVP VSQLTR	79 105	[69]
Channa striatus	Thermolysin	VPAAPPK NGTWFEPP	0.45 0.63	[70]
Microalgae				
Chlorella vulgaris	Pepsin	IVVE FAL AEL VVPPA AFL	315 26.3 57.1 79.5 63.8	[40]
Chlorella ellipsoidea	Alcalase	VEGY	128.4	[71]
Spirulina platensis	Pepsin	IAE IAPG VAF	34.7 11.4 35.8	[40]
Molluscs				
Sea cucumber (*Acaudina molpadioidea*)	Bromelain and Alcalase	MEGAQEAQGD	15.9	[72]
Cuttlefish (*Sepia officinalis*)	Cuttlefish hepatopancreas	VYAP VIIF MAW	6.1 8.7 16.32	[73]
		GIHETTY EKSYELP VELYP	25.66 14.41 5.22	[74]
Squid (*Dosidicus gigas*) skin collagen	Esperase	GRGSVPAPGP	47.78	[75]
Corbicula fluminea	Protamex + Flavourzyme	VKP	3.7	[76]
		VKK	1045	

The potency of peptides derived from marine organisms is expressed as the half maximal inhibitory concentration (IC_{50}), which indicates the ACE inhibitor concentration that leads to 50% inhibition of ACE activity. Moreover, Lineweaver–Burk plots are usually used to determine the inhibition mode of

ACE inhibitory peptides. Most of the reported peptides act as competitive inhibitors of ACE. In the competitive inhibition mode, the inhibitor competes with the substrate and binds to the active site of ACE. In the non-competitive inhibition mode, the inhibitor binds to a site other than the active site. The binding of inhibitor to ACE alters the conformation of ACE, which prevents the substrate from binding to the active site of ACE. The enzyme, substrate, and inhibitor cannot form a complex; thus, the enzyme–substrate complex or enzyme–inhibitor complex is formed. In the uncompetitive inhibition mode, the inhibitor binds to only the substrate–enzyme complex. The C-terminal end of the inhibitory peptide associates with the active site pockets of ACE. ACE harbors three sub-sites: antepenultimate position (S1), penultimate position (S1'), and ultimate position (S2'). In the substrate, the amino acids Pro, Ala, Val, and Leu are the most favorable for S1; Ile is the most favorable for S1'; and Pro and Leu are the most favorable for S2' [77]. The S1 sub-site includes Ala354, Glu384, and Tyr523 residues; S1' pocket contains Glu162; and S2' pocket includes Gln281, His353, His513, Lys511, and Tyr520 [78,79]. Many studies have shown that peptides with high ACE inhibitory activity contain Trp, Phe, Tyr, or Pro at the C-terminus and branched aliphatic amino acids at the N-terminus [49].

In China, soft-shelled turtle eggs have been used as a tonic food for a long time. Low-molecular weight peptides (<3 kDa) have been isolated from soft-shelled turtle egg by ultrafiltration and fractionated by reversed-phase high-performance liquid chromatography (RP-HPLC). In vitro screening of the resulting fractions for ACE inhibitory activity has revealed an IC_{50} value of 4.39 μM for the peptide IVRDPNGMGAW isolated from soft-shelled turtle egg white. This peptide has been identified as a competitive inhibitor of ACE [26]. The peptide AKLPSW, isolated from soft-shelled turtle egg yolk, has also been shown to exhibit potent ACE inhibitory activity, with an IC_{50} value of 15.3 μM, and inhibition kinetics has indicated that this peptide is a non-competitive inhibitor of ACE. The AKLPSW peptide significantly reduces the systolic blood pressure by approximately 13 mm Hg after 6 h of oral administration, thus confirming its antihypertensive effect [25]. In another study, Sardinella protein hydrolysates (SPHs) were obtained from fermentation with *Bacillus subtilis* (SPH-A26) and *Bacillus amyloliquefaciens* (SPH-An6). Approximately 800 peptides have been identified in SPH-A26 and SPH-An6 using nano electrospray ionization liquid chromatography tandem mass spectrometry. Of these 800 peptides, eight isolated from SPH-A26 and seven from SPH-An6 have been selected based on homologies with previously characterized peptides (Biopep data bank), as well as peptide length. Among the synthesized peptides, NVPVYEGY and ITALAPSTM show ACE inhibitory activity with IC_{50} values of 210 and 229 μM, respectively. Fermented SPHs have a potential for use as hypotensive nutraceutical ingredients [80]. The popular freshwater tilapia also reported the potential antihypertensive peptides from hydrolysate by using papain, bromelain, and pepsin. In order to enhance the activity, the hydrolysate was fractionated into four fractions (<1 kDa, 1–3 kDa, 3–5 kDa, and 5–10 kDa). The pepsin-hydrolyzed FPH (FPHPe) with the highest DH (23%) possessed the strongest ACE-inhibitory activity (IC_{50} of 0.57 mg/mL). Its <1 kDa ultrafiltration fraction (FPHPe1) suppressed both ACE (IC_{50} of 0.41 mg/mL). In addition, FPHPe1 significantly reduced SBP (maximum −33 mmHg), DBP (maximum −24 mmHg), mean arterial pressure (MAP) (maximum −28 mmHg), and hearth rates (HR) (maximum −58 beats) in SHRs [81].

The production of peptides with ACE inhibitory activity must consider the amino acid composition and molecular weight of hydrolysates. Purification is carried out to obtain a single peptide with a specific amino acid residues which is in accordance with characterized sequence of bioactive peptide inhibiting ACE. The pure peptide could be easily observed its activity and stability, as well as the dosage of peptide administration in the patients with hypertension symptom would be validly determined. Total hydrolysates with high molecular weight revealed lower activity for inhibiting the ACE rather than single peptide. The shorter amino acid residues is more visible to reach the target site when through the digestive tract and they can be absorbed easily. Then, lower-molecular weight peptides also have a higher probability of passing through the intestinal barrier and exerting biological function [65]. The C-terminal residue in tripeptides or dipeptides plays an important role in binding to sub-sites S1, S1', and S2' sub-sites within the active site of ACE [82]. Aromatic or hydrophobic amino acid residues, such as Trp, Phe, Tyr, and Pro, are more active if present at positions in the C-terminal end that bind to each of the three sub-sites of ACE. In addition, tripeptides or dipeptides with a branched aliphatic amino acid at the N-terminus show potent ACE inhibition. Basic amino acid residues, such as Lys and Arg, at the C-terminus also contribute to potent inhibition against ACE [83]. Many studies have shown that the C-terminal residue of potent ACE inhibitory peptides is usually a hydrophobic amino acid [39,70,74,84,85].

There is no correlation between competitive inhibitor with high ACE inhibitory activity. Several non-competitive inhibitors show high ACE inhibitory activity. The peptide Ala-Lys-Leu-Pro-Ser-Trp derived from soft-shelled turtle egg yolk exhibits a low IC_{50} value of 13.7 μM [25], whereas the peptide Val-Glu-Leu-Tyr-Pro isolated from cuttlefish muscle protein exhibits an even lower IC_{50} value of 5.22 μM [74]; both these peptides are considered non-competitive inhibitors. Moreover, some peptides inhibit ACE activity by the uncompetitive mode of inhibition. For example, the peptides Ile-Trp and Phe-Tyr have been ientified as uncompetitive inhibitors [86]; similarly, the peptides Tyr-Ley-Tyr-Glu-Ile-Ala and Tyr-Leu-Tyr-Glu-Ile-Ala-Arg-Arg have been identified as uncompetitive inhibitors [87]. Depending on the results of pre-incubation of the peptide with ACE, the ACE inhibitory peptides are divided into three categories: true inhibitors, prodrugs, and real substrates. A true inhibitor shows no significant difference in the IC_{50} value before and after pre-incubation with ACE, whereas a prodrug shows dramatic reduction in the IC_{50} value after pre-incubation with ACE. On the other hand, a real substrate shows an increase in the IC_{50} value after pre-incubation with ACE, suggesting a reduction in its inhibitory activity against ACE. Generally, the prodrug- and true inhibitor-type peptides are expected to exhibit long-lasting antihypertensive activity in spontaneously hypertensive rats used as a model to study hypertension in humans [88,89].

3. Generation of Bioactive Peptides

Protein hydrolysates have an excellent amino acid balance, are readily digestible, show rapid uptake, and contain bioactive peptides [90]. Bioactive peptides act as therapeutic agents and are characterized by high biological specificity, low toxicity, high structural diversity, high and wide spectrum of activity, and small size, which implies that they have a low likelihood of triggering undesirable immune responses [91]. Bioactive peptides are defined as protein fragments with beneficial effects on bodily functions and human health. Peptides isolated from food sources are structurally similar to endogenous peptides and therefore interact with the same receptors and play a prominent role as immune regulators, growth factors, and modifiers of food intake [92]. Depending on the sequence of amino acids, these peptides can exhibit diverse activities, including antimicrobial [93], antioxidant [94], antithrombotic [95], and antihypertensive [25].

Bioactive peptides are generally produced via enzymatic hydrolysis using digestive enzymes, fermentation using proteolytic starter cultures, or proteolysis using microorganism-or plant-derived enzymes. To generate short-chain functional peptides, enzymatic hydrolysis is used in combination with fermentation or proteolysis [96]. During growth, microorganisms release the protease enzyme into the extracellular medium, leading to proteolysis and peptide generation. Microorganisms are typically used for fermentation for several hours to several days, depending on the desired peptide and the type of fermentation [97]. During fermentation, microorganisms break down complex compounds into smaller molecules with various physiological functions [98]. Fermented marine food products are rich sources of bioactive compounds, including amino acids and peptides [99]. Digestive enzymes, such as trypsin, chymotrypsin, and pepsin, release the bioactive peptides for gastrointestinal digestion in vivo. To stimulate gastrointestinal digestion, several proteolytic enzymes, such as alcalase and thermolysin, engage with trypsin and pepsin. In addition, recombinant DNA technology and chemical synthesis have been used to produce bioactive peptides [92]. The physicochemical properties, such as molecular weight, isoelectric point, and hydrophilic or hydrophobic indices of the resulting peptides, change after enzymatic hydrolysis. Prominent amino acids, such as Pro and Val, play key roles in most antihypertensive peptides [91].

In the digestive system, bioactive peptides are absorbed through the intestine and enter the blood stream to exert systemic effects or local effects in the gastrointestinal tract. Dipeptides and tripeptides are easily absorbed in the intestine. To exert antihypertensive effects, bioactive peptides must reach the target cells after absorption through the intestine. Common bioactive peptides with antihypertensive effects include Val-Pro-Pro (VPP) and Ile-Pro-Pro (IPP); they are produced via fermentation using *Lactobacillus helveticus* and *Saccharomyces cerevisiae*. These two peptides have been detected in the aortal tissue using HPLC, and their effect on ACE activity was lower in the aorta in the study group than in the control group (saline) [14].

4. Screening Approach

The search for peptides capable of inhibiting ACE activity has been intensified. The pursue of ACE inhibitory peptides from marine, as well as other sources, has been substantiated. A reliable assay to determine the ability of peptides to inhibit ACE activity is of paramount concern. In vitro determination of ACE inhibitory peptides is preceded by enzymatic digestion or microbial fermentation, followed by the analysis of structure and chemical synthesis of active peptides. Most assays evaluating the ACE inhibitory activity of peptides have been performed as described previously [100]. The technique used to evaluate the ACE inhibitory activity of peptides must be simple, sensitive, and reliable. Several such methods have been developed, such as spectrophotometry, HPLC, fluorometric capillary electrophoresis, and radiochemistry. Among these, spectrophotometry is the most commonly used method to measure ACE inhibitory activity. This method involves the hydrolysis of hippuryl-histidyl-leucine (HHL) by ACE to hippuric acid (HA). The amount of HA produced from HHL is directly correlated with ACE activity [101]. The amount of HA formed is determined by measuring the absorbance at 228 nm (absorption maximum of HA) [102]. Although the spectrophotometry is useful, it is time consuming, complicated, and is unable to detect trace amounts of the sample.

In practice, results of different assays may vary because of the use of different substrates, such as the synthetic peptides HHL and furanacryloyl-L-phenylalanylglycyl-glycine (FAPGG), which are the most commonly used substrates, and the fluorescent molecule o-aminobenzoylglycyl-p-nitrophenylalanylproline for specific detection and quantification [103]. Results may also vary within the same assay because of the use of different test conditions or the use of ACE from different origins. Thus, ACE activity levels must be carefully controlled to obtain comparable and reproducible results [83,104].

HPLC is a common method to determine ACE inhibitory activity of peptides as it generates reproducible results. Although HPLC has been used for decades, it requires the extraction of the product from the reaction mixture using an organic solvent, which limits the number of samples that can be analyzed per day and is also a source of error [105]. Moreover, HPLC analysis shows peculiar results from samples with added inhibitor, which exhibit high HA release than samples without the added inhibitor. This occurs if the enzyme or the substrate (HHL) is unstable in solution. The evaluation of ACE inhibition is depends on the comparison between the concentration of HA in the presence or absence of an inhibitor (inhibitor blank). The occurrence of autolysis of HHL to give HA was evaluated by a reaction blank, i.e., a sample with the higher inhibitor concentration and without the enzyme [24]. Another substrate, FAPGG, has also been used for HPLC [106,107]; FAPGG releases 2-furylacryloyl-L-phenylalanine (FAP) as a product. This method is used to quantitate the levels and can be used a model of inhibition according to the sigmoid character of the response curve. The slope of the curve, describing absorbance versus time, is thus a direct measure of ACE activity. It is based on the combination of enzymatic reaction with HPLC detection of the inhibition of enzyme activity by measuring the levels of the substrate and product formed. The amount of FAP formed is determined by measuring the absorbance at 305 nm. This method is beneficial, as it does not require sophisticated equipment or radiolabelled compounds [108]. Because the price of the two substrates, HHL and FAPGG, is similar, the HPLC method is advantageous over spectrophotometry, as it requires less labor and has a higher throughput than spectrophotometry [103].

The determination of ACE activity also utilizes fluorescent tripeptides, such as o-aminobenzoylglycyl-p-nitro-L-phenylalanyl-L-proline [Abz-Gly-Phe(NO$_2$)-Pro]. The hydrolysis of this substrate by ACE generates o-aminobenzoylglycine (Abz-Gly) as a product, which is easily quantified fluorometrically using appropriate excitation and emission wavelengths. Fluorescence detection of the reaction products is highly sensitive and precise. Moreover, commercial availability of all reagents is a major advantage, allowing easy introduction of the assay in laboratories [109].

To obtain ACE inhibitory peptides, slight modification of the assay is crucial. Orthogonal bioassay-guided fractionation is considered as a potential method to obtain ACE inhibitory peptides. This method involves the separation of the potential peptides using two ways of fractionation: Strong cation exchange (SCX) and RP-HPLC (Figure 2). SCX separates peptides based on their charge, whereas RP-HPLC separates peptides based on their hydrophobicity [110]. Although both SCX and RP-HPLC separate peptides using different mechanisms, peptides are regarded as potential ACE inhibitors because they remain in the most active fraction using both methods. Pujiastuti et al. [25] revealed the identification of overlapping peptides using SCX and RP-HPLC.

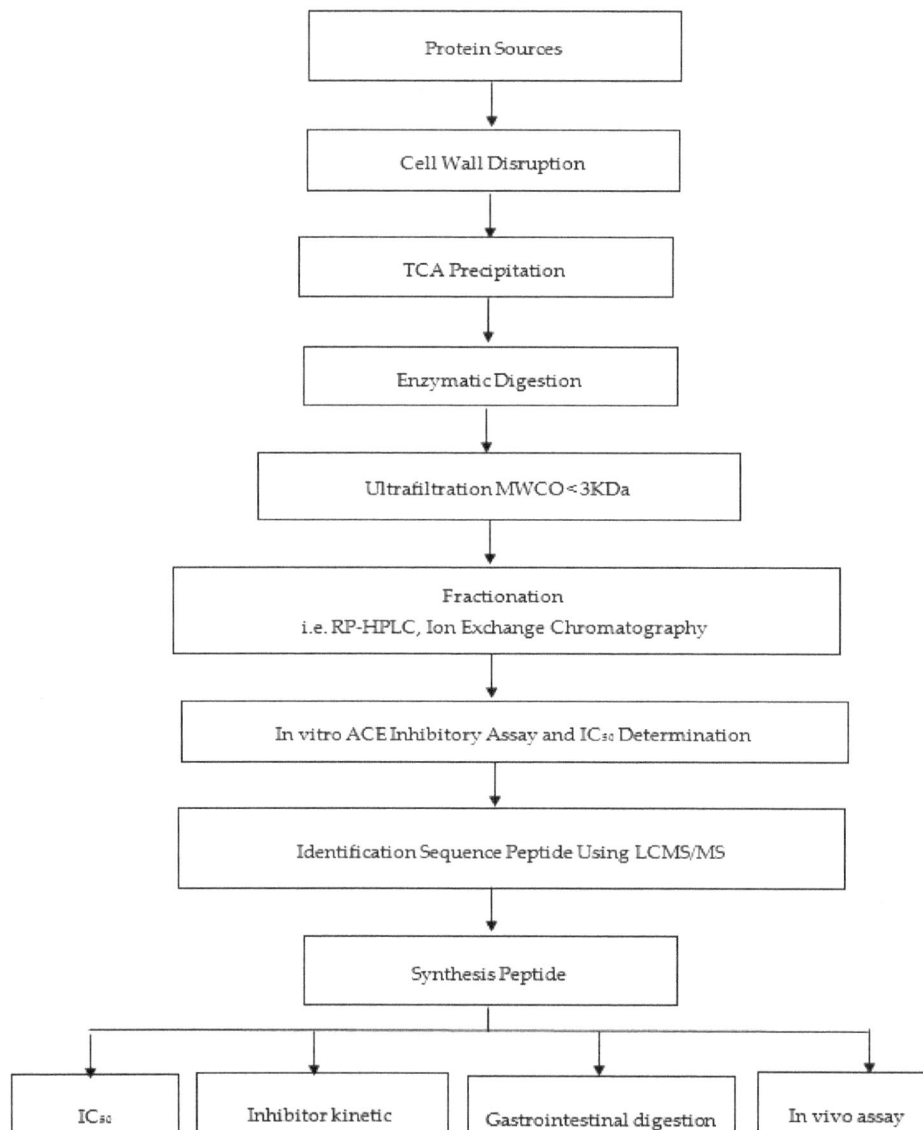

Figure 2. Flowchart showing the production of bioactive peptides for angiotensin I-converting enzyme (ACE) inhibitory assay [25].

A new method used to measure ACE activity is ultra-performance liquid chromatography (UPLC). The UPLC-mass spectrometry method has been developed to determine ACE activity using HHL as the substrate and purified rabbit ACE. This method is rapid, accurate, and reproducible, and is used to determine trace amounts of compounds. In addition, this method requires a short analysis time and small reaction volume and is highly selective compared to conventional methods. It is also suitable for high-throughput screening of potential ACE inhibitors and candidate compounds isolated from herbal medicines [111].

The in vitro gastrointestinal digestion approach provides a straightforward approach to imitate peptide function by incubating the peptide with ACE before in vivo oral administration. Oral administration of ACE inhibitory peptides in hypertensive patients requires these peptides to pass through the digestive tract and be absorbed through the intestinal epithelium. Pepsin is widely used to represent gastrointestinal enzymes that function at acidic pH. Polypeptides are further truncated by pancreatic proteases, including trypsin, α-chymotrypsin, elastase, and carboxypeptidases A and B at alkaline pH. In vivo testing of peptides is frequently performed in spontaneously hypertensive rats as they mimic hypertension in humans. This animal model has been used to evaluate the effects of both short-and long-term administration of antihypertensive peptides. In human studies, food-derived peptides have been used to establish whether peptides exhibit an antihypertensive effect in humans with high-to-normal blood pressure. For example, the antihypertensive effect of the peptides IPP and VPP isolated from the commercial fermented milk show antihypertensive effects after long-term administration. The sour milk product Calpis from Japan has been examined in mildly hypertensive patients [112]. In some cases, ACE inhibitory peptides fail to show hypotensive activity after oral administration in vivo, possibly because of the hydrolysis of these peptides by ACE or gastrointestinal proteases [74,113]. It is difficult to evaluate a direct correlation between in vitro ACE inhibitory activity and in vivo antihypertensive activity because the bioavailability of these peptides after oral administration varies. ACE inhibitory peptides must remain active during gastrointestinal digestion and reach the specific organ. However, it is possible that ACE inhibitory peptides are degraded before reaching the specific organ. The antihypertensive mechanism of ACE inhibitory peptides, rather than the ACE inhibition mechanism, may be of greater interest [77,114].

In silico methods are used to predict the structure of ACE inhibitory peptides based on similarity between sequences available in databases. The molecular docking approach is widely used to predict and characterize the binding site of target proteins according to ligand conformation and binding affinity score [115]. The most convenient approach to elucidate the accuracy of molecular docking is to determine the distance of binding conformation using the scoring function in the docking program [116]. Several scoring functions are used to evaluate the docking procedure, such as CDocker Energy, CDocker Interaction Energy, LibDockScore, PLP1, PLP2, LigScore1, LigScore2, Jain, PMF, and PMFO4. Besides, BIOPEP-UWM and BLAST database is increasingly popular to be in silico approaches for investigating biological activities from tilapia and chickpea [117]. BIOPEP-UWM database is used to predict bioactive peptides composed in protein sequences. This method has benefits such as time and cost reduction, as well as being a rapid method to identify and characterize proteins. Briefly, the bioactivities, sequences, number, and location of the peptides were obtained from the sequences of the identified proteins analyzed using the "profiles of potential bioactivity" tool. Moreover, the sequences of the identified proteins were examined using the "enzyme action" tool to simulate enzymatic hydrolysis [118]. Knowing the position of the binding site before docking significantly increases the docking efficiency. Moreover, knowledge of the structure and activity relationship is important to explore potential ACE inhibitory peptides. The ACE structure contains a Zn site, which usually coordinates with oxygen, nitrogen, and sulphur donors of Asp, Cys, and His, respectively, wherein His is the most regularly encountered in the sphere of Zn^{2+} ion. The other Zn ligand in catalytic sites is water; it is activated for polarization, ionization, and arrangement of ligands in coordination with Zn [11]. The Zn^{2+} ion is also important for the binding strength between ACE and its inhibitors [119]. Generally, ACE inhibitors contain one or more molecular functionalities, such as Zn-binding ligand, a hydrogen bond donor, and a carboxyl-terminal group [120]. The ability of a protein to interact with small molecules plays a major role in the dynamics of that protein, which may enhance or inhibit its biological function. Studies on the catalytic mechanism of ACE have revealed that the 19 amino acid residues in the active site of ACE, including His353, Ala354, Ser355, Ala356, His383, Glu384, His387, Phe391, Pro407, His410, Glu411, Phe512, His513, Ser516, Ser517, Val518, Pro519, Arg522, and Tyr523, bind to small molecules or to protein (ligand).

5. Conclusions

Bioactive peptides derived from marine resources have potential ACE inhibitory activity and are considered as therapeutic agents to combat hypertension. The main characteristic of ACE inhibitory peptides is the position of the hydrophobic residue, usually Pro, at the C-terminus. In vitro and in vivo testing are the most challenging tasks in antihypertensive research as their results do not always show direct correlation, although gastrointestinal digestion is suggested to mimic peptide release in human body. Marine organisms represent sustainable sources of ACE inhibitory peptides for the production of pharmaceuticals and nutraceuticals at an industrial scale. Due to the importance of pure peptide inhibiting ACE for future pharmaceutical and nutraceutical industry, the purification techniques of identified peptide is highly crucial. Therefore, upscaling research on bioactive peptide purification should trigger biotechnologists to perform the research.

Highlights:

- Angiotensin I-converting enzyme (ACE) is a key target for treating hypertension.
- Food-derived bioactive peptides inhibit ACE activity, decreasing blood pressure.
- These peptides improve bodily functions and human health, without adverse effects.
- Marine organisms are sustainable sources of ACE inhibitory peptides.
- Various methods for their industrial production and testing are available.

Author Contributions: Conceptualization, writing original draft preparation, D.Y.P.; English editing, M.N.G.A. and J.-L.H.; project administration, M.A.A.; supervision J.-L.H., review, editing, read and approve the final manuscript, ALL.

Acknowledgments: We gratefully acknowledge to Scientific Publication and Journal Development Center (PPJPI) Universitas Airlangga for English proofreading and Department of Biological Science and Technology, NPUST, Taiwan for supporting this work.

References

1. World Health Organization. World Health Day 2013. *Glob. Brief Hypertens.* **2013**, 9–11.
2. Ferreira-Santos, P.; Carrón, R.; Recio, I.; Sevilla, M.Á.; Montero, M.J. Effects of milk casein hydrolyzate supplemented with phytosterols on hypertension and lipid profile in hypercholesterolemic hypertensive rats. *J. Funct. Foods* **2017**, *28*, 168–176. [CrossRef]
3. Volpe, M.; Battistoni, A.; Chin, D.; Rubattu, S.D.; Tocci, G. Renin as a biomarker of cardiovascular disease in clinical practice. *Nutr. Metab. Cardiovasc. Dis.* **2012**, *22*, 312–317. [CrossRef]
4. Meng, Q.C.; Oparil, S. Purification and assay methods for angiotensin-converting enzyme. *J. Chromatogr. A* **1996**, *743*, 105–122. [CrossRef]
5. Bhullar, K.S.; Lassalle-Claux, G.; Touaibia, M.; Rupasinghe, H.V. Antihypertensive effect of caffeic acid and its analogs through dual renin–angiotensin–aldosterone system inhibition. *Eur. J. Pharmacol.* **2014**, *730*, 125–132. [CrossRef]
6. Guang, C.; Phillips, R.D.; Jiang, B.; Milani, F. Three key proteases–angiotensin-I-converting enzyme (ACE), ACE2 and renin–within and beyond the renin-angiotensin system. *Arch. Cardiovasc. Dis.* **2012**, *105*, 373–385. [CrossRef]

7. Lee, S.-H.; Qian, Z.-J.; Kim, S.-K. A novel angiotensin I converting enzyme inhibitory peptide from tuna frame protein hydrolysate and its antihypertensive effect in spontaneously hypertensive rats. *Food Chem.* **2010**, *118*, 96–102. [CrossRef]

8. Shahidi, F.; Zhong, Y. Bioactive Peptides. *J. AOAC Int.* **2008**, *91*, 914–931.

9. Bougatef, A.; Nedjar-Arroume, N.; Ravallec-Plé, R.; Leroy, Y.; Guillochon, D.; Barkia, A.; Nasri, M. Angiotensin I-converting enzyme (ACE) inhibitory activities of sardinelle (Sardinella aurita) by-products protein hydrolysates obtained by treatment with microbial and visceral fish serine proteases. *Food Chem.* **2008**, *111*, 350–356. [CrossRef]

10. Chen, J.; Wang, Y.; Zhong, Q.; Wu, Y.; Xia, W. Purification and characterization of a novel angiotensin-I converting enzyme (ACE) inhibitory peptide derived from enzymatic hydrolysate of grass carp protein. *Peptides* **2012**, *33*, 52–58. [CrossRef]

11. Spyroulias, G.; Galanis, A.; Pairas, G.; Manessi-Zoupa, E.; Cordopatis, P. Structural Features of Angiotensin-I Converting Enzyme Catalytic Sites: Conformational Studies in Solution, Homology Models and Comparison with Other Zinc Metallopeptidases. *Curr. Top. Med. Chem.* **2004**, *4*, 403–429. [CrossRef]

12. Lee, J.K.; Hong, S.; Jeon, J.-K.; Kim, S.-K.; Byun, H.-G. Purification and characterization of angiotensin I converting enzyme inhibitory peptides from the rotifer, Brachionus rotundiformis. *Bioresour. Technol.* **2009**, *100*, 5255–5259. [CrossRef]

13. Udenigwe, C.C. Bioinformatics approaches, prospects and challenges of food bioactive peptide research. *Trends Food Sci. Technol.* **2014**, *36*, 137–143. [CrossRef]

14. Möller, N.P.; Scholz-Ahrens, K.E.; Roos, N.; Schrezenmeir, J. Bioactive peptides and proteins from foods: Indication for health effects. *Eur. J. Nutr.* **2008**, *47*, 171–182. [CrossRef]

15. Li, G.-H.; Le, G.-W.; Shi, Y.-H.; Shrestha, S. Angiotensin I–converting enzyme inhibitory peptides derived from food proteins and their physiological and pharmacological effects. *Nutr. Res.* **2004**, *24*, 469–486. [CrossRef]

16. Wilson, J.; Hayes, M.; Carney, B. Angiotensin-I-converting enzyme and prolyl endopeptidase inhibitory peptides from natural sources with a focus on marine processing by-products. *Food Chem.* **2011**, *129*, 235–244. [CrossRef]

17. Li, Y.; Sadiq, F.A.; Liu, T.; Chen, J.; He, G. Purification and identification of novel peptides with inhibitory effect against angiotensin I-converting enzyme and optimization of process conditions in milk fermented with the yeast Kluyveromyces marxianus. *J. Funct. Foods* **2015**, *16*, 278–288. [CrossRef]

18. Yu, Y.; Hu, J.; Miyaguchi, Y.; Bai, X.; Du, Y.; Lin, B. Isolation and characterization of angiotensin I-converting enzyme inhibitory peptides derived from porcine hemoglobin. *Peptides* **2006**, *27*, 2950–2956. [CrossRef]

19. Chen, Y.; Wang, Z.; Chen, X.; Liu, Y.; Zhang, H.; Sun, T. Identification of angiotensin I-converting enzyme inhibitory peptides from koumiss, a traditional fermented mare's milk. *J. Dairy Sci.* **2010**, *93*, 884–892. [CrossRef]

20. Yang, Y.; Tao, G.; Liu, P.; Liu, J. Peptide with Angiotensin I-Converting Enzyme Inhibitory Activity from Hydrolyzed Corn Gluten Meal. *J. Agric. Food Chem.* **2007**, *55*, 7891–7895. [CrossRef]

21. Suh, H.J.; Whang, J.H.; Kim, Y.S.; Bae, S.H.; Noh, D.O. Preparation od angiotensin I converting enzyme inhibitor from corn gluten. *Process Biochem.* **2003**, *38*, 1239–1244. [CrossRef]

22. Huang, Q.; Li, S.-G.; Teng, H.; Jin, Y.-G.; Ma, M.-H.; Song, H.-B. Optimizing preparation conditions for Angiotensin-I-converting enzyme inhibitory peptides derived from enzymatic hydrolysates of ovalbumin. *Food Sci. Biotechnol.* **2015**, *24*, 2193–2198. [CrossRef]

23. Zhang, Y.; Pechan, T.; Chang, S.K. Antioxidant and angiotensin-I converting enzyme inhibitory activities of phenolic extracts and fractions derived from three phenolic-rich legume varieties. *J. Funct. Foods* **2018**, *42*, 289–297. [CrossRef]

24. Boschin, G.; Scigliuolo, G.M.; Resta, D.; Arnoldi, A. ACE-inhibitory activity of enzymatic protein hydrolysates from lupin and other legumes. *Food Chem.* **2014**, *145*, 34–40. [CrossRef]

25. Pujiastuti, D.Y.; Shih, Y.-H.; Chen, W.-L.; Hsu, J.-L. Screening of angiotensin-I converting enzyme inhibitory peptides derived from soft-shelled turtle yolk using two orthogonal bioassay-guided fractionations. *J. Funct. Foods* **2017**, *28*, 36–47. [CrossRef]

26. Rawendra, R.D.; Chang, C.-I.; Chen, H.-H.; Huang, T.-C.; Hsu, J.-L. A novel angiotensin converting enzyme inhibitory peptide derived from proteolytic digest of Chinese soft-shelled turtle egg white proteins. *J. Proteom.* **2013**, *94*, 359–369. [CrossRef]

27. Priyanto, A.D.; Doerksen, R.J.; Chang, C.-I.; Sung, W.-C.; Widjanarko, S.B.; Kusnadi, J.; Lin, Y.-C.; Wang, T.-C.; Hsu, J.-L. Screening, discovery, and characterization of angiotensin-I converting enzyme inhibitory peptides derived from proteolytic hydrolysate of bitter melon seed proteins. *J. Proteom.* **2015**, *128*, 424–435. [CrossRef]

28. Lu, Y.; Govindasamy-Lucey, S.; Lucey, J.A. Angiotensin-I-converting enzyme-inhibitory peptides in commercial Wisconsin Cheddar cheeses of different ages. *J. Dairy Sci.* **2016**, *99*, 41–52. [CrossRef]

29. Sieber, R.; Bütikofer, U.; Egger, C.; Portmann, R.; Walther, B.; Wechsler, D. ACE-inhibitory activity and ACE-inhibiting peptides in different cheese varieties. *Dairy Sci. Technol.* **2010**, *90*, 47–73. [CrossRef]

30. Majumder, K.; Chakrabarti, S.; Morton, J.S.; Panahi, S.; Kaufman, S.; Davidge, S.T.; Wu, J. Egg-derived ACE-inhibitory peptides IQW and LKP reduce blood pressure in spontaneously hypertensive rats. *J. Funct. Foods* **2015**, *13*, 50–60. [CrossRef]

31. Miguel, M.; Alonso, M.J.; Salaices, M.; Aleixandre, A.; López-Fandiño, R. Antihypertensive, ACE-inhibitory and vasodilator properties of an egg white hydrolysate: Effect of a simulated intestinal digestion. *Food Chem.* **2007**, *104*, 163–168. [CrossRef]

32. Yoshii, H.; Tachi, N.; Ohba, R.; Sakamura, O.; Takeyama, H.; Itani, T. Antihypertensive effect of ACE inhibitory oligopeptides from chicken egg yolks. *Comp. Biochem. Physiol. Part C Toxicol. Pharmacol.* **2001**, *128*, 27–33. [CrossRef]

33. Yu, Z.; Liu, B.; Zhao, W.; Yin, Y.; Liu, J.; Chen, F. Primary and secondary structure of novel ACE-inhibitory peptides from egg white protein. *Food Chem.* **2012**, *133*, 315–322. [CrossRef] [PubMed]

34. Jiang, Z.; Wang, L.; Che, H.; Tian, B. Effects of temperature and pH on angiotensin-I-converting enzyme inhibitory activity and physicochemical properties of bovine casein peptide in aqueous Maillard reaction system. *LWT* **2014**, *59*, 35–42. [CrossRef]

35. Lin, K.; Zhang, L.-W.; Han, X.; Cheng, D.-Y. Novel angiotensin I-converting enzyme inhibitory peptides from protease hydrolysates of Qula casein: Quantitative structure-activity relationship modeling and molecular docking study. *J. Funct. Foods* **2017**, *32*, 266–277. [CrossRef]

36. Yamada, A.; Sakurai, T.; Ochi, D.; Mitsuyama, E.; Yamauchi, K.; Abe, F. Novel angiotensin I-converting enzyme inhibitory peptide derived from bovine casein. *Food Chem.* **2013**, *141*, 3781–3789. [CrossRef] [PubMed]

37. Hayes, M.; Mora, L.; Hussey, K.; Aluko, R.E.; Soler, L.M. Boarfish protein recovery using the pH-shift process and generation of protein hydrolysates with ACE-I and antihypertensive bioactivities in spontaneously hypertensive rats. *Innov. Food Sci. Emerg. Technol.* **2016**, *37*, 253–260. [CrossRef]

38. Itou, K.; Akahane, Y. Antihypertensive effect of heshiko, a fermented mackerel product, on spontaneously hypertensive rats. *Fish. Sci.* **2004**, *70*, 1121–1129. [CrossRef]

39. Neves, A.C.; Harnedy, P.A.; O'Keeffe, M.B.; Fitzgerald, R.J. Bioactive peptides from Atlantic salmon (Salmo salar) with angiotensin converting enzyme and dipeptidyl peptidase IV inhibitory, and antioxidant activities. *Food Chem.* **2017**, *218*, 396–405. [CrossRef] [PubMed]

40. Suetsuna, K.; Chen, J.-R. Identification of Antihypertensive Peptides from Peptic Digest of Two Microalgae, Chlorella vulgaris and Spirulina platensis. *Mar. Biotechnol.* **2001**, *3*, 305–309. [CrossRef] [PubMed]

41. Suetsuna, K.; Maekawa, K.; Chen, J.-R. Antihypertensive effects of Undaria pinnatifida (wakame) peptide on blood pressure in spontaneously hypertensive rats. *J. Nutr. Biochem.* **2004**, *15*, 267–272. [CrossRef] [PubMed]

42. Aneiros, A.; Garateix, A. Bioactive peptides from marine sources: Pharmacological properties and isolation procedures. *J. Chromatogr. B* **2004**, *803*, 41–53. [CrossRef] [PubMed]

43. Suleria, H.A.R.; Gobe, G.; Masci, P.; Osborne, S.A. Marine bioactive compounds and health promoting perspectives; innovation pathways for drug discovery. *Trends Food Sci. Technol.* **2016**, *50*, 44–55. [CrossRef]

44. Yasuhara-Bell, J.; Lu, Y. Marine compounds and their antiviral activities. *Antivir. Res.* **2010**, *86*, 231–240. [CrossRef] [PubMed]

45. Harnedy, P.A.; Fitzgerald, R.J. Bioactive peptides from marine processing waste and shellfish: A review. *J. Funct. Foods* **2012**, *4*, 6–24. [CrossRef]

46. Ryan, J.T.; Ross, R.P.; Bolton, D.; Fitzgerald, G.F.; Stanton, C. Bioactive Peptides from Muscle Sources: Meat and Fish. *Nutrients* **2011**, *3*, 765–791. [CrossRef]

47. Ngo, D.-H.; Ryu, B.; Kim, S.-K. Active peptides from skate (Okamejei kenojei) skin gelatin diminish angiotensin-I converting enzyme activity and intracellular free radical-mediated oxidation. *Food Chem.* **2014**, *143*, 246–255. [CrossRef]

48. Martins, A.; Vieira, H.M.; Gaspar, H.; Santos, S. Marketed Marine Natural Products in the Pharmaceutical and Cosmeceutical Industries: Tips for Success. *Mar. Drugs* **2014**, *12*, 1066–1101. [CrossRef]

49. Ni, H.; Li, L.; Liu, G.; Hu, S.-Q. Inhibition Mechanism and Model of an Angiotensin I-Converting Enzyme (ACE)-Inhibitory Hexapeptide from Yeast (Saccharomyces cerevisiae). *PLoS ONE* **2012**, *7*, e37077. [CrossRef]

50. Shi, L.; Mao, C.; Xu, Z.; Zhang, L. Angiotensin-converting enzymes and drug discovery in cardiovascular diseases. *Drug Discov. Today* **2010**, *15*, 332–341. [CrossRef]

51. Gavras, H. Angiotensin converting enzyme inhibition and its impact on cardiovascular disease. *Circulation* **1990**, *81*, 381–388. [CrossRef] [PubMed]

52. Wijesekara, I.; Kim, S.-K. Angiotensin-I-Converting Enzyme (ACE) Inhibitors from Marine Resources: Prospects in the Pharmaceutical Industry. *Mar. Drugs* **2010**, *8*, 1080–1093. [CrossRef] [PubMed]

53. Wijesinghe, W.A.J.P.; Ko, S.C.; Jeon, Y.J. Effect of phlorotannins isolated from Ecklonia cava on angiotensin I-converting enzyme (ACE) inhibitory activity. *Nutr. Res. Pract.* **2011**, *5*, 93–100. [CrossRef] [PubMed]

54. Wu, H.; He, H.-L.; Chen, X.-L.; Sun, C.-Y.; Zhang, Y.-Z.; Zhou, B.-C. Purification and identification of novel angiotensin-I-converting enzyme inhibitory peptides from shark meat hydrolysate. *Process Biochem.* **2008**, *43*, 457–461. [CrossRef]

55. Byun, H.-G.; Kim, S.-K. Purification and characterization of angiotensin I converting enzyme (ACE) inhibitory peptides from Alaska pollack (Theragra chalcogramma) skin. *Process Biochem.* **2001**, *36*, 1155–1162. [CrossRef]

56. Wang, Y.-K.; He, H.-L.; Chen, X.-L.; Sun, C.-Y.; Zhang, Y.-Z.; Zhou, B.-C. Production of novel angiotensin I-converting enzyme inhibitory peptides by fermentation of marine shrimp Acetes chinensis with Lactobacillus fermentum SM 605. *Appl. Microbiol. Biotechnol.* **2008**, *79*, 785–791. [CrossRef]

57. Ono, S.; Hosokawa, M.; Miyashita, K.; Takahashi, K. Inhibition properties of dipeptides from salmon muscle hydrolysate on angiotensin I-converting enzyme. *Int. J. Food Sci. Technol.* **2006**, *41*, 383–386. [CrossRef]

58. Liu, J.; Yu, Z.; Zhao, W.; Lin, S.; Wang, E.; Zhang, Y.; Hao, H.; Wang, Z.; Chen, F. Isolation and identification of angiotensin-converting enzyme inhibitory peptides from egg white protein hydrolysates. *Food Chem.* **2010**, *122*, 1159–1163. [CrossRef]

59. Wu, J.; Aluko, R.E.; Nakai, S. Structural Requirements of Angiotensin I-Converting Enzyme Inhibitory Peptides: Quantitative Structure–Activity Relationship Study of Di- and Tripeptides. *J. Agric. Food Chem.* **2006**, *54*, 732–738. [CrossRef]

60. Robert, M.-C.; Razaname, A.; Mutter, M.; Juillerat, M.A. Identification of Angiotensin-I-Converting Enzyme Inhibitory Peptides Derived from Sodium Caseinate Hydrolysates Produced byLactobacillus helveticusNCC 2765. *J. Agric. Food Chem.* **2004**, *52*, 6923–6931. [CrossRef]

61. Saito, T.; Nakamura, T.; Kitazawa, H.; Kawai, Y.; Itoh, T. Isolation and Structural Analysis of Antihypertensive Peptides That Exist Naturally in Gouda Cheese. *J. Dairy Sci.* **2000**, *83*, 1434–1440. [CrossRef]

62. Fahmi, A.; Morimura, S.; Guo, H.; Shigematsu, T.; Kida, K.; Uemura, Y. Production of angiotensin I converting enzyme inhibitory peptides from sea bream scales. *Process Biochem.* **2004**, *39*, 1195–1200. [CrossRef]

63. Lan, X.; Liao, D.; Wu, S.; Wang, F.; Sun, J.; Tong, Z.; Wu, S. Rapid purification and characterization of angiotensin converting enzyme inhibitory peptides from lizard fish protein hydrolysates with magnetic affinity separation. *Food Chem.* **2015**, *182*, 136–142. [CrossRef] [PubMed]

64. Wu, S.; Feng, X.; Lan, X.; Xu, Y.; Liao, D. Purification and identification of Angiotensin-I Converting Enzyme (ACE) inhibitory peptide from lizard fish (Saurida elongata) hydrolysate. *J. Funct. Foods* **2015**, *13*, 295–299. [CrossRef]

65. Gu, R.-Z.; Li, C.-Y.; Liu, W.-Y.; Yi, W.-X.; Cai, M.-Y. Angiotensin I-converting enzyme inhibitory activity of low-molecular-weight peptides from Atlantic salmon (*Salmo salar* L.) skin. *Food Res. Int.* **2011**, *44*, 1536–1540. [CrossRef]

66. Intarasirisawat, R.; Benjakul, S.; Wu, J.; Visessanguan, W. Isolation of antioxidative and ACE inhibitory peptides from protein hydrolysate of skipjack (Katsuwana pelamis) roe. *J. Funct. Foods* **2013**, *5*, 1854–1862. [CrossRef]

67. Jung, W.-K.; Mendis, E.; Je, J.-Y.; Park, P.-J.; Son, B.W.; Kim, H.C.; Choi, Y.K.; Kim, S.-K. Angiotensin I-converting enzyme inhibitory peptide from yellowfin sole (Limanda aspera) frame protein and its antihypertensive effect in spontaneously hypertensive rats. *Food Chem.* **2006**, *94*, 26–32. [CrossRef]

68. Ngo, D.-H.; Vo, T.-S.; Ryu, B.; Kim, S.-K. Angiotensin-I-converting enzyme (ACE) inhibitory peptides from Pacific cod skin gelatin using ultrafiltration membranes. *Process Biochem.* **2016**, *51*, 1622–1628. [CrossRef]

69. Ko, J.-Y.; Kang, N.; Lee, J.-H.; Kim, J.-S.; Kim, W.-S.; Park, S.-J.; Kim, Y.-T.; Jeon, Y.-J. Angiotensin I-converting enzyme inhibitory peptides from an enzymatic hydrolysate of flounder fish (Paralichthys olivaceus) muscle as a potent anti-hypertensive agent. *Process Biochem.* **2016**, *51*, 535–541. [CrossRef]

70. Ghassem, M.; Arihara, K.; Babji, A.S.; Said, M.; Ibrahim, S. Purification and identification of ACE inhibitory peptides from Haruan (Channa striatus) myofibrillar protein hydrolysate using HPLC–ESI-TOF MS/MS. *Food Chem.* **2011**, *129*, 1770–1777. [CrossRef]

71. Ko, S.-C.; Kang, N.; Kim, E.-A.; Kang, M.C.; Lee, S.-H.; Kang, S.-M.; Lee, J.-B.; Jeon, B.-T.; Kim, S.-K.; Park, S.-J.; et al. A novel angiotensin I-converting enzyme (ACE) inhibitory peptide from a marine Chlorella ellipsoidea and its antihypertensive effect in spontaneously hypertensive rats. *Process Biochem.* **2012**, *47*, 2005–2011. [CrossRef]

72. Zhao, Y.; Li, B.; Dong, S.; Liu, Z.; Zhao, X.; Wang, J.; Zeng, M. A novel ACE inhibitory peptide isolated from Acaudina molpadioidea hydrolysate. *Peptides* **2009**, *30*, 1028–1033. [CrossRef] [PubMed]

73. Balti, R.; Nedjar-Arroume, N.; Bougatef, A.; Guillochon, D.; Nasri, M. Three novel angiotensin I-converting enzyme (ACE) inhibitory peptides from cuttlefish (Sepia officinalis) using digestive proteases. *Food Res. Int.* **2010**, *43*, 1136–1143. [CrossRef]

74. Balti, R.; Bougatef, A.; Sila, A.; Guillochon, D.; Dhulster, P.; Nedjar-Arroume, N. Nine novel angiotensin I-converting enzyme (ACE) inhibitory peptides from cuttlefish (Sepia officinalis) muscle protein hydrolysates and antihypertensive effect of the potent active peptide in spontaneously hypertensive rats. *Food Chem.* **2015**, *170*, 519–525. [CrossRef]

75. Alemán, A.; Gómez-Guillén, M.C.; Montero, P. Identification of ace-inhibitory peptides from squid skin collagen after in vitro gastrointestinal digestion. *Food Res. Int.* **2013**, *54*, 790–795. [CrossRef]

76. Tsai, J.; Lin, T.; Chen, J.; Pan, B. The inhibitory effects of freshwater clam (Corbicula fluminea, Muller) muscle protein hydrolysates on angiotensin I converting enzyme. *Process Biochem.* **2006**, *41*, 2276–2281. [CrossRef]

77. Jao, C.-L.; Huang, S.-L.; Hsu, K.-C. Angiotensin I-converting enzyme inhibitory peptides: Inhibition mode, bioavailability, and antihypertensive effects. *BioMedicine* **2012**, *2*, 130–136. [CrossRef]

78. Ko, S.-C.; Jang, J.; Ye, B.-R.; Kim, M.-S.; Choi, I.-W.; Park, W.-S.; Jung, W.-K. Purification and molecular docking study of angiotensin I-converting enzyme (ACE) inhibitory peptides from hydrolysates of marine sponge Stylotella aurantium. *Process Biochem.* **2016**, *54*, 180–187. [CrossRef]

79. Wu, Q.; Jia, J.; Yan, H.; Du, J.; Gui, Z. A novel angiotensin-I converting enzyme (ACE) inhibitory peptide from gastrointestinal protease hydrolysate of silkworm pupa (Bombyx mori) protein: Biochemical characterization and molecular docking study. *Peptides* **2015**, *68*, 17–24. [CrossRef]

80. Jemil, I.; Mora, L.; Nasri, R.; Abdelhedi, O.; Aristoy, M.-C.; Hajji, M.; Nasri, M.; Toldrá, F.; Soler, L.M. A peptidomic approach for the identification of antioxidant and ACE-inhibitory peptides in sardinelle protein hydrolysates fermented by Bacillus subtilis A26 and Bacillus amyloliquefaciens An6. *Food Res. Int.* **2016**, *89*, 347–358. [CrossRef]

81. Lin, H.-C.; Alashi, A.M.; Aluko, R.E.; Pan, B.S.; Chang, Y.-W. Antihypertensive properties of tilapia (Oreochromis spp.) frame and skin enzymatic protein hydrolysates. *Food Nutr. Res.* **2017**, *61*, 1391666. [CrossRef] [PubMed]

82. Ondetti, M.A.; Cushman, D.W. Enzymes of the renin-angiotensin system and their inhibitors. *Annu. Rev. Biochem.* **1982**, *51*, 283–308. [CrossRef] [PubMed]

83. López-Fandiño, R.; Otte, J.; Van Camp, J. Physiological, chemical and technological aspects of milk-protein-derived peptides with antihypertensive and ACE-inhibitory activity. *Int. Dairy J.* **2006**, *16*, 1277–1293. [CrossRef]

84. Lassoued, I.; Mora, L.; Barkia, A.; Aristoy, M.C.; Nasri, M.; Toldrá, F. Bioactive peptides identified in thornback ray skin's gelatin hydrolysates by proteases from Bacillus subtilis and Bacillus amyloliquefaciens. *J. Proteom.* **2015**, *128*, 8–17. [CrossRef]

85. So, P.B.T.; Rubio, P.; Lirio, S.; Macabeo, A.P.; Huang, H.-Y.; Corpuz, M.J.-A.T.; Villaflores, O.B. In vitro angiotensin I converting enzyme inhibition by a peptide isolated from Chiropsalmus quadrigatus Haeckel (box jellyfish) venom hydrolysate. *Toxicon* **2016**, *119*, 77–83. [CrossRef] [PubMed]

86. Sato, M.; Hosokawa, T.; Yamaguchi, T.; Nakano, T.; Muramoto, K.; Kahara, T.; Funayama, K.; Kobayashi, A.; Nakano, T. Angiotensin I-Converting Enzyme Inhibitory Peptides Derived from Wakame (Undaria pinnatifida) and Their Antihypertensive Effect in Spontaneously Hypertensive Rats. *J. Agric. Food Chem.* **2002**, *50*, 6245–6252. [CrossRef] [PubMed]

87. Nakagomi, K.; Fujimura, A.; Ebisu, H.; Sakai, T.; Sadakane, Y.; Fujii, N.; Tanimura, T. Acein-1, a novel angiotensin-I-converting enzyme inhibitory peptide isolated from tryptic hydrolysate of human plasma. *FEBS Lett.* **1998**, *438*, 255–257. [CrossRef]

88. Fujita, H.; Yoshikawa, M. LKPNM: A prodrug-type ACE-inhibitory peptide derived from fish protein. *Immunopharmacol.* **1999**, *44*, 123–127. [CrossRef]

89. Vercruysse, L.; Van Camp, J.; Morel, N.; Rougé, P.; Herregods, G.; Smagghe, G. Ala-Val-Phe and Val-Phe: ACE inhibitory peptides derived from insect protein with antihypertensive activity in spontaneously hypertensive rats. *Peptides* **2010**, *31*, 482–488. [CrossRef]

90. Shahidi, F.; Ambigaipalan, P. Novel functional food ingredients from marine sources. *Curr. Opin. Food Sci.* **2015**, *2*, 123–129. [CrossRef]

91. Agyei, D.; Ongkudon, C.M.; Wei, C.Y.; Chan, A.S.; Danquah, M.K. Bioprocess challenges to the isolation and purification of bioactive peptides. *Food Bioprod. Process.* **2016**, *98*, 244–256. [CrossRef]

92. Sánchez-Rivera, L.; Martínez-Maqueda, D.; Cruz-Huerta, E.; Miralles, B.; Recio, I. Peptidomics for discovery, bioavailability and monitoring of dairy bioactive peptides. *Food Res. Int.* **2014**, *63*, 170–181. [CrossRef]

93. Jemil, I.; Abdelhedi, O.; Mora, L.; Nasri, R.; Aristoy, M.-C.; Jridi, M.; Hajji, M.; Toldrá, F.; Nasri, M.; Soler, L.M. Peptidomic analysis of bioactive peptides in zebra blenny (Salaria basilisca) muscle protein hydrolysate exhibiting antimicrobial activity obtained by fermentation with Bacillus mojavensis A21. *Process Biochem.* **2016**, *51*, 2186–2197. [CrossRef]

94. Sheih, I.-C.; Wu, T.-K.; Fang, T.J. Antioxidant properties of a new antioxidative peptide from algae protein waste hydrolysate in different oxidation systems. *Bioresour. Technol.* **2009**, *100*, 3419–3425. [CrossRef] [PubMed]

95. Ustyuzhanina, N.E.; Ushakova, N.A.; Zyuzina, K.A.; Bilan, M.I.; Elizarova, A.L.; Somonova, O.V.; Madzhuga, A.V.; Krylov, V.B.; Preobrazhenskaya, M.E.; Usov, A.I.; et al. Influence of Fucoidans on Hemostatic System. *Mar. Drugs* **2013**, *11*, 2444–2458. [CrossRef]

96. Korhonen, H.; Pihlanto, A. Bioactive peptides: Production and functionality. *Int. Dairy J.* **2006**, *16*, 945–960. [CrossRef]

97. Rizzello, C.G.; Tagliazucchi, D.; Babini, E.; Rutella, G.S.; Saa, D.L.T.; Gianotti, A. Bioactive peptides from vegetable food matrices: Research trends and novel biotechnologies for synthesis and recovery. *J. Funct. Foods* **2016**, *27*, 549–569. [CrossRef]

98. Sanjukta, S.; Rai, A.K. Production of bioactive peptides during soybean fermentation and their potential health benefits. *Trends Food Sci. Technol.* **2016**, *50*, 1–10. [CrossRef]

99. Kleekayai, T.; Harnedy, P.A.; O'Keeffe, M.B.; Poyarkov, A.A.; CunhaNeves, A.; Suntornsuk, W.; Fitzgerald, R.J. Extraction of antioxidant and ACE inhibitory peptides from Thai traditional fermented shrimp pastes. *Food Chem.* **2015**, *176*, 441–447. [CrossRef]

100. Cushman, D.; Cheung, H. Spectrophotometric assay and properties of the angiotensin-converting enzyme of rabbit lung. *Biochem. Pharmacol.* **1971**, *20*, 1637–1648. [CrossRef]

101. Li, G.-H.; Liu, H.; Shi, Y.-H.; Le, G.-W. Direct spectrophotometric measurement of angiotensin I-converting enzyme inhibitory activity for screening bioactive peptides. *J. Pharm. Biomed. Anal.* **2005**, *37*, 219–224. [CrossRef] [PubMed]

102. Belović, M.M.; Ilić, N.M.; Tepić, A.N.; Šumić, Z. Selection of conditions for angiotensin-converting enzyme inhibition assay: Influence of sample preparation and buffer. *Food Feed Res.* **2013**, *40*, 11–16.

103. Shalaby, S.M.; Zakora, M.; Otte, J. Performance of two commonly used angiotensin-converting enzyme inhibition assays using FA-PGG and HHL as substrates. *J. Dairy Res.* **2006**, *73*, 178–186. [CrossRef] [PubMed]

104. Murray, B.; Walsh, D.; Fitzgerald, R. Modification of the furanacryloyl-l-phenylalanylglycylglycine assay for determination of angiotensin-I-converting enzyme inhibitory activity. *J. Biochem. Biophys. Methods* **2004**, *59*, 127–137. [CrossRef] [PubMed]

105. Sentandreu, M.Á.; Toldrá, F. A rapid, simple and sensitive fluorescence method for the assay of angiotensin-I converting enzyme. *Food Chem.* **2006**, *97*, 546–554. [CrossRef]

106. Van Der Ven, C.; Gruppen, H.; De Bont, D.B.; Voragen, A.G. Optimisation of the angiotensin converting enzyme inhibition by whey protein hydrolysates using response surface methodology. *Int. Dairy J.* **2002**, *12*, 813–820. [CrossRef]

107. Vermeirssen, V.; Van Camp, J.; Verstraete, W. Optimisation and validation of an angiotensin-converting enzyme inhibition assay for the screening of bioactive peptides. *J. Biochem. Biophys. Methods* **2002**, *51*, 75–87. [CrossRef]

108. Anzenbacherová, E.; Anzenbacher, P.; Macek, K.; Květina, J. Determination of enzyme (angiotensin convertase) inhibitors based on enzymatic reaction followed by HPLC. *J. Pharm. Biomed. Anal.* **2001**, *24*, 1151–1156. [CrossRef]

109. Sentandreu, M.A.; Toldrá, F. A fluorescence-based protocol for quantifying angiotensin-converting enzyme activity. *Nat. Protoc.* **2006**, *1*, 2423–2427. [CrossRef]

110. Betancourt, L.H.; De Bock, P.-J.; Staes, A.; Timmerman, E.; Perez-Riverol, Y.; Sánchez, A.; Besada, V.; González, L.J.; Vandekerckhove, J.; Gevaert, K. SCX charge state selective separation of tryptic peptides combined with 2D-RP-HPLC allows for detailed proteome mapping. *J. Proteom.* **2013**, *91*, 164–171. [CrossRef]

111. Geng, F.; He, Y.; Yang, L.; Wang, Z. A rapid assay for angiotensin-converting enzyme activity using ultra-performance liquid chromatography-mass spectrometry. *Biomed. Chromatogr. BMC* **2010**, *24*, 312–317. [CrossRef] [PubMed]

112. Hernández-Ledesma, B.; Contreras, M.D.M.; Recio, I. Antihypertensive peptides: Production, bioavailability and incorporation into foods. *Adv. Colloid Interface Sci.* **2011**, *165*, 23–35. [CrossRef] [PubMed]

113. Wu, J.; Ding, X. Characterization of inhibition and stability of soy-protein-derived angiotensin I-converting enzyme inhibitory peptides. *Food Res. Int.* **2002**, *35*, 367–375. [CrossRef]

114. Vermeirssen, V.; Van Camp, J.; Verstraete, W. Bioavailability of angiotensin I converting enzyme inhibitory peptides. *Br. J. Nutr.* **2004**, *92*, 357–366. [CrossRef] [PubMed]

115. Meng, X.-Y.; Zhang, H.-X.; Mezei, M.; Cui, M. Molecular Docking: A Powerful Approach for Structure-Based Drug Discovery. *Curr. Comput. Drug Des.* **2011**, *7*, 146–157. [CrossRef]

116. Politi, A.; Durdagi, S.; Moutevelis-Minakakis, P.; Kokotos, G.; Mavromoustakos, T. Development of accurate binding affinity predictions of novel renin inhibitors through molecular docking studies. *J. Mol. Graph. Model.* **2010**, *29*, 425–435. [CrossRef]

117. Panjaitan, F.C.A.; Gomez, H.L.R.; Chang, Y.-W. In Silico Analysis of Bioactive Peptides Released from Giant Grouper (Epinephelus lanceolatus) Roe Proteins Identified by Proteomics Approach. *Molecules* **2018**, *23*, 2910. [CrossRef]

118. Tejano, L.A.; Peralta, J.P.; Yap, E.E.S.; Panjaitan, F.C.A.; Chang, Y.-W. Prediction of Bioactive Peptides from Chlorella sorokiniana Proteins Using Proteomic Techniques in Combination with Bioinformatics Analyses. *Int. J. Mol. Sci.* **2019**, *20*, 1786. [CrossRef]

119. Pan, D.; Guo, H.; Zhao, B.; Cao, J. The molecular mechanisms of interactions between bioactive peptides and angiotensin-converting enzyme. *Bioorganic Med. Chem. Lett.* **2011**, *21*, 3898–3904. [CrossRef]

120. Andrews, P.R.; Carson, J.M.; Caselli, A.; Spark, M.J.; Woods, R. Conformational analysis and active site modeling of angiotensin-converting enzyme inhibitors. *J. Med. Chem.* **1985**, *28*, 393–399. [CrossRef]

3

Enzyme Properties of a Laccase Obtained from the Transcriptome of the Marine-Derived Fungus *Stemphylium lucomagnoense*

Wissal Ben Ali [1,2,*][ID], Amal Ben Ayed [1,2], Annick Turbé-Doan [1], Emmanuel Bertrand [1][ID], Yann Mathieu [3][ID], Craig B. Faulds [1], Anne Lomascolo [1][ID], Giuliano Sciara [1], Eric Record [1][ID] and Tahar Mechichi [2][ID]

[1] Biodiversité et Biotechnologie Fongiques, Aix-Marseille Université, INRAE, UMR1163 Marseille, France; amal.benayed@enis.tn (A.B.A.); annick.doan@inrae.fr (A.T.-D.); Emmanuel.Bertrand@Univ-Amu.Fr (E.B.); Craig.Faulds@Univ-Amu.Fr (C.B.F.); anne.lomascolo@univ-amu.fr (A.L.); giuliano.sciara@inrae.fr (G.S.); eric.record@inrae.fr (E.R.)

[2] Laboratoire de Biochimie et de Génie Enzymatique des Lipases, Ecole Nationale d'Ingénieurs de Sfax, Université de Sfax, Sfax 3029, Tunisia; tahar.mechichi@enis.rnu.tn

[3] Michael Smith Laboratories, University of British Columbia, Vancouver, BC V6T 1Z4, Canada; yann.mathieu85@gmail.com

* Correspondence: wissal.BEN-ALI@etu.univ-amu.fr

Abstract: Only a few studies have examined how marine-derived fungi and their enzymes adapt to salinity and plant biomass degradation. This work concerns the production and characterisation of an oxidative enzyme identified from the transcriptome of marine-derived fungus *Stemphylium lucomagnoense*. The laccase-encoding gene *Sl*Lac2 from *S. lucomagnoense* was cloned for heterologous expression in *Aspergillus niger* D15#26 for protein production in the extracellular medium of around 30 mg L^{-1}. The extracellular recombinant enzyme *Sl*Lac2 was successfully produced and purified in three steps protocol: ultrafiltration, anion-exchange chromatography, and size exclusion chromatography, with a final recovery yield of 24%. *Sl*Lac2 was characterised by physicochemical properties, kinetic parameters, and ability to oxidise diverse phenolic substrates. We also studied its activity in the presence and absence of sea salt. The molecular mass of *Sl*Lac2 was about 75 kDa, consistent with that of most ascomycete fungal laccases. With syringaldazine as substrate, *Sl*Lac2 showed an optimal activity at pH 6 and retained nearly 100% of its activity when incubated at 50°C for 180 min. *Sl*Lac2 exhibited more than 50% of its activity with 5% wt/vol of sea salt.

Keywords: laccase; *Stemphylium*; heterologous expression; enzyme properties; alkaline; salt tolerance

1. Introduction

The laccases (benzenediol:oxygen oxidoreductase, EC 1.10.3.2) belong to a small group of enzymes called the blue copper protein or copper oxidases [1] that catalyse the one-electron oxidation of four reducing-substrate molecules concomitant with the four-electron reduction of molecular oxygen to water [2]. Laccases are almost ubiquitous enzymes, widely distributed among plants, fungi, prokaryotes, and arthropods [3]. Before 2011, more than 100 laccases from Basidiomycota and Ascomycota were purified and characterised [4]. They are characterised by a molecular mass and an isoelectric point (pI) that range from about 50 to 100 kDa and 3 to 7, respectively [5,6]. Laccases have active sites containing four copper atoms (Cu) bound to three redox sites (Type 1, Type 2, and Type 3 Cu pair) involved in the catalytic mechanisms of these cuproteins [7,8]. Christopher et al. [9] identified three types of copper using UV/visible and electronic paramagnetic resonance (EPR) spectroscopy. Type 1 Cu at its oxidised resting state is responsible for the blue colour of the protein and is EPR-detectable. Type 2 Cu is only EPR-detectable [9]. Type 3 Cu is composed of a pair of Cu atoms in a binuclear conformation, which is not detectable by EPR. The Type 2 and Type 3 copper sites form a trinuclear centre directly involved in the enzyme catalytic mechanism [9]. Laccases exhibit broad substrate ranges that vary from one laccase to another [10]. They are known as *p*-diphenol:oxygen oxidoreductases, preferentially oxidising monophenols such as 2,6-dimethoxyphenol or guaiacol. Laccase-catalysed reactions are strongly dependent on redox potential, temperature and reaction medium [11]. Laccases fall into two groups depending on their redox potential. Low-redox-potential enzymes occur in bacteria and plants, and high-redox-potential laccases are widely distributed in fungi [12,13]. The low-redox laccases (400–600 mV), unlike high-redox potential laccases (700–800 mV) and ligninolytic peroxidases (>1 V), enable only direct degradation of phenolic compounds of low redox potential and not oxidation of more recalcitrant aromatic compounds, such as some industrial dyes [14]. In the presence of a suitable redox mediator, for example 1-hydroxybenzotriazole (HBT), laccases are able to oxidise non-phenolic structures [15,16]. Many industrial applications for laccases have been proposed, including in pulp and paper [17], organic synthesis, environment, food, pharmaceuticals, and textile dye decolourisation [18].

Synthetic dyes are chemicals widely used in many industries including textiles, paper, printing, cosmetics, and pharmaceuticals [19]. The discharge of their effluents into water systems is a serious environmental concern [20,21]. Enzymatic decolourisation of dyes has shown several advantages over the physico-chemical methods, including low energy costs, ease of control, and eco-friendly impact on ecosystems [22]. However, biological treatments of dye wastewater usually involve an environment with high salinity and numerous organic solvents. Most laccases lose activity under these extreme conditions [23]. It is therefore important to identify new laccases with high tolerance to salt, organic solvents, and high temperature [24].

According to Bonugli-Santos et al. [25], studies of laccases from marine-derived fungi are still limited, and these enzymes may have different properties from those produced by terrestrial microorganisms, owing to different environmental conditions, such as salinity, temperature, pH and pressure. *Stemphylium lucomagnoense* was previously selected by screening fungi isolated from Tunisian coastal waters for its capacity to exert a laccase-like activity under saline conditions [26]. In addition, its secretome analysis confirmed the presence of laccase-like activities in the presence of sea salts [27]. In this study, we expressed a laccase encoding gene from *S. lucomagnoense* in *Aspergillus niger* and purified and characterised the corresponding protein to study the main properties of this marine-derived enzyme.

2. Results

2.1. Target Selection, Aspergillus Niger Transformation, and Screening

Among the seven putative full-length laccase-encoding genes annotated from the transcriptome of *S. lucomagnoense* [27] (NCBI BioSample accession ID:191 SAMN15897915 with corresponding NCBI BioProject accession ID: PRJNA659110, assembled contig sequences available at GitHub URL as 203 https://github.com/drabhishekkumar/Stemphylium-lucomagnoense-transcriptomics), one was selected for heterologous expression in *A. niger*, based on the divergence (i) of the predicted protein sequence and (ii) of the predicted pI compared with those of known laccases, and on (iii) the presence of the coordination site for Type 1, 2, and 3 coppers, as found in already-characterised laccases (Table 1 and Figure S1). The full-length gene encoding *Sl*Lac2 consisting of 1839 bp (613 amino acids including 21 amino acid for the signal peptide) was selected for the following reasons. *Sl*Lac2 shared only 44.1, 44.8, and 44.6% identity with the closest laccases *Melanocarpus albomyces* (*Ma*Lac), and *Pestalotiopsis* sp. *Ps*Lac1 and *Ps*Lac2, respectively. The calculated molecular weight was 64.6 kDa with a theoretical pI of 8.07, 3.4 to 2 points higher than the lowest and highest calculated pI related to APMZ2_prot2393 (pI 4.66) and APMZ2_7477 (pI 6.03) proteins. In comparison with the closest characterised laccases, *Ps*Lac1, *Ps*Lac 2, and *Ma*Lac have a calculated pI of 6.17, 4.13, and 5.13, respectively (accession numbers

KY55480, KY554801, and Q70KY3, respectively). The theoretical extinction coefficients of SlLac2 at 280 nm was 111,645 M^{-1} cm^{-1}.

Table 1. Properties of the putative laccases of *Stemphylium lucomagnoense* and the closest characterised laccases from the terrestrial fungus, *Melanocarpus albomyces* (*Ma*Lac1: Q70KY3), and the marine-derived fungus *Pestalotiopsis* sp. KF079 (*Ps*Lac1: KY554800 and *Ps*Lac 2: KY554801), including the number of amino acids of the signal peptide and the mature protein, the calculated molecular mass and pI, and the amino acids involved in the enzyme coordination sites for the Type 1, 2 and 3 coppers. APMZ2_prot14771 * encoding SlLac2 was deposited in the nucleotide sequence database (GenBank) under the accession code MT470191.

Accession Number	Signal Peptide (Amino Acids)	Mature Protein (Amino Acids)	Molecular Mass (Da)	PI	Coordination Sites	
Q70KY3	22	601	68958	5.18	IHWHG HPMHLH	WYHSH HCHIAWH
KY554800	21	555	60904	6.09	IHWHG HPMHLH	WYHSH HCHIAWH
KY554801	18	543	59813	4.10	IHWHG HPMHLH	WYHSH HCHIAWH
APMZ2_prot14771 *	21	592	64581	8.07	IHWHG HPIHLH	WYHSH HCHIAFH
APMZ2_prot8345	23	578	63456	5.64	IHWHG HPIHLH	WYHSH HCHIAWH
APMZ2_prot7177	21	615	68221	6.03	IHFHG HPIHKH	WYHSH HCHINNH
APMZ2_prot10330	No	754	83453	5.40	IHFHG HPFHLH	WYHAH HCHNMWH
APMZ2_prot15523	No	729	80156	5.37	LHAHG HPMHLH	WYHSH HCHLAWH
APMZ2_prot2393	No	554	60438	4.66	LHFHG HPFHLH	WYHSH HCHIEWH
APMZ2_prot9198	19	576	64532	5.08	MHWHG HPFHLH	FYHSH HCHVLQH

To produce the recombinant SlLac2, protoplasts from *A. niger* D15#26 were co-transformed with a mixture of plasmid pAB4-1 and an expression vector containing the corresponding gene. Transformants were selected for their ability to grow without uridine supplementation, and co-transformants containing the laccase-encoding cDNA were screened for laccase expression by growth on minimum medium plates supplemented with 2,2'-azino-bis (3-ethylbenzothiazoline-6-sulphonic acid) (ABTS). Laccase-producing transformants were identified by the appearance of a green zone around the colonies after 3 to 6 days of culture. Coloured zones on plates were not observed for control transformants lacking the laccase-encoding cDNA (transformed only with pAB4-1). A total of 27 positive clones were cultured in standard liquid media and checked daily for protein production by sodium dodecyl sulfate-polyacrylamide gel electrophoresis (SDS-PAGE) and for enzymatic activity by spectrophotometric assays. Approximately 90% of the tested transformants exhibited laccase-like activity in the culture medium and the transformant SlLac2 reached a peak of laccase activity (85.9 nkat mL^{-1}) on Day 9.

2.2. Purification of the Recombinant Laccase and Study of Physico-Chemical Properties

The recombinant laccase was purified from the culture medium of *A. niger* in a three-step procedure (Table 2): ultrafiltration, anion-exchange chromatography, and size exclusion chromatography. 930 mL of the filtered culture media (containing 176 mg of protein) was concentrated (1.9-fold) and separated from most impurities, which included a brown pigment absorbing strongly at 280 nm [28], by ultrafiltration through a polyethersulphone membrane (10 kDa molecular mass cut-off), with a resulting purification factor of 1.4-fold. The resulting concentrate was then loaded onto a carboxymethyl (CM)-Sepharose column to be further purified with a purification factor of 3-fold. In the last step, the resulting sample was concentrated through a 10 kDa molecular mass cut-off Amicon membrane and loaded onto a Sephacryl S-200HR size exclusion chromatography column. The resulting purification factor was 5.9-fold, yielding 7.2 mg of protein. The final recovery yield was 24.1%. The molecular mass of the purified laccase by SDS/PAGE was about 75 kDa (Figure 1A).

Table 2. Purification for the recombinant SlLac2 produced in *Aspergillus niger* D15#26. CM-Sepharose: carboxymethyl-Sepharose.

Purification Steps	Volume (mL)	Total Activity (nKat)	Protein (mg)	Specific Activity (nkat mg^{-1})	Activity Yield (%)	Purification Factor (fold)
Culture medium	930	78,027	176	443	100	1
Ultrafiltration	500	99,000	156	635	126.9	1.4
CM-Sepharose	325	20,775	15.4	1349	26.6	3.0
Sephacryl S-200HR	40	18,782	7.2	2609	24.1	5.9

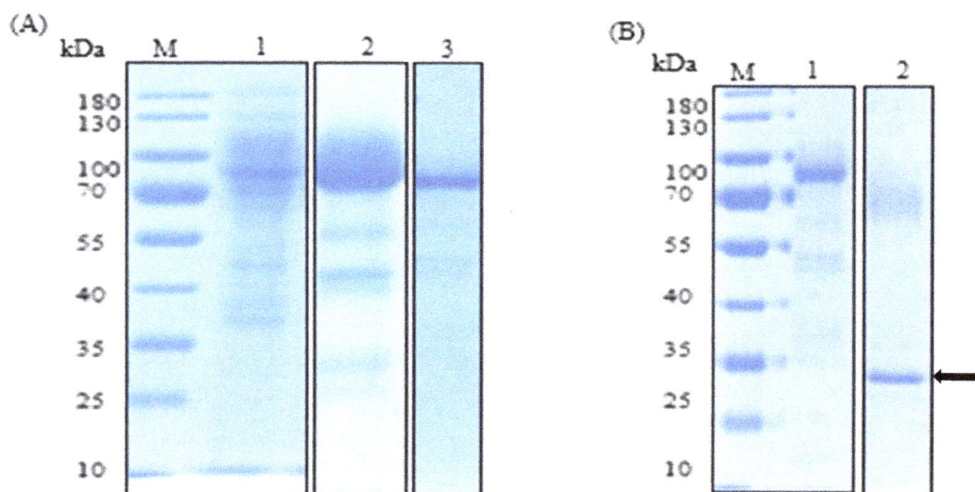

Figure 1. Sodium dodecyl sulfate-polyacrylamide gel electrophoresis (SDS-PAGE) (12% polyacrylamide gels) and N-glycosylation analysis of the purified SlLac2. Three-step purification procedure of SlLac2 (**A**). Lanes 1, culture supernatant concentrated by ultrafiltration; 2, purified SlLac2 after carboxymethyl (CM)-Sepharose column; 3, purified SlLac2 obtained after Sephacryl S-200HR. N-glycosylation analysis by SDS-PAGE of the purified SlLac2 (**B**). Lanes 1, purified SlLac2; 2; purified SlLac2 deglycosylated by N-glycosidase PNGase F (marked with an arrow). M are molecular mass standards. Proteins were stained with Coomassie blue.

To check the N-terminal protein processing, the first five amino acids of the recombinant laccase were sequenced and aligned with the deduced mature protein. They showed 100% identity. This result demonstrated that the 24-amino-acid glucoamylase (GLA) prepropeptide from *A. niger* was correctly processed. For N-glycosylation, five sites (71, 115, 232, 404, and 424) were predicted. In addition, SlLac2 was deglycosylated using N-glycosidase PNGase (Figure 1B). On SDS-PAGE, the deglycosylated protein showed an apparent difference in its molecular mass from the untreated protein. The molecular masses of the purified and digested laccases were 75 kDa and <70 kDa, respectively. The molecular mass of the deglycosylated SlLac2 and the corresponding calculated molecular mass (64.6 kDa) confirmed the presence of N-glycosylations of approximately 10% of the total protein.

2.3. Kinetic Parameters

The kinetic constants (K_M and V_{max}) of SlLac2 were determined using various substrates such as 2,2'-azino-bis (3-ethylbenzothiazoline-6-sulphonic acid) (ABTS), 2,6-dimethoxyphenol (DMP) and syringaldazine under the optimal reaction conditions at 30 °C and pH (4.0, 5.0 and 6.0 respectively) (Table 3). Of the six substrates, SlLac2 exhibited the highest affinity for syringaldazine ($K_M = 0.0035$ mM) followed by ABTS ($K_M = 0.0206$ mμM) and DMP ($K_M = 0.024$ mμM). The highest catalytic efficiency was found for ABTS (7.42 s^{-1} mM^{-1}), followed by syringaldazine (2.11 s^{-1} mM^{-1}) and DMP (0.24 s^{-1} mM^{-1}).

Table 3. Kinetic parameters of *Sl*Lac2.

Substrate		K_M (mM)	K_{cat} (s^{-1})	K_{cat}/K_M (s^{-1} mM^{-1})
ABTS		0.0206 +/− 0.0039	0.153 +/− 0.0026	7.42
DMP		0.0240 +/− 0.0039	0.0059 +/− 0.0003	0.24
Syringaldazine		0.0035 +/− 0.00057	0.0074 +/− 0.0002	2.11

2.4. Enzyme Activity and Stability at Different pH and Temperature Values

*Sl*Lac2 activity increased gradually with increasing temperature up to 60 °C, the optimal temperature value (Figure 2A). Regarding thermal stability, *Sl*Lac2 was stable at temperatures ranging from 25 to 50 °C, but the activity was significantly lost above 60 °C (Figure 2B). The half-life value for *Sl*Lac2 was 95 min and 30 min, at 60 °C and 70 °C, respectively. The highest laccase activity level was recorded at pH 6.0 (Figure 2C). The enzyme possessed significant activity at pH 4.0 and 7.0 (60% and 75%, respectively). A lower activity could be measured at alkaline pH, with 30% of its initial activity at pH 8.0, and the activity almost disappeared above pH 8.0. When incubated at pH 5.0, 6.0 and 7.0 (Figure 2D), laccase activity of the purified enzyme *Sl*Lac2 retained around 80% initial activity after 4 h of incubation, about 70% after 24 h and 60% after 48 h. When incubated at pH 3.0 or 4, the laccase activity retained over 65% residual activity after 4 h, about 50–65% after 24 h and about 50–60% after 48 h. These results demonstrate that the enzyme endures at neutral to weak-acidic pH but has lower resistance to strong-acidic pH values.

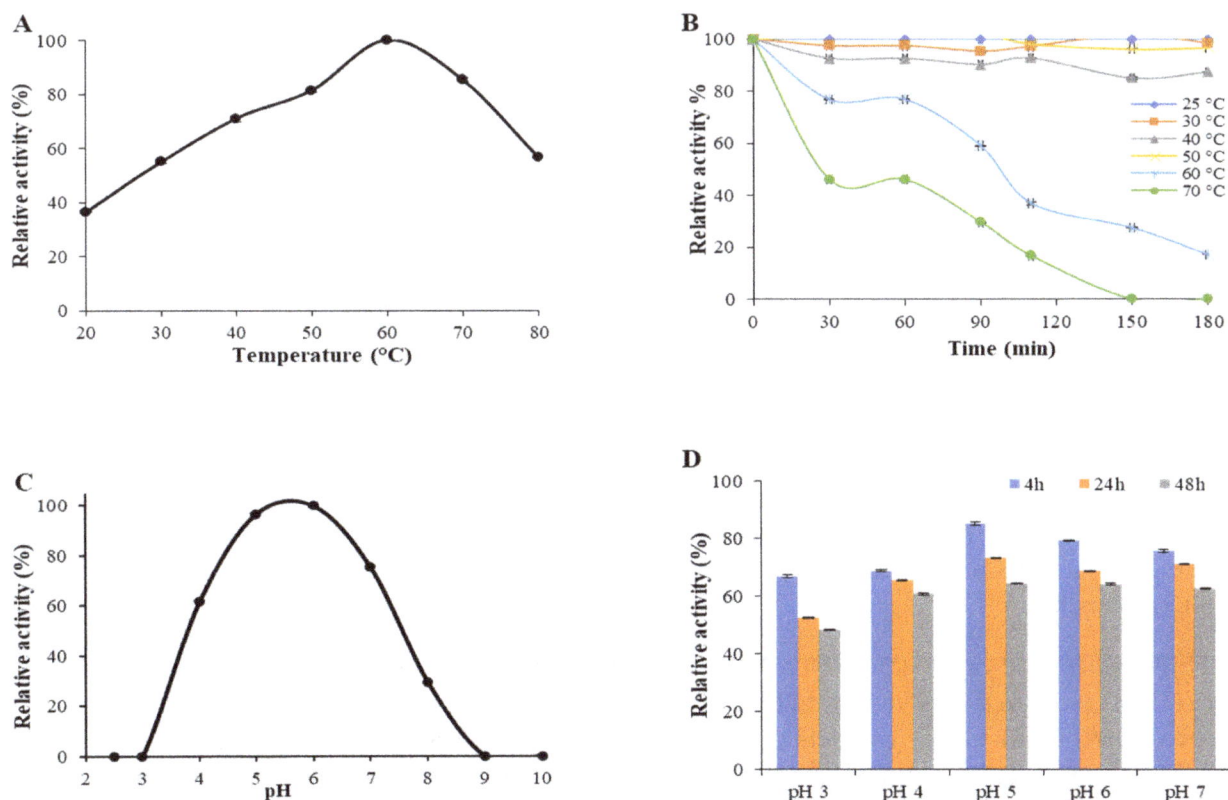

Figure 2. Enzyme activity and stability of *Sl*Lac2 at different pH and temperature. (**A**) The oxidation of syringaldazine (1 mM) was determined for the temperature curve in a sodium acetate buffer (50 mM,

pH 6.0), and (**C**) for the pH curve at 30 °C. Values were calculated as a percentage of maximum activity (set at 100%) at optimum temperature and pH. (**B**) Residual activities were estimated after 30, 60, 90, 120, and 180 min of incubation at six different temperatures ranging from 25 to 70 °C or (**D**) after 4, 24, and 48 h at five different pH values ranging from 3.0 to 7.0. Residual activities are expressed as a percentage of the initial activity (point at time 0, measured immediately after adding the enzyme), which was set at 100%. Assays were performed using syringaldazine as a substrate in standard conditions. Each data point (mean +/− standard deviation) is the result of triplicate experiments.

2.5. Effect of Metal Ions, Inhibitors and Solvents on Laccase Activity

Various metal ions (Mo^{2+}, Ag^{2+}, AsO_4^{3-}, Fe^{2+}, Cd^{2+}, Zn^{2+}, and Cu^{2+}) at 10 mM concentration were tested on the activity of *Sl*Lac2 (Figure 3A). Mo^{2+} and Ag^{2+} caused the disappearance of 30% and 60% of *Sl*Lac2 activity, respectively. AsO_4^{3}, Fe^{2+}, Cd^{2+}, and Zn^{2+} led to an even greater decrease in laccase activity (about 90% inhibition), and Cu^{2+} completely inhibited *Sl*Lac2 activity (98%).

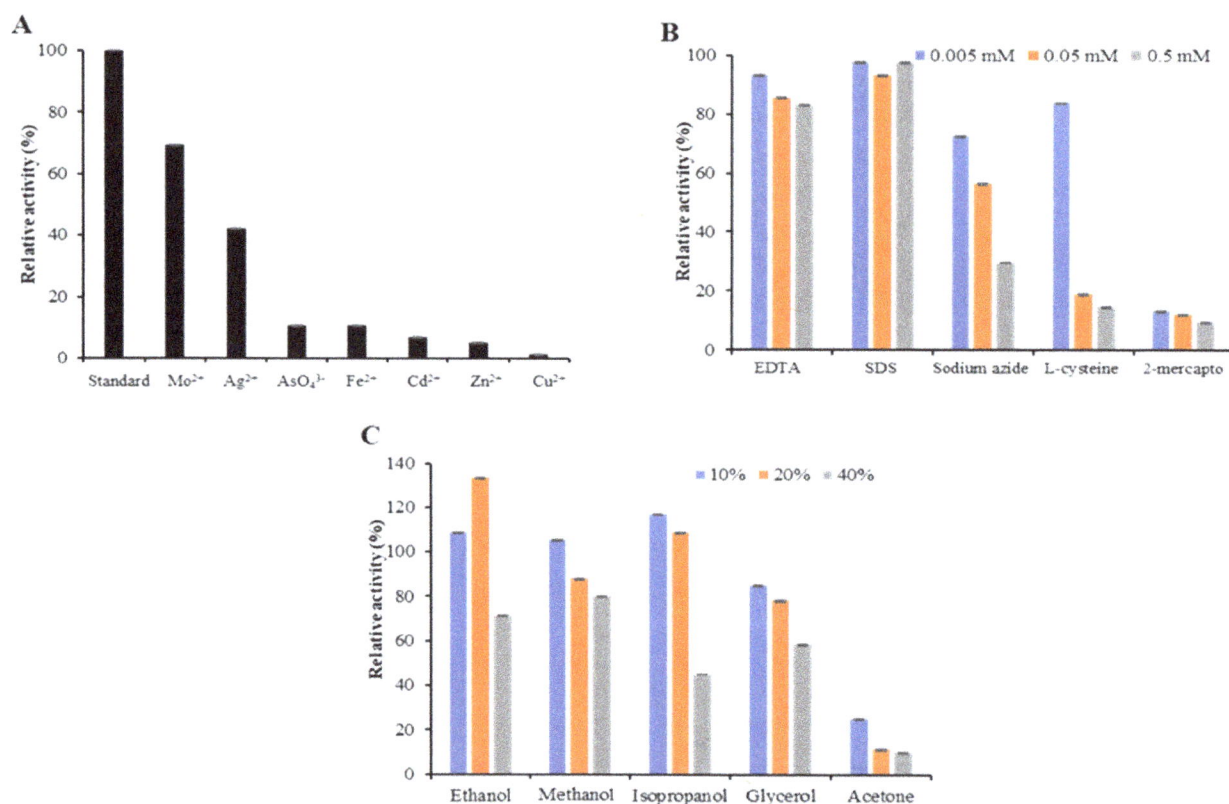

Figure 3. Effect of metal ions, inhibitors, and solvents on *Sl*Lac2 activity. (**A**) Effects of various metal ions (Mo^{2+}, Ag^{2+}, AsO_4^{3-}, Fe^{2+}, Cd^{2+}, Zn^{2+}, or Cu^{2+}) at 10 mM on the purified recombinant enzyme. (**B**) Effect of various inhibitors (EDTA, SDS, 2-mercaptoethanol, L-cysteine, and sodium azide) on *Sl*Lac2 activity at different concentrations (0.5, 0.05 and 0.005 mM). (**C**) Effect of solvents (ethanol, methanol, isopropanol, glycerol, and acetone) at different concentrations (10, 20, and 40% vol/vol) on *Sl*Lac2 activity.

Three concentrations (0.05 to 0.5 mM) of potential laccase inhibitors, ethylenediamine tetra-acetic acid (EDTA), sodium dodecyl sulfate (SDS), 2-mercaptoethanol, cysteine, and sodium azide (NaN_3) were evaluated to check their effects on the *Sl*Lac2 activity (Figure 3B). In the tested concentration range, more than 85% of enzyme activity remained in the presence of EDTA and SDS. Sodium azide inhibited *Sl*Lac2 with a loss of 25–70% of its initial activity, while cysteine acted as a stronger inhibitor (15–85% of inhibition). The most efficient inhibitor was found to be 2-mercaptoethanol, which reduced activity by 88–91%.

The effect of different concentrations of solvents ethanol, methanol, isopropanol, glycerol and acetone on *Sl*Lac2 activity in the range 10–40% (vol/vol) was studied (Figure 3C). *Sl*Lac2 was relatively stable, with 10–20% (vol/vol) of ethanol, methanol and isopropanol, but its activity was increased to more than 130% of its initial activity with 20% (vol/vol) ethanol. However, the recombinant laccase was less stable towards 40% (vol/vol) of these solvents, with a decrease of about 20% of its activity (methanol) to 60% (isopropanol). Of all the organic solvents tested, acetone showed a markedly negative effect on *Sl*Lac2 activity, causing more than 80% activity inhibition at a concentration of 10% (vol/vol) and near-inactivation with 20% and 40% (vol/vol) of acetone.

2.6. Effect of Sea Salt on Laccase Activity and Surface Charge of SlLac2

The effect of different concentrations of sea salt (1–5% wt/vol) on the purified laccase was analysed (Figure 4). The results showed that the activity gradually decreased as the salt concentration increased from 10% to almost half of its initial activity (Figure 3A). The enzyme could thus tolerate the presence of sea salt, retaining more than 50% of its activity with 5% wt/vol of sea salt. In parallel, analysis of the primary sequence and the overall surface charges of *St*Lac2 were carried out and compared with those of the terrestrial-derived laccase, *Ma*Lac1, and of the marine-derived laccases, *Ps*Lac1 and *Ps*Lac2. The primary sequence of *St*Lac2 exhibited a similar recurrence of negatively charged (D + E) over positively charged (R + K) amino acids, with the exception of *Ps*Lac2, which had 4.15 times higher (D + E)/(R + K) ratio (Figure 4B). *Ma*Lac1 and *Ps*Lac1 presented intermediate values, i.e., 1.55 and 1.20, respectively. The three-dimensional models of *Ma*Lac1 and *Ps*Lac1 showed an even charge distribution in their surface plots, whereas for *Ps*Lac2, the surface charge was strongly negative (Figure 4C). In contrast, *St*Lac2 exhibited a slightly higher positive charge distribution at the surface.

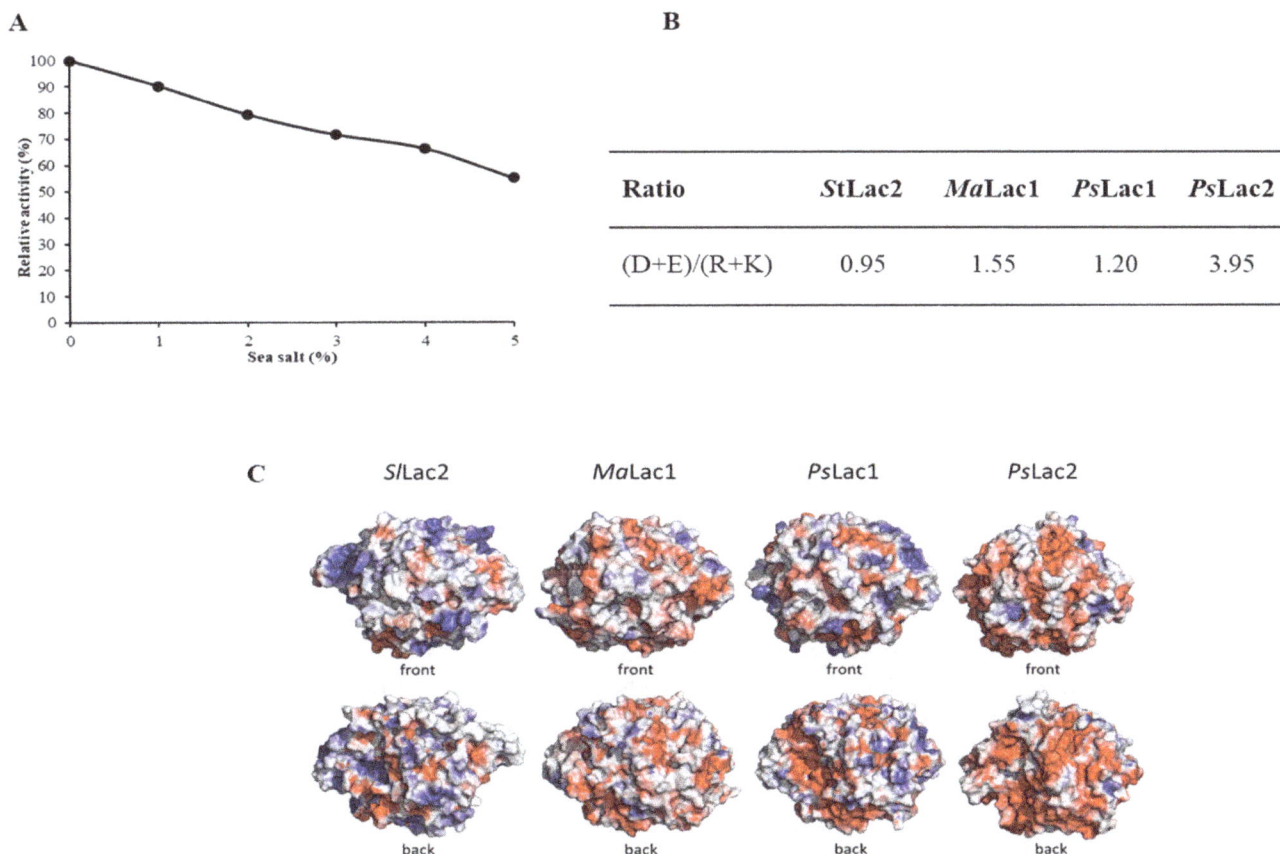

A

B

Ratio	StLac2	*Ma*Lac1	*Ps*Lac1	*Ps*Lac2
(D+E)/(R+K)	0.95	1.55	1.20	3.95

C

Figure 4. Effect of sea salt and surface charge of *Sl*Lac2. (**A**) Effect of different concentrations (1, 2, 3, 4 and 5% wt/vol) of sea salt on *Sl*Lac2 activity. (**B**) (D + E)/(R + K) amino acids ratio of *Sl*Lac2 compared with those from *Melanocarpus albomyces* laccase 1, *Ma*Lac1 (Q70KY3) and *Pestalotiopsis* sp. laccases 1 and 2,

*Ps*Lac1 (KY554800) and *Ps*Lac2 (KY5548001). (**C**) Surface charge plots (negative and positive charges are in red and blue, respectively) of *Sl*Lac2 compared with those from *M. albomyces* laccase 1, *Ma*Lac1 and *Pestalotiopsis* sp. laccases 1 and 2, *Ps*Lac1 and *Ps*Lac2. The surface potentials were calculated using the vacuum electrostatics function of the PyMOL molecular graphics system (Schrödinger, New York, NY, USA).

3. Discussion

The objective of our research project was to seek new insights into the physiology of the marine-derived fungi, more specifically their lignocellulose enzyme machinery, and how they adapt in their marine environment. In previous work, we screened five fungal strains isolated from Tunisian coastal waters [26] and characterised for their laccase-like activities. Based on a microbial approach and on enzyme activity screening, one of these marine-derived strains, *S. lucomagnoense*, was selected for its adapted growth on xylan in saline conditions, and its improved laccase (seagrass-containing cultures) and cellulase (wheat straw-containing cultures) activities in the presence of sea salt [27]. To further study the selected marine-derived fungus, its transcriptome was sequenced and its proteome analysed when grown on wheat-straw and sea grass in the absence or presence of sea salt. From this study, we were unable to identify the laccases involved in the improved laccase activity in the presence of salts, possibly owing to the sensitivity limit of our combined transcriptomic-proteomic approach, because of undetected transcript domains. However, seven genes were identified in *S. lucomagnoense* transcriptome that encoded putative laccases, and one was selected for heterologous expression in *A. niger* and characterisation of the corresponding laccase, *Sl*Lac2.

Only a few biochemical characterisations of laccases from marine organisms are available [28–31]. Our aim was accordingly to obtain additional data on these enzymes and their potential for biotechnological applications. Marine-derived fungi are considered as a source of original enzymes active in saline conditions and pH ranging from 3.0 to 11.0 [24], which are properties of interest for many industrial applications. *Sl*Lac2 was produced in *A. niger* with a yield of about 30 mg L^{-1} in the culture medium, which is in the range of protein production obtained for the laccase of the terrestrial fungus *Pycnoporus cinnabarinus*, heterogously produced in *A. niger* (70 mg L^{-1}) [32]. Recently, two laccase-encoding genes isolated from the marine-derived fungus *Pestalotiopsis* sp., were expressed in *A. niger* and lower yields were obtained: 2.6 and 6.2 mg L^{-1} for *Ps*Lac 1 and *Ps*Lac2, respectively [28]. The main physico-chemical properties of *Sl*Lac2 were determined to compare them with literature reports on laccases from marine and terrestrial fungi. The purified laccase was active on all typical laccase substrates including ABTS, DMP, and syringaldazine, and showed the highest affinity (lowest K_M value) for syringaldazine (3.5 μM). However, the highest catalytic efficiency (K_{cat}/K_M) was found for ABTS (7.42 s^{-1} mM^{-1}) as its turnover value was much higher (K_{cat} 0.153 s^{-1}) compared with that determined for syringaldazine (0.0074 s^{-1}). Laccases are widespread in the natural environment, and very well-characterised for their kinetic parameters. Many previous studies found that syringaldazine and ABTS were their preferred substrates [33]. Laccase affinity for syringaldazine (tens of μM) is generally higher than that measured for ABTS (hundreds of μM), whereas K_M constants for other phenolic compounds are considerably higher [24,34]. In our case, *Sl*Lac2 exhibited a preference (affinity 6-fold) for syringaldazine. By comparison, *Ps*Lac1 from the marine-derived fungus *Pestalotiopsis* sp., showed K_M values of 4 μM and 24 μM for syringaldazine and ABTS, respectively, while *Ps*Lac2, did not exhibit a specific preference for either substrate, with higher K_M values (hundreds of μM) [28]. Two other marine-derived fungal laccases from *Trematosphaeria mangrovei* [29] and *Cerrena unicolor* [31] showed different behaviours, with K_M values of 1.42 mM and 54 μM against ABTS. Even though the affinity was stronger for ABTS, the catalytic efficiency (K_{cat}/K_M) of *Sl*Lac2 was slightly higher for syringaldazine (7.42 s^{-1} mM^{-1}). This value is very close to the K_{cat}/K_M measured for *Pestalotiopsis* sp. Lac2 (5.67 s^{-1} mM^{-1}) [28]. *Ps*Lac2 was described as a relatively versatile laccase, exhibiting a similar catalytic efficiency for syringaldazine, ABTS and DMP (5.67, 4.33 and 9.1 s^{-1} mM^{-1}, respectively). By contrast, the second laccase of *Pestalotiopsis* sp., *Ps*Lac1 showed the highest catalytic efficiency for syringaldazine at 82.93 s^{-1} mM^{-1}, and a 3-times lower K_{cat}/K_M for the other two substrates ABTS and DMP (29.25 and 24.53 s^{-1} mM^{-1}, respectively). Higher catalytic efficiency for syringaldazine can be found in the literature, such as for the laccase-like enzyme cloned form a marine library with 29 s^{-1} mM^{-1} [23]. In conclusion for the kinetic parameters, *Sl*Lac2 was shown to possess the general characteristics of fungal laccases, with the highest efficiency for syringaldazine.

The physical and biochemical properties of the purified recombinant *Sl*Lac2 were tested for further comparisons with fungal laccases. As reported by Baldrian, fungal laccases typically exhibit an optimal pH in the acidic pH range (3.0–5.0), using ABTS as substrate [33], whereas the optimal pH for the oxidation of syringaldazine is higher (3.5–6.0) [33]. The purified recombinant *Sl*Lac2 was tested to determine its optimal pH in the range 3.0–9.0 using syringaldazine as a substrate, and it showed that the optimal pH value driving the maximum activity of the recombinant enzyme was pH 6.0, with a measurable activity at pH 8.0. For the laccases obtained from *Pestalotiopsis sp.* and *T. mangrovei*, the optimal pH was determined with ABTS as substrate: 5.0 and 4.0, respectively [28,29]. Usually, the stability of most fungal laccases is higher at acidic pH and decreases when close to neutral pH conditions, which is the case for the terrestrial *P. cinnabarinus* laccase [34]. In our work, we demonstrated that the purified recombinant *Sl*Lac2 was relatively resistance in several pH conditions tested (3.0 to 7.0). In addition, *Sl*Lac2 was more stable in the tested range of pH than *Pestalotiopis* sp. laccases, i.e., *Ps*Lac1 was less stable at alkaline pH. It showed 30% residual activity after 100 h of incubation at pH 6.0, whereas *Ps*Lac2, was instead more sensitive to low pH; its stability decreased with time between pH 3.0 and 6.0, and abruptly at pH 2.0 (no activity left after 40 min of incubation) [28].

Concerning the behaviour of fungal laccases with temperature, previous studies found that the optimal temperature was between 50 and 60 °C for most fungal laccases [32,33]. In our study, we also showed that the optimum temperature of the purified laccase *Sl*Lac2 was in this range (60 °C). Compared with marine laccases, *T. mangrovei* and *Ps*Lac 1 had slightly higher optimal temperatures: 65 °C (syringaldazine. 80 °C with ABTS) and 70 °C (ABTS), respectively [28,29]. Regarding thermal stability, *Sl*Lac2 was stable for 180 min at temperatures ranging from 25 to 50 °C, whereas the activity was significantly lost above 60 °C, as demonstrated for *Ps*Lac2 [28]. In addition, *T. mangrovei* laccase and *Ps*Lac1 showed curves with more pronounced thermal deactivation at temperatures above 60 °C [28–30]. We conclude that *Sl*Lac2 presents novel, useful properties due to its activity relative to temperature and pH, and to its stability compared with its marine laccase homologs.

Enzyme activity is often affected positively or negatively by metal ions present in the mixtures [35–38]. Because laccases are used in several processes containing metal ions, it is useful to characterise the effect of metal ions on their activities. It has been shown in many studies that laccase activities vary with the type of metal ion present and the enzyme source [38–41]. *Sl*Lac2 showed that adding a 10 mM concentration of Mo^{2+} and Ag^{2+} to the purified enzyme left 70% and 40% of the activity, respectively, but the presence of AsO4^{3-}, Fe^{2+}, Cd^{2+}, Zn^{2+}, and especially Cu^{2+}, caused the loss of all or most activity. For instance, Cu^{2+} has already been shown to inhibit *Scytalidium thermophilum* laccase activity [42]. Among several metal ions tested, Xu et al. showed that Ag$^+$, Ag^{2+}, Li$^+$, and Pb^{2+} could inhibit the laccase of *Cerenia* sp. reversibly. Hg$^+$ was shown to reduce pH and thermal stability, and dynamic simulation showed the presence of binding sites for Hg close to copper binding sites on the laccase molecule [43]. For *T. mangrovei*, the effect of metal ions was analysed at 1 mM, and only Fe^{2+} showed a complete inactivation of its activity, whereas Cu^{2+} was not tested [29]. At the same concentration, Fe^{3+} showed a 30% inactivation of the marine-derived fungus, *C. unicolor* laccase, whereas Cu^{2+} addition produced a slight effect with a loss of less than 10% of activity [31]. The most common laccase inhibitors tested include cysteine, EDTA, sodium fluoride, sodium azide, dithiothreitol, thioglycolic acid, and diethyldithiocarbamic acid. These inhibitors are not laccase-specific and their applications on phenoloxidases was because they could inhibit metalloenzymes [44,45]. They can act on the active site: EDTA binds to copper atoms and stops the transfer of electrons. Cysteine, dithiothreitol, thioglycolic acid, and diethyldithiocarbamic acid are reducing substances described as potential laccase inhibitors sequestering dioxygen and stopping the oxidation of the phenolic substrates.

However, Johannes and Majcherczyk (2002) have tested several potential laccase inhibitors including dithiothreitol, thioglycolic acid (TGA), diethyldithiocarbamic acid (DDC), cysteine, and sodium azide (NaN$_3$) [46]. They found that only NaN$_3$ stopped the substrate oxidation by no oxygen uptake, whereas the sulfydryl compounds did not affect the oxygen consumption and even increased it through an autoxidation reaction. SlLac2 was significantly inactivated by NaN$_3$, L-cysteine and 2-mercaptoethanol, whereas *C. unicolor* laccase was inhibited by 1 mM of NaN$_3$ (95% inhibition) and 2-mercaptoethanol (67% inhibition), but not by L-cysteine (0.1 mM) [28]. For *T. mangrovei* laccase, a 50% inhibition was demonstrated for 1 mM of NaN$_3$ and the most efficient inhibitor was shown to be NaCN (80% imbibition) [29].

Biochemical reactions involving laccases in compatible organic solvents give access to some insoluble substrates, which may help in the detoxification of several sparingly water-soluble persistent organic pollutants. However, several studies have shown negative effects of organic solvents on laccase activity [47,48]. For instance, Robles et al. [49] studied a laccase of the hyphomycete *Chalara* (syn. *Thielaviopsis*) *paradoxa* CH32 and showed an activity decrease of 27% and 36% in the presence of 25% (vol/vol) ethanol and methanol, respectively. In our study, we showed that with 40% (vol/vol) of ethanol and methanol, there was a similar decrease in the activity from about 30% and 20%, respectively. However, of the solvents tested in our experiments, acetone was the most damaging, probably by precipitating the proteins. Both free and immobilised laccases decreased their activity with an increase in water-miscible organic solvent concentrations, and in most reported cases, immobilised laccase is less strongly affected than the free enzyme by the organic solvent concentrations, which could offer an alternative, depending on the solvents needed to implement for the laccase application.

As SlLac2 was isolated from a marine-derived fungus, this enzyme behaviour was studied in relation to saline conditions. In the presence of sea salt, SlLac2 was affected, with a continuous decrease in its activity following the sea salt concentration increase. For the two laccases of the marine-derived fungus *Pestalotiopsis* sp., PsLac1 and 2, completely different behaviour was observed, as their activities were enhanced in the presence of increasing concentrations of sea salts [28]. PsLac2 activity was clearly stronger (1.6 times higher compared with PsLac2 at 5% sea salt). In addition, *T. mangrovei* laccase lost half of its activity in only 1 mM NaCl [29] while *Pestalotiopsis* sp. lytic polysaccharide monoxidases A (LMPOA) could act on cellulose for up to 6% sea salt [50]. A few studies have been carried out on enzymes originating from saline environments to identify the molecular determinants responsible for their salt tolerance. Recently, Li et al. [51] demonstrated that two amino acid sites were involved in the salt activation, as Cl$^-$ ion could bind to specific local sites to interfere with substrate binding and/or electron transfer. In addition, salt-adapted enzymes were shown to be strongly negatively charged on their surface, a characteristic contributing to protein stability and to enzyme activities adapted to extreme osmotic conditions [52–54]. SlLac2 was shown to possess a slightly higher positive charge distribution at the protein surface. In contrast, MaLac1 and PsLac1 exhibited a relatively even balance of negative charges over positive charges (D + E/R + K ratio), whereas PsLac2 presented a very high ratio of 3.92. In a previous study on LMPOs identified from *Pestalotiopsis* sp., the calculated ratio also exhibited very high ratios greater than 4.0 [50] and an excess of negatively charged residues at their surface, typical of marine enzymes. Altogether, these results show that the resistance to salt is not yet fully understood, and that further research is still needed to elucidate the adaptation mechanisms.

4. Materials and Methods

4.1. Strains and Culture Conditions

Escherichia coli strain TOP 10 was used for vector storage and propagation. *Aspergillus niger* strain D15#26 (pyrG deficient) [55] was used for the heterologous expression of the SlLac2 encoding synthetic gene. After co-transformation with vectors respectively containing the pyrG gene and the laccase cDNA, transformants of *A. niger* were grown for selection on solid minimal medium without uridine and containing 70 mM of NaNO$_3$, 7 mM of KCl, 11 mM of KH$_2$PO$_4$, 2 mM of MgSO$_4$, glucose 1% (wt/vol), and trace elements (1000 × stock; 76 mM ZnSO$_4$, 178 mM H$_3$BO$_3$, 25 mM MnCl$_2$, 18 mM FeSO$_4$,

7.1 mM $CoCl_2$, 6.4 mM $CuSO_4$, 6.2 mM Na_2MoO_4, and 174 mM EDTA). For the screening procedure of the positive transformants, 100 mL of culture medium containing 70 mM $NaNO_3$, 7 mM KCl, 200 mM Na_2HPO_4, 2 mM $MgSO_4$, glucose 5% (wt/vol), and trace elements was inoculated with 2×10^6 spores mL^{-1} in a 250 mL baffled flask.

4.2. Cloning and Expression of SlLac2-Encoding Gene

The open reading frame sequence encoding *Sl*Lac2 was synthesised and codon bias optimised for *A. niger* (GeneArt, Regensburg, Germany), with some modifications. The amino acids of the signal peptide, which was predicted with the program SignalP hosted on the ExPASy Proteomics server (http://www.expasy.ch), were replaced by the 24-amino-acid glucoamylase (GLA) preprosequence from *A. niger* (MGFRSLLALSGLVCNGLANVISKR). Two restriction sites (*Bss*HII and *Hind*III) were respectively added at the 5′ and 3′ ends of the sequence for cloning into the expression vector pAN52.4 (GenBank/EMBL accession number Z32699). In the final expression cassette, the *Aspergillus nidulans* glyceraldehyde-3-phosphate dehydrogenase-encoding gene (*gpdA*) promoter, the 5′ untranslated region of the *gpdA* mRNA, and the *A. nidulans trpC* terminator were used to drive the expression of the inserted coding sequences. The co-transformation was carried out as described by Punt and van den Hondel [25] using both the expression vector containing the expression cassette and pAB4-1 [56] containing the *pyrG* selection marker in a 10:1 ratio. Transformants were selected for uridine prototrophy by growth on selective solid minimal medium (without uridine).

4.3. Screening of Transformants and Laccase Activity Assay

To screen for the best clones for enzyme production in liquid medium, 100 mL of minimal medium (adjusted to pH 5.5 with 1 M of citric acid) was inoculated with 2×10^6 spores mL^{-1} in a 250 mL flask. The cultures were incubated for 10 days at 30°C in a shaker incubator (110 rpm), and pH was adjusted daily to 5.5 with 1 M citric acid. From these liquid cultures, aliquots (2 mL) were withdrawn daily, and mycelia were pelleted (20 min at $15,000 \times g$). Laccase activity of the resulting supernatant was assayed spectrophotometrically by monitoring the oxidation of syringaldazine (1 mM) as substrate at 436 nm ($\varepsilon_{436} = 29,300$ M^{-1} cm^{-1}) in a sodium acetate buffer (50 mM, pH 6.0). The reaction was monitored for 1 min at 30 °C in an Uvikon XS spectrophotometer (BioTek Instruments, Colmar, France). Activity is expressed in nkat mL^{-1}, 1 nkat corresponding to the oxidation of 1 nanomole of substrate per second. Measurements in all the experiments were performed in triplicate.

4.4. Production and Purification of Recombinant SlLac2

For protein production, the best positive clone corresponding to the clone with the highest laccase activity was selected and cultured. 930 mL of culture medium containing 70 mM $NaNO_3$, 7 mM KCl, 200 mM K_2HPO_4, 2 mM $MgSO_4$, glucose 10% (wt/vol), trace elements and adjusted to pH 5.0 with a 1 M citric acid solution, inoculated with 2×10^6 spores mL^{-1}, was prepared to initiate a large-scale protein production (1 L). The culture was harvested after 11 days. The culture medium was clarified by filtration through GF/D, GF/A and GF/F glass fibre filters (Whatman, Maidstone, UK), followed by filtrations through 0.45 μm and 0.22 μm polyethersulphone membranes (Express Plus, Merck Millipore). The collected filtrate was concentrated by ultrafiltration through a polyethersulphone membrane with a 10 kDa molecular mass cut-off (Vivaflow crossflow cassette, Sartorius, Les Ulis, France). After dialysing overnight at 4 °C against 100 mM Tris buffer pH 8.0, 150 mM NaCl and 1 mM EDTA, the sample was loaded onto on a CM-Sepharose fast flow column (GE Healthcare Life Science, Velizy-Villacoublay, France) using an AKTA purifier (GE Healthcare Life Science). The sample was loaded onto the column previously equilibrated with a binding buffer (0.2 mM sodium tartrate buffer pH 5.0, containing 25 mM NaCl). Protein elution was performed using a linear gradient of 0–100% of the elution buffer (0.2 mM of sodium tartrate buffer pH 5.0, containing 1 mM NaCl). Activity was determined in the 10 mL collected fractions, and protein production was evaluated by SDS-PAGE. The active fractions were concentrated through a 10 kDa molecular mass cut-off Amicon membrane (Millipore). The concentrated samples were loaded onto a Sephacryl S-200HR size exclusion chromatography column (GE Healthcare Life Science) equilibrated with a 50 mM sodium acetate buffer pH 5.0, containing 50 mM NaCl. 5 mL fractions were collected and assayed for laccase activity as described above, and protein homogeneity was tested by SDS-PAGE. For the different purification steps, the total protein concentration was determined with the Bradford assay using the BioRad Protein Assay Kit (BioRad, Marnes-la-Coquette, France) and bovine serum albumin (BSA) as a standard. Final protein concentration was determined spectrophotometrically at 280 nm using a NanoDrop 2000 (Thermo Fisher Scientific, Illkirch, France) and the theoretical molar extinction coefficients SlLac2 at 280 nm, 125,290 M^{-1} cm^{-1}. The molecular mass of the purified protein was determined by 12% sodium dodecyl sulphate-polyacrylamide gel electrophoresis (SDS-PAGE).

4.5. Bioinformatic Analysis

The molecular mass, theoretical pI, and molar extinction coefficient of enzymes were predicted by the ProtParam tool (http://web.expasy.org/protparam/). Protein sequences were aligned using MUSCLE [57,58] and CLUSTAL W [59] at http://www.ebi.ac.uk/Tools/msa/. Signal peptides were predicted using Signal P [60] at http://www.cbs.dtu.dk/services/SignalP/. N-glycosylation sites [60] were predicted at http://www.cbs.dtu.dk/services/NetNGlyc/. The automated protein structure homology modelling online tool Phyre 2 [61] was used to predict the three-dimensional models of SlLac2 using the closest homologs of known structure available in the Protein Data Bank (PDB). Protein models c2q9oA (*Melanocarpus albomyces* laccase, 44% identity), c5lwxA (*Aspergillus niger* laccase McoG H253D variant, 37% identity), c3ppsD (*Thielavia arenaria* laccase, 45% identity) and c3sqrA (*Botrytis aclada* laccase, 44% identity) were used as templates for SlLac2. Surface charges were calculated, and all the figures were prepared with PyMOL.

4.6. Determination of N-Terminal Amino Acid Sequence and Glycosylation Level

The N-terminal sequence was determined according to Edman degradation. Analysis was carried out on an Applied Biosystem 476Asequencer by the proteomic platform of the Institut de Microbiologie de la Méditerrranée, CNRS, Aix-Marseille Université (France). Matrix-assisted laser desorption ionisation—time-of-flight mass spectrometry of samples was carried out on a Microflex II time-of-flight mass spectrometer (Bruker Daltonik, Germany).

To determine the level of glycosylation of the purified laccase SlLac2, a deglycosylation reaction was carried out using the PNGase F (New England Biolabs, France). 15 μg of purified laccase was mixed with 4 μL of the denaturation solution (5% SDS and 40 mM dithiothreitol (DTT)) denatured at 100 °C for 10 min. The denatured sample was deglycosylated in a final volume of 40 μL containing 1% Nonidet P-40, 50 mM sodium phosphate buffer (pH 7.5) and 2 μL of the deglycosylation enzyme PNgase F solution (500,000 units mL^{-1}). The reactions were incubated at 37 °C for 2 h. After the deglycosylation, 10 μL of the digested enzyme was loaded onto a 12% homogeneous SDS-PAGE gel.

4.7. Substrate Specificity and Kinetics

Enzyme activities were measured using a UVIKONxs spectrophotometer (Bio-TEK Instruments) at 30 °C by following the oxidation of different aromatic substrates: 2,2'-azino-bis(3-ethylbenzothiazoline-6-sulphonic acid) (ABTS), 2,6-dimethoxyphenol (DMP) and syringaldazine. The absorbance increases at 436 nm (ε_{436} = 29,300 M^{-1} cm^{-1}), 469 nm (ε_{469} = 27,500 M^{-1} cm^{-1}) and 530 nm (ε_{530} = 65,000 M^{-1} cm^{-1}) were followed for ABTS, DMP, and syringaldazine, respectively. Kinetic constants (enzyme concentration 0.18 mg mL^{-1}) were determined for each of these compounds (substrate range of 0.05 to 1.1 mM for ABTS, and 0.025 to 0.1 mM for DMP and syringaldazine) with sodium acetate buffer (50 mM, pH 4.0, 5.0, and 6.0, respectively). All assays were performed in triplicate. Mean apparent affinity constant (Michaelis constant, K_m) and enzyme turnover (catalytic constant, k_{cat}) values and standard errors were obtained by nonlinear least-squares fitting to the Michaelis-Menten model. Fitting of these constants to the normalized Michaelis-Menten equation $\upsilon = (k_{cat}/K_m)[S]/(1 + [S]/K_m)$ yielded enzyme efficiency values (k_{cat}/K_m) with their standard errors

4.8. Effect of pH and Temperature on the Activity and Stability of the Laccase

The effect of pH on the purified enzyme was determined by measuring the enzyme activity at 30 °C in various buffers at different pH (2.5–10.0) using the following buffers: 50 mM sodium acetate buffer (pH 2.5–7.0), phosphate buffer (pH 7.0–9.0), and sodium bicarbonate (pH 9.0–10.0), against 1 mM syringaldazine as substrate at 30°C. The pH stability of the purified enzyme was determined by incubating the enzyme in the different buffer solutions (50 mM) with different pH values (3.0–7.0) and allowing the mixture to stand for 4 h, 24 h and 48 h incubation at room temperature. The residual enzyme activity was determined under standard assay conditions after centrifugation to remove the denatured proteins.

To determine the effect of temperature on laccase, the activity of the purified enzyme was measured in a sodium acetate buffer (50 mM, pH 6.0) at different temperatures ranging from 20 to 80°C. The stability to thermal treatment of purified enzyme was determined by measuring residual activity after incubation at different temperatures (25–70°C) and pH 6.0 for 30 min to 3 h. All the reactions were run in triplicate.

4.9. Effect of Metal Ions, Inhibitors, Solvents, and Sea Salt on Laccase Activity.

To study the effect of metal ions, potential inhibitors, different solvents, and sea salt on the purified enzyme, the activity was determined in the presence of metal ions in standard laccase activity conditions. The following compounds were used as the source of metal ions: $CuSO_4.5H_2O$, Na_2MoO_4, $FeSO_4.7H_2O$, $CdCl_2$, $Na_3AsO_4.12H_2O$, and $ZnSO_4.7H_2O$. They were added at a final concentration of 10 mM. The potential inhibitors were L-cysteine, 2-mercaptoethanol, EDTA, SDS, and sodium azide (final concentrations of 0.5, 0.05 and 0.05 mM). The different organic solvents were ethanol, methanol, iso-propanol, glycerol, and acetone at different concentrations: 10, 20, and 40%. Different concentrations of sea salt (Sigma-Aldricht, Lyon, France) (1–5% wt/vol) were added to the reaction mixture and the activity was determined in standard conditions. All the reactions were run in triplicate.

4.10. Nucleotide Sequence Accession Number

The gene sequences encoding SlLac2 were deposited in the nucleotide sequence database (GenBank) under the accession code MT470191.

5. Conclusions

The present work carried out on the S. lucomagnoense gives new insights into the marine-derived laccases, which are catalysts of great interest for many biotechnological applications. In addition, SlLac2 produced in A. niger presented potential properties related to pH and some tolerance to sea salt, which could be exploited in biotechnological applications, such as the pulp and paper sector and enzymatic degradation of industrial dyes, for which alkaline-active enzymes and high tolerance to salt are needed, respectively

Supplementary Materials:
Supplementary Materials, Figure S1: Alignments of the closest characterized laccases from *Melanocarpus albomyces* (MaLac1, Q70KY3) and *Pestalotiopsis* sp. KF079 (PsLac1: KY554800 and PsLac 2: KY554801), using CLUSTAL W sequence alignment algorithm. Figure S2: Kinetics of SlLac2 on the three substrates.

Author Contributions: W.B.A., A.B.A., Y.M., and A.T.-D. provided experiment support; W.B.A., A.B.A., and A.T.-D. provided technical support and contributed to the discussion; W.B.A., E.R., T.M. designed the experiments; W.B.A., E.B., C.B.F., G.S., A.L., E.R., T.M. wrote the manuscript. All authors have read and agreed to the published version of the manuscript.

References

1. Sivakumar, R.; Rajendran, R.; Balakumar, C.; Tamilvendan, M. Isolation, screening and optimization of production medium for thermostable laccase production from *Ganoderma* sp. *Int. J. Eng. Sci. Technol.* **2010**, *2*, 7133–7141.
2. Kumar, R.; Kaur, J.; Jain, S.; Kumar, A. Optimization of laccase production from *Aspergillus flavus* by design of experiment technique: Partial purification and characterization. *J. Genet. Eng. Biotechnol.* **2016**, *14*, 125–131. [CrossRef] [PubMed]
3. Giardina, P.; Faroca, V.; Pezzela, P.C.; Piscitelli, A.; Vanhilla, S.; Sannia, G. Laccase: A never-ending story. *Cell. Mol. Life Sci.* **2010**, *62*, 369–386. [CrossRef] [PubMed]
4. Durán, N.; Rosa, M.A.; D'Annibale, A.; Gianfreda, L. Applications of laccases and tyrosinases (phenoloxidases) immobilized on different supports: A review. *Enzym. Microb. Technol.* **2002**, *31*, 907–931. [CrossRef]
5. Arora, D.S.; Sharma, R.K. Ligninolytic fungal laccases and their biotechnological applications. *Appl. Biochem. Biotechnol.* **2010**, *160*, 1760–1788. [CrossRef]
6. Xu, F. Oxidation of phenols, anilines, and benzenethiols by fungal laccases: Correlation between activity and redox potentials as well as halide inhibition. *Biochemistry* **1996**, *35*, 7608–7614. [CrossRef]
7. Minussi, R.C.; Pastore, G.M.; Durán, N. Enzima de interés en enología: Lacasas. *Aliment* **1999**, *304*, 145–150.
8. Christopher, F. The structure and function of fungal laccase. *Microbiology* **1994**, *140*, 19.
9. Baldrian, P. Purification and characterization of laccase from the white-rot fungus *Daedalea quercina* and decolorization of synthetic dyes by the enzyme. *Appl. Microbiol. Biotechnol.* **2004**, *63*, 560–563. [CrossRef]
10. Li, K.; Xu, F.; Eriksson, K.E. Comparison of fungal laccases and redox mediators in oxidation of a nonphenolic lignin model compound. *Appl. Microbiol. Biotechnol.* **1999**, *65*, 2654–2660. [CrossRef]
11. Munk, L.; Sitarz, A.K.; Kalyani, D.C.; Mikkelsen, J.D.; Meyer, A.S. Can laccases catalyze bond cleavage in lignin? *Biotechnol. Adv.* **2015**, *33*, 13–24. [CrossRef] [PubMed]
12. Janusz, G.; Pawlik, A.; Świderska-Burek, U.; Polak, J.; Sulej, J.; Jarosz-Wilkołazka, A.; Paszczyński, A. Laccase Properties, Physiological Functions, and Evolution. *Int. J. Mol. Sci.* **2020**, *21*, 966. [CrossRef]
13. Xu, F.; Shin, W.; Brown, S.H.; Wahleithner, J.A.; Sundaram, U.M.; Solomon, E.I. A study of a series of recombinant fungal laccases and bilirubin oxidase that exhibit significant differences in redox potential, substrate specificity, and stability. *Biochim. Biophys. Acta Protein Struct. Mol. Enzym.* **1996**, *1292*, 303–311. [CrossRef]
14. Bourbonnais, R.; Paice, M.G. Oxidation of non-phenolic substrates: An expanded role for laccase in lignin biodegradation. *FEBS Lett.* **1990**, *267*, 99–102. [CrossRef]
15. Call, H.P.; Mücke, I. History, overview and applications of mediated lignolytic systems, especially laccase-mediator-systems (Lignozym®-process). *J. Biotechnol.* **1997**, *53*, 163–202. [CrossRef]
16. Bajpai, P.; Kondo, R. *Biotechnology for Environmental Protection in the Pulp and Paper Industry*; Springer Science & Business Media: Berlin/Heidelberg, Germany, 2012; ISBN 3-642-60136-7.
17. Wesenberg, D.; Kyriakides, I.; Agathos, S.N. White-rot fungi and their enzymes for the treatment of industrial dye effluents. *Biotechnol. Adv.* **2003**, *22*, 161–187. [CrossRef]
18. Vinodhkumar, T.; Thiripurasundari, N.; Ramanathan, G.; Karthik, G. Screening of dye degrading bacteria from textile effluents. *J. Chem. Pharm. Res.* **2013**, *3*, 848–857.
19. Michaels, G.B.; Lewis, D.L. Sorption and toxicity of azo and triphenylmethane dyes to aquatic microbial populations. *Environ. Toxicol. Chem.* **1985**, *4*, 45–50. [CrossRef]
20. Chung, K.-T.; Stevens, S.E., Jr. Degradation azo dyes by environmental microorganisms and helminths. *Environ. Toxicol. Chem.* **1993**, *12*, 2121–2132.
21. Ghoreishi, S.M.; Haghighi, R. Chemical catalytic reaction and biological oxidation for treatment of non-biodegradable textile effluent. *Chem. Eng. J.* **2003**, *95*, 163–169. [CrossRef]
22. Santhanam, N.; Vivanco, J.M.; Decker, S.R.; Reardon, K.F. Expression of industrially relevant laccases: Prokaryotic style. *Trends Biotechnol.* **2011**, *29*, 480–489. [CrossRef]
23. Yang, Q.; Zhang, M.; Zhang, M.; Wang, C.; Liu, Y.; Fan, X.; Li, H. Characterization of a novel, cold-adapted, and thermostable laccase-like enzyme with high tolerance for organic solvents and salt and potent dye decolorization ability, derived from a marine metagenomic library. *Front. Microbiol.* **2018**, *9*, 2998. [CrossRef]
24. Bonugli-Santos, R.C.; Dos Santos Vasconcelos, M.R.; Passarini, M.R.Z.; Vieira, G.A.L.; Lopes, V.C.P.; Mainardi, P.H.; Dos Santos, J.A.; de Azevedo Duarte, L.; Otero, I.V.R.; da Silva Yoshida, A.M.; et al. Marine-derived fungi: Diversity of enzymes and biotechnological applications. *Front. Microbiol.* **2015**, *6*, 269. [CrossRef] [PubMed]

25. Punt, P.J.; van den Hondel, C.A.M.J.J. Transformation of filamentous fungi based on hygromycin b and phleomycin resistance markers. *Methods Enzymol.* **1992**, *216*, 447–457.

26. Ben Ali, W.; Chaduli, D.; Navarro, D.; Lechat, C.; Turbé-Doan, A.; Bertrand, E.; Faulds, C.B.; Sciara, G.; Lesage-Meessen, L.; Record, E.; et al. Screening of five marine-derived fungal strains for their potential to produce oxidases with laccase activities suitable for biotechnological applications. *BMC Biotechnol.* **2020**, *20*, 27. [CrossRef] [PubMed]

27. Ben Ali, W.; Navarro, D.; Kumar, A.; Drula, E.; Turbé-Doan, A.; Correia, L.O.; Baumberger, S.; Bertrand, E.; Faulds, C.B.; Henrissat, B.; et al. Characterization of the CAZy Repertoire from the Marine-Derived Fungus *Stemphylium lucomagnoense* in Relation to Saline Conditions. *Mar. Drugs* **2020**, *18*, 461. [CrossRef]

28. Wikee, S.; Hatton, J.; Turbé-Doan, A.; Mathieu, Y.; Daou, M.; Lomascolo, A.; Kumar, A.; Lumyong, S.; Sciara, G.; Faulds, C.B.; et al. Characterization and Dye Decolorization Potential of Two Laccases from the Marine-Derived Fungus *Pestalotiopsis* sp. *Int. J. Mol. Sci.* **2019**, *20*, 1864. [CrossRef]

29. Atalla, M.M.; Zeinab, H.K.; Eman, R.H.; Amani, A.Y.; Abeer, A.A.E. Characterization and kinetic properties of the purified *Trematosphaeria mangrovei* laccase enzyme. *Saudi J. Biol. Sci.* **2013**, *20*, 373–381. [CrossRef]

30. Mainardi, P.H.; Feitosa, V.A.; de Paiva, L.B.B.; Bonugli-Santos, R.C.; Squina, F.M.; Pessoa, A., Jr.; Sette, L.D. Laccase production in bioreactor scale under saline condition by the marine-derived basidiomycete *Peniophora* sp. CBMAI 1063. *Fungal Biol.* **2018**, *122*, 302–309. [CrossRef]

31. D'Souza-Ticlo, D.; Sharma, D.; Raghukumar, C. A Thermostable Metal-Tolerant Laccase with Bioremediation Potential from a Marine-Derived Fungus. *Mar. Biotechnol.* **2009**, *11*, 725–737. [CrossRef]

32. Record, E.; Punt, P.J.; Chamkha, M.; Labat, M.; van den Hondel, C.A.; Asther, M. Expression of the *Pycnoporus cinnabarinus* laccase gene in *Aspergillus niger* and characterization of the recombinant enzyme. *Eur. J. Biochem.* **2002**, *269*, 602–609. [CrossRef] [PubMed]

33. Baldrian, P. Fungal laccases-occurrence and properties. *FEMS Microbiol. Rev.* **2006**, *30*, 215–242. [CrossRef]

34. Eggert, C.; Temp, U.; Eriksson, K.-E. The ligninolytic system of the white rot fungus *Pycnoporus cinnabarinus*: Purification and characterization of the laccase. *Appl. Environ. Microbiol.* **1996**, *62*, 1151–1158. [CrossRef] [PubMed]

35. Sun, K.; Kang, F.; Waigi, M.G.; Gao, Y.; Huang, Q. Laccase-mediated transformation of triclosan in aqueous solution with metal cations and humic acid. *Environ. Pollut.* **2017**, *220*, 105–111. [CrossRef]

36. Murugesan, K.; Kim, Y.-M.; Jeon, J.-R.; Chang, Y.-S. Effect of metal ions on reactive dye decolorization by laccase from *Ganoderma lucidum*. *J. Hazard. Mater.* **2009**, *168*, 523–529. [CrossRef]

37. Lu, L.; Zhao, M.; Wang, T.-N.; Zhao, L.-Y.; Du, M.-H.; Li, T.-L.; Li, D.-B. Characterization and dye decolorization ability of an alkaline resistant and organic solvents tolerant laccase from *Bacillus licheniformis* LS04. *Bioresour. Technol.* **2012**, *115*, 35–40. [CrossRef]

38. Si, J.; Peng, F.; Cui, B. Purification, biochemical characterization and dye decolorization capacity of an alkali-resistant and metal-tolerant laccase from *Trametes pubescens*. *Bioresour. Technol.* **2013**, *128*, 49–57. [CrossRef] [PubMed]

39. Siroosi, M.; Amoozegar, M.A.; Khajeh, K. Purification and characterization of an alkaline chloride-tolerant laccase from a halotolerant bacterium, *Bacillus* sp. strain WT. *J. Mol. Catal. B Enzym.* **2016**, *134*, 89–97. [CrossRef]

40. Wang, S.-S.; Ning, Y.-J.; Wang, S.-N.; Zhang, J.; Zhang, G.-Q.; Chen, Q.-J. Purification, characterization, and cloning of an extracellular laccase with potent dye decolorizing ability from white rot fungus *Cerrena unicolor* GSM. *Int. J. Biol. Macromol.* **2017**, *95*, 920–927. [CrossRef]

41. Yan, J.; Chen, D.; Yang, E.; Niu, J.; Chen, Y.; Chagan, I. Purification and characterization of a thermotolerant laccase isoform in *Trametes trogii* strain and its potential in dye decolorization. *Int. Biodeter. Biodegr.* **2014**, *93*, 186–194. [CrossRef]

42. Younes, S.B.; Sayadi, S. Purification and characterization of a novel trimeric and thermotolerant laccase produced from the ascomycete *Scytalidium thermophilum* strain. *J. Mol. Catal. B Enzym.* **2011**, *73*, 35–42. [CrossRef]

43. Xu, X.; Huang, X.; Liu, D.; Lin, J.; Ye, X.; Yang, J. Inhibition of metal ions on *Cerrena* sp. laccase: Kinetic, decolorization and fluorescence studies. *J. Taiwan Inst. Chem. E* **2018**, *84*, 1–10. [CrossRef]

44. Bollag, J.-M.; Leonowicz, A. Comparative studies of extracellular fungal laccases. *Appl. Environ. Microbiol.* **1984**, *48*, 849–854. [CrossRef]

45.   Slomczynski, D.; Nakas, J.P.; Tanenbaum, S.W. Production and Characterization of Laccase from *Botrytis cinerea* 61. *Appl. Environ. Microbiol.* **1995**, *61*, 907–912. [CrossRef]
46.   Johannes, C.; Majcherczyk, A. Laccase activity tests and laccase inhibitors. *J. Biotechnol.* **2000**, *78*, 193–199. [CrossRef]
47.   Milstein, O.; Hüttermann, A.; Majcherczyk, A.; Schulze, K.; Fründ, R.; Lüdemann, H.-D. Transformation of lignin-related compounds with laccase in organic solvents. *J. Biotechnol.* **1993**, *30*, 37–48. [CrossRef]
48.   Luterek, J.; Gianfreda, L.; Wojtaś-Wasilewska, M.; Cho, N.S.; Rogalski, J.; Jaszek, M.; Malarczyk, E.; Staszczak, M.; Fink-Boots, M.; Leonowicz, A. Activity of free and immobilized extracellular *Cerrena unicolor* laccase in water miscible organic solvents. *Holzforsch. Int. J. Biol. Chem. Phys. Technol. Wood* **1998**, *52*, 589–595.
49.   Robles, A.; Lucas, R.; Martínez-Cañamero, M.; Ben Omar, N.; Pérez, R.; Gálvez, A. Characterization of laccase activity produced by the hyphomycete *Chalara* (*syn. Thielaviopsis*) *paradoxa* CH. *Enzym. Microb. Technol.* **2002**, *31*, 516–522. [CrossRef]
50.   Patel, I.; Kracher, D.; Ma, S.; Garajova, S.; Haon, M.; Faulds, C.B.; Berrin, J.-G.; Ludwig, R.; Record, E. Salt-responsive lytic polysaccharide monooxygenases from the mangrove fungus *Pestalotiopsis* sp. NCi. *Biotechnol. Biofuels* **2016**, *9*, 108. [CrossRef]
51.   Li, Z.; Jiang, S.; Xie, Y.; Fang, Z.; Xiao, Y.; Fang, W.; Zhang, X. Mechanism of the salt activation of laccase Lac. *Biochem. Biophys. Res. Commun.* **2020**, *521*, 997–1002. [CrossRef]
52.   Kern, M.; McGeehan, J.E.; Streeter, S.D.; Martin, R.N.A.; Besser, K.; Elias, L.; Eborall, W.; Malyon, G.P.; Payne, C.M.; Himmel, M.E.; et al. Structural characterization of a unique marine animal family 7 cellobiohydrolase suggests a mechanism of cellulase salt tolerance. *Proc. Natl. Acad. Sci. USA* **2013**, *110*, 10189–10194. [CrossRef]
53.   Paul, S.; Bag, S.K.; Das, S.; Harvill, E.T.; Dutta, C. Molecular signature of hypersaline adaptation: Insights from genome and proteome composition of halophilic prokaryotes. *Genome Biol.* **2008**, *9*, R70. [CrossRef]
54.   Lanyi, J.K. Salt-dependent properties of proteins from extremely halophilic bacteria. *Bacteriol. Rev.* **1974**, *38*, 272. [CrossRef]
55.   Gordon, C.L.; Khalaj, V.; Ram, A.F.; Archer, D.B.; Brookman, J.L.; Trinci, A.P.; Jeenes, D.J.; Doonan, J.H.; Wells, B.; Punt, P.J. Glucoamylase: Green fluorescent protein fusions to monitor protein secretion in *Aspergillus niger*. *Microbiology* **2000**, *146*, 415–426. [CrossRef]
56.   Van Hartingsveldt, W.; Mattern, I.E.; van Zeijl, C.M.; Pouwels, P.H.; van den Hondel, C.A. Development of a homologous transformation system for *Aspergillus niger* based on the pyrG gene. *Mol. Gen. Genet. MGG* **1987**, *206*, 71–75. [CrossRef]
57.   Edgar, R.C. MUSCLE: A multiple sequence alignment method with reduced time and space complexity. *BMC Bioinform.* **2004**, *5*, 113. [CrossRef]
58.   Chojnacki, S.; Cowley, A.; Lee, J.; Foix, A.; Lopez, R. Programmatic access to bioinformatics tools from EMBL-EBI update: 2017. *Nucleic Acids Res.* **2017**, *45*, W550–W553. [CrossRef]
59.   Armenteros, J.J.A.; Tsirigos, K.D.; Sønderby, C.K.; Petersen, T.N.; Winther, O.; Brunak, S.; von Heijne, G.; Nielsen, H. SignalP 5.0 improves signal peptide predictions using deep neural networks. *Nat. Biotechnol.* **2019**, *37*, 420–423. [CrossRef]
60.   Steentoft, C.; Vakhrushev, S.Y.; Joshi, H.J.; Kong, Y.; Vester-Christensen, M.B.; Schjoldager, K.T.-B.; Lavrsen, K.; Dabelsteen, S.; Pedersen, N.B.; Marcos-Silva, L. Precision mapping of the human O-GalNAc glycoproteome through SimpleCell technology. *EMBO J.* **2013**, *32*, 1478–1488. [CrossRef]
61.   Kelley, L.A.; Mezulis, S.; Yates, C.M.; Wass, M.N.; Sternberg, M.J. The Phyre2 web portal for protein modeling, prediction and analysis. *Nat. Protoc.* **2015**, *10*, 845–858. [CrossRef] [PubMed]

4

AlgM4: A New Salt-Activated Alginate Lyase of the PL7 Family with Endolytic Activity

Guiyuan Huang [1], **Qiaozhen Wang** [1], **Mingqian Lu** [1], **Chao Xu** [1], **Fei Li** [1], **Rongcan Zhang** [1], **Wei Liao** [1,2] **and Shushi Huang** [1,*]

[1] Guangxi Key Laboratory of Marine Natural Products and Combinatorial Biosynethesis Chemistry, Guangxi Academy of Sciences, Nanning 530007, China; guiyuan1105@163.com (G.H.); wqzh-333@163.com (Q.W.); lumq520@126.com (M.L.); 18174662721@163.com (C.X.); lifei@gxas.cn (F.L.); zhangrongcan@126.com (R.Z.); laoweimail@163.com (W.L.)

[2] The Food and Biotechnology, Guangxi Vocational and Technical College, Nanning 530226, China

* Correspondence: hshushi@gxas.cn.

Abstract: Alginate lyases are a group of enzymes that catalyze the depolymerization of alginates into oligosaccharides or monosaccharides. These enzymes have been widely used for a variety of purposes, such as producing bioactive oligosaccharides, controlling the rheological properties of polysaccharides, and performing structural analyses of polysaccharides. The *algM4* gene of the marine bacterium *Vibrio weizhoudaoensis* M0101 encodes an alginate lyase that belongs to the polysaccharide lyase family 7 (PL7). In this study, the kinetic constants V_{max} (maximum reaction rate) and K_m (Michaelis constant) of AlgM4 activity were determined as 2.75 nmol/s and 2.72 mg/mL, respectively. The optimum temperature for AlgM4 activity was 30 °C, and at 70 °C, AlgM4 activity dropped to 11% of the maximum observed activity. The optimum pH for AlgM4 activity was 8.5, and AlgM4 was completely inactive at pH 11. The addition of 1 mol/L NaCl resulted in a more than sevenfold increase in the relative activity of AlgM4. The secondary structure of AlgM4 was altered in the presence of NaCl, which caused the α-helical content to decrease from 12.4 to 10.8% and the β-sheet content to decrease by 1.7%. In addition, NaCl enhanced the thermal stability of AlgM4 and increased the midpoint of thermal denaturation (Tm) by 4.9 °C. AlgM4 exhibited an ability to degrade sodium alginate, poly-mannuronic acid (polyM), and poly-guluronic acid (polyG), resulting in the production of oligosaccharides with a degree of polymerization (DP) of 2–9. AlgM4 possessed broader substrate, indicating that it is a bifunctional alginate lyase. Thus, AlgM4 is a novel salt-activated and bifunctional alginate lyase of the PL7 family with endolytic activity.

Keywords: *Vibrio weizhoudaoensis*; alginate lyase; PL7 family; salt-activated enzyme

1. Introduction

Alginic acid has the ability to form viscous solutions and gels in aqueous media and is nontoxic to living organisms. Therefore, it has been widely used in the pharmaceutical, cosmetic, food, and biotech industries [1]. In addition, the degradation products of alginic acid—alginate oligosaccharides—have a wide range of biological activities, such as the promotion of growth and the alleviation of abiotic stress in plants; antitumour, antibacterial, anti-inflammatory, anticoagulant, antioxidative, and immunomodulatory activities; and the reduction of free radicals and blood glucose and lipids. Alginate oligosaccharides have broad application prospects in the green agriculture, medical, food, and household chemical industries, among others [2].

Alginic acid, also known as algin or alginate, is a straight-chain polysaccharide composed of β-D-mannuronic acid (M) and its C5 stereoisomer α-L-guluronic acid (G), which are randomly linked via α-1,4-glycosidic bonds. Alginate molecules can have the sugar monomers M and G arranged in three ways: M monomers can be linked in succession, forming poly-mannuronic acid (polyM); G monomers can be linked in succession, forming poly-guluronic acid (polyG); and M and G monomers can be randomly and alternately linked, forming poly-guluronic acid -mannuronic acid (polyMG) [3]. The relative proportions of M and G vary among the alginic acids derived from different organisms [4,5].

Alginate lyases catalyze the cleavage of the 1,4-glycosidic bonds between the uronic acid monomers in alginate, resulting in the production of oligouronic acids or uronic acid monomers. Alginate lyases are present in a wide range of organisms and can be isolated from marine algae, molluscs, and microorganisms, as well as soil microorganisms. Currently, the primary sources of alginate lyases are marine bacteria, including members of the genera *Pseudomonas* and *Vibrio* [6].

The catalytic mechanisms of alginate lyases are well understood. Alginate lyases catalyze the cleavage of 1,4-glycosidic bonds in alginic acid via a β-elimination reaction. In addition, a double bond is formed between C4 and C5 of the saccharide ring containing the 4-*O* glycosidic bond, generating an oligomer with 4-deoxy-L-erythro-hex-4-enepyranosyluronate at the nonreducing end [7]. Based on their catalytic characteristics, alginate lyases are divided into endolytic and exolytic alginate lyases [8,9]. According to the substrate specificity, endolytic alginate lyases can be divided into mannuronate lyases (polyM lyase, EC 4.2.2.3) and guluronate lyases (polyG lyase, EC 4.2.2.11). Bifunctional enzymes that exhibit activities towards both polyG and polyM have also been identified [10].

While numerous alginate lyases have been discovered, few studies have focused on the enzymatic properties of alginate lyases. In recent years, a number of alginate lyases have been discovered and reported, including the cold-adapted alginate lyases [11–13], thermostable alginate lyases [14], high-alkaline alginate lyases [15], and salt-activated alginate lyases [12,15–17]. According to the evolution and homology of amino acid sequences, most alginate lyases belong to seven families of polysaccharide lyases (PL-5, PL-7, PL-14, PL-15, PL-17, and PL-18) 18]. Most of the reported alginate lyases have endolytic activity [15,19–21] and hydrolyze sodium alginate (SA) to produce oligosaccharides. Alginate lyase A1-IV, produced by the bacterial strain *Sphingomonas* sp. A1 [22], and alginate lyase Atu3025, produced by *Agrobacterium tumefaciens* C58 23], possess exolytic activity and hydrolyze SA into mannuronate or guluronate. Alginate lyases that degrade alginate into monosaccharides are part of the PL15 family [22,23], whereas alginate lyases with endolytic activity belong to the PL-5, 6, 7, 14, 16, 17, and 18 families [24].

The marine bacterium *Vibrio weizhoudaoensis* M0101 harbors the *algM4* gene, which encodes an alginate lyase belonging to polysaccharide lyase family 7 (PL7). In this study, we purified exogenously expressed AlgM4 and observed it to exhibit high salt tolerance, as AlgM4 activity increased more than sevenfold in the presence of 1 mol/L NaCl. The result is different from those for other salt-activated alginate lyases for which enzyme activity is decreased at 1 mol/L NaCl [15,25]. In the depolymerization of a high content of sodium alginate, Alg2A generated equal total molar amounts of oligosaccharides, but the amounts of oligosaccharides with DP of 5–10 were higher than those for both of the two commercial enzymes [26]. AlgM4 showed activities toward both polyM and polyG, which may degrade alginate more effectively. Moreover, AlgM4 catalyzing polyM released oligosaccharides with DP 7–9 from the polyM, which was different from the previously reported endolytic alginate lyases, despite their diverse substrate specificities [27,28]. Therefore, the unique endolytic reaction mode of AlgM4 gives it a distinct advantage in facilitating uronic acid oligosaccharides with high DPs. AlgM4 could be a good tool for the preparation of alginate oligosaccharides. AlgM4 not only functions as a key enzyme in the preparation and functional study of oligosaccharides but also plays an important role in utilization of alginate for ethanol fermentation.

2. Results and Discussion

2.1. Analysis of the AlgM4 Sequence

The alginate lyase gene *algM4* is 1563 bp in length and encodes a 520-amino-acid protein. A signal peptide analysis of AlgM4 using SignalP 4.0 (http://www.cbs.dtu.dk/services/SignalP/) predicted that the N-terminus of AlgM4 contains a 24-amino-acid signal peptide (Figure 1). A sequence alignment using the National Center for Biotechnology Information (NCBI) BLAST (https://blast.ncbi.nlm.nih.gov/) engine revealed that AlgM4 is a dual-domain protease, containing an N-terminal F5/8 type C domain and a C-terminal alginate-lyase 2 domain. The alginate lyases of the PL7 family contain three highly conserved domains: SA3 (RXEXR), SA4 (YXKAGXYXQ), and SA5 (QXH). The BLAST sequence analysis showed that AlgM4 contains the conserved amino acid sequences of the SA3 domain (RTEMR), the SA4 domain (YFKAGVYNQ), and the SA5 catalytic domain (QIH) (Figure 1). Therefore, AlgM4 belongs to the PL7 family of alginate lyases. The amino acid sequence of AlgM4 was compared with the sequences of other alginate lyases of the PL7 family selected in the CAZy database (http://www.cazy.org/), and a phylogenetic tree was constructed using the neighbor joining method (Figure 2). AlgM4 was most closely related to an alginate lyase from *Vibrio litoralis* BZM-2 (ALP75562.1), with an amino acid sequence similarity of 74% observed between AlgM4 and ALP75562.1, without annotation by genome analysis. These results suggest that AlgM4 is a new alginate lyase of the PL7 family.

MKHKFLKTLLASSVLFAVGCTSNGNDTSQLHPQSDTGAPVLTPVAIEASSHDGNGPDRLFDQDINTR
　　　　　　Signal peptide
WSASGDGEWAVLDYGSVHEFDAIRAAFSKGNERQSIFDILVSTDGETWTPVLENVQSSGGVIGYERFE
FAPVQARYVKYVGHGNTANGWNSVTELAAIKCGVNACPSNHIITPEVIAAEQGLIAKQKAEEAARK
EARKDLRKGNFGAPAVYPCQTTVKCGKTALPVPTGLPATPKAGNAPSENFDLTSWYLSQPFDHNND
NRPDDVSEWDLANGYSHPDVFYTADDGGLVFKSFVKGVRTSPNTKYARTEMREMLRRGDTSIPTKG
　　　　　　　　　　　　　　　　　　　　　　　　　　　　SA3
VNKNNWVFSSAPVADQKAAGGVDGVMEATLKIDHTTTTGEAGEVGRFIIGQIHDQDDEPIRLYYRKL
　　　　　　　　　　　　　　　　　　　　　　　SA5
PNQDKGTVYFAHENTLKGTDQYFDLVGGMTGEIGDDGIALGEKFSYRIEVKGNTLTVTVMREGKPDV
KQVVDMSESGYDVGGKYMYFKAGVYNQNISGEMDDYVQATFYKLEKSHGSYKG-
　　　　　SA4

Figure 1. The deduced amino acid sequences of AlgM4. The signal peptide is underlined; SA3, SA5, and SA4 are indicated with blue symbols.

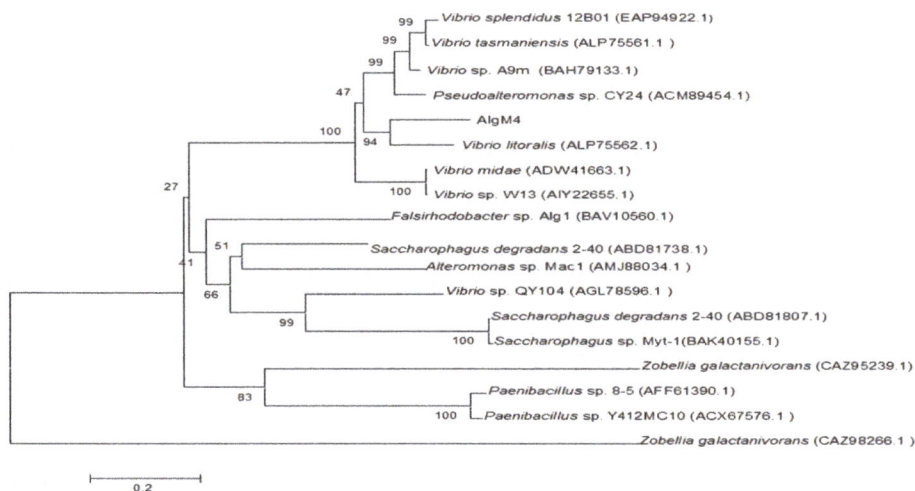

Figure 2. Neighbor-joining phylogenetic tree of *V. weizhoudaoensis* strain M0101 based on putative AlgM4 protein sequences.

2.2. Purification and Enzymatic Activity of AlgM4

Using *V. weizhoudaoensis* M0101 genomic DNA as a template, the *algM4* gene was PCR amplified without the N-terminal signal peptide sequence or the stop codon and ligated into the pET30a(+) plasmid. The plasmid was then transformed into *Escherichia coli* BL21 (DE3) cells for AlgM4 expression. Purified AlgM4 protein with a C-terminal 6×histidine (His) tag was obtained by Ni^{2+} affinity chromatography. Sodium dodecyl sulfate polyacrylamide gel electrophoresis (SDS-PAGE) showed a single protein band with a molecular weight of approximately 55 kDa (Figure 3). The specific activity of the purified AlgM4 was 4638 U/mg (Table 1), exhibiting a 36.7% increase in specific activity compared with the crude enzyme extract.

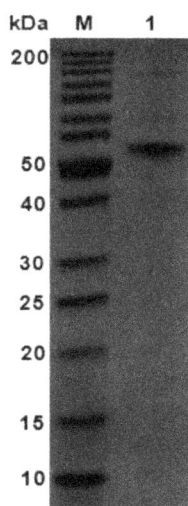

Figure 3. SDS-PAGE analysis of purified AlgM4. Lane M, molecular weight markers; Lane 1, purified AlgM4.

Table 1. Summary of AlgM4 purification.

Steps	Total Protein (mg)	Total Activity (U)	Specific Activity (U/mg)	Recover (%)	Purification (Fold)
Crude enzyme AlgM4	58.352	73878	1266	100	1
Purified AlgM4	11.55	53570	4638	72.5	3.7

2.3. V_{max} (Maximum Reaction Rate) and K_m (Michaelis Constant) of AlgM4

AlgM4 activity was measured when incubated with various concentrations of substrate (SA) in a water bath at 30 °C for 10 min, and the resulting values were used to calculate the kinetic constants of AlgM4 activity. The V_{max} and K_{cat} values of AlgM4 were 2.75 nmol/s and 30.25 S^{-1}, respectively, and the K_m value was 2.72 mg/mL. Two other NaCl-activated enzymes—AlyPM from *Pseudoalteromonas* sp. SM0524 and AlySY08 from *Nitratiruptor* sp. SB155-2—were previously shown to have K_m values of 74.39 [17] and 0.36 mg/mL [14], respectively. The oligoalginate lyase Alg17C derived from *Saccharophagus degradans* 2-40 had an observed K_m value of 35.2 mg/mL [18]. *Vibrio splendidus* 12B01 expresses three oligoalginate lyases—OalA, OalB, and OalC—which had observed K_m values of 3.25, 0.76, and 0.53 mg/mL, respectively [29].

2.4. Enzymatic Characteristics of AlgM4

AlgM4 exhibited high enzymatic activity at 20–40 °C, which decreased to 11% of the maximal observed activity when incubated at 70 °C. Similar to other NaCl-activated alginate lyases in the PL7 family (such as A1m [15], rA9mT [16], and AlyPM [17]), AlgM4 had highest enzymatic activity at 30 °C (Figure 4A).

Many alginate lyases have optimal pH values between pH 7 and 8, in contrast to the optimal high-alkaline pH values of pectate lyases [20]. The optimum pH for AlgM4 activity was assessed, and the results are shown in Figure 4B. The optimum reaction pH was 8.5, indicating that AlgM4 was weakly basophilic. In slightly alkaline environments (pH 7.5–9.0), AlgM4 retained 80% activity, whereas AlgM4 was completely inactive at pH 11.

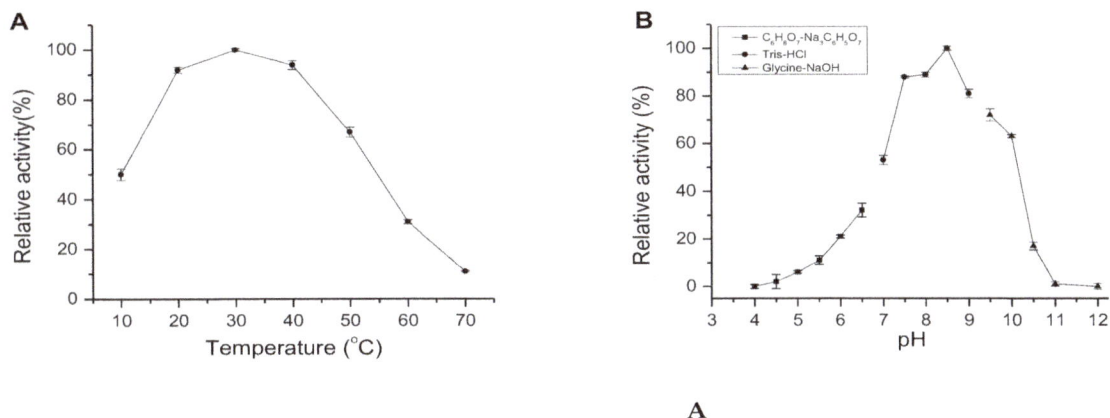

Figure 4. Effect of temperature and pH on AlgM4 activity. ( ) The optimal temperature for AlgM4 activity. The activity of AlgM4 at 30 °C was completely retained; (**B**) The optimal pH for AlgM4 activity. The activity of AlgM4 at pH 8.5 was completely retained.

The activity of AlgM4 remained stable for 30 min at 25 °C, regardless of the presence or absence of 1 mol/L NaCl (Figure 5A). After a 30 min incubation at 30 or 35 °C in the presence of 1 mol/L NaCl, AlgM4 retained 92% of its initial activity, which decreased rapidly at temperatures exceeding 40 °C and decreased by 63% at 45 °C. In the absence of NaCl, the enzymatic activity of AlgM4 was reduced by 94% at 45 °C. An examination of temperature tolerance showed that NaCl improved the thermal stability of AlgM4.

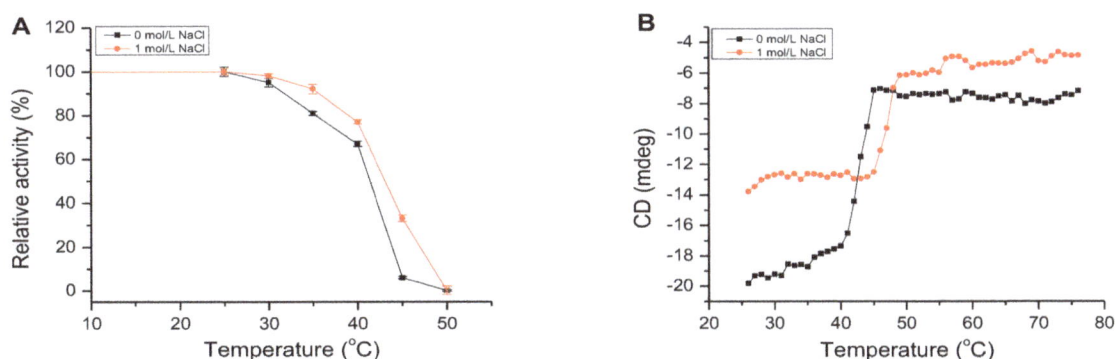

Figure 5. The thermal stability and melting temperature (Tm) of AlgM4. (**A**) The thermal stability of AlgM4. The residual activity of AlgM4 at 25 °C was completely retained; (**B**) Circular dichroism signals at 218 nm were used for analysis of the Tm value.

The effects of metal ions and surfactants on AlgM4 activity are shown in Table 2. Ca^{2+} did not influence enzyme activity, unlike for other alginate lyases [12,17,20]. While Mg^{2+} promoted the activity of AlgM4, Cu^{2+}, Mn^{2+}, and Zn^{2+} inhibited AlgM4 activity. The most significant inhibitory effect on AlgM4 was observed by Zn^{2+}, which caused an 82% reduction in AlgM4 activity. Ethylene diamine tetraacetic acid (EDTA) and SDS suppressed AlgM4 activity to varying degrees. The anionic surfactant SDS strongly inhibited the enzymatic activity of AlgM4, causing a 97% reduction in AlgM4 activity, while EDTA reduced AlgM4 activity by 35%.

Table 2. Effect of chemical reagents on AlgM4 activity.

Reagent	Concentration (mmol/L)	Relative Activity (%)
None	-	100 ± 0.3
$CaCl_2$	1	97 ± 0.1
$MgCl_2$	1	114 ± 1.1
KCl	1	91 ± 0.9
$CuCl_2$	1	75 ± 1.7
$MnCl_2$	1	77 ± 2.4
$ZnCl_2$	1	18 ± 4.5
EDTA	1	65 ± 6.2
SDS	1	3 ± 3.1

The data are expressed as the means $\pm$ SD, n = 3. The activity of AlgM4 in the absence of chemical reagents was completely retained.

2.5. Effect of NaCl on AlgM4 Activity

Different concentrations of NaCl were added to 0.1 mg/mL purified AlgM4, and the enzymatic activity of AlgM4 was then examined. AlgM4 was promoted at NaCl concentrations of 0.1–1.4 mol/L and was greatest in the presence of 1.0 mol/L NaCl, exhibiting more than 7 times the activity observed in the absence of NaCl (Table 3). The activity of AlyPM in the presence of 0.5–1.2 mol/L NaCl was 6 times that observed in the absence of NaCl [17]. Under optimum reaction conditions, the activity of A1m in the presence of 0.6–0.8 mol/L NaCl was 20 times that observed in the absence of NaCl [15]. The activity of rA9mT in the presence of 0.4 mol/L NaCl was 24 times that observed in the absence of NaCl [16].

Table 3. Effect of NaCl on AlgM4 activity.

NaCl Concentration (mol/L)	Relative Activity (%)
0	100 ± 0.2
0.1	229 ± 5.6
0.2	286 ± 6.4
0.3	356 ± 3.8
0.4	447 ± 1.4
0.5	481 ± 0.8
0.6	528 ± 3.1
0.7	572 ± 0.4
0.8	628 ± 3.4
0.9	664 ± 1.9
1.0	741 ± 3.2
1.2	462 ± 5.8
1.4	236 ± 4.5

The data are expressed as the means $\pm$ SD, n = 3. The activity of AlgM4 in the absence of NaCl was completely retained.

2.6. Determination of the Secondary Structure and Thermal Denaturation Temperature of AlgM4

The thermal denaturation temperature of AlgM4 was determined by circular dichroism (CD) spectroscopy at 25–75 °C. In addition, the secondary structure of AlgM4 was determined by ultraviolet–visible (UV–Vis) spectroscopy and CD spectroscopy. The CD absorption values of AlgM4 protein at 218 nm in different temperatures are shown in Figure 5B, and the CD values were used for analysis of the Tm value. The CD values of AlgM4 relatively remained stable at 25–35 °C in the absence of NaCl; subsequently, CD values slowly increased with the increase of temperature and the denaturation of AlgM4, then increased dramatically at 40–45 °C and remained stable at temperatures exceeding 45 °C. The AlgM4 protein was denatured completely when the temperature was higher than 45 °C. However, in the presence of 1 mol/L NaCl, the CD values increased rapidly at 45–50 °C and remained stable at temperatures exceeding 50 °C. The results showed that NaCl enhanced the ability of AlgM4 to resist thermal denaturation.

As shown in Figure 6A, the UV absorbance of AlgM4 at 220–240 nm was significantly reduced in the presence of 1 mol /L NaCl. CD spectroscopy was used to determine the secondary structure of the purified AlgM4, and the results are shown in Figure 6B. The absorption spectrum of AlgM4 was characterized by positive and negative peaks at approximately 198 and 218 nm, respectively, and the intensity of these peaks was reduced by the addition of 1 mol/L NaCl. The results showed that the α-helix and β-sheet contents in the secondary structure of AlgM4 were decreased after addition of 1 mol/L NaCl. Specifically, the α-helix and β-sheet contents were reduced from 12.4% and 38.2% to 10.8% and 36.5%, respectively. The changes in the contents of secondary structural elements are summarized in Table 4. The secondary structure of AlgM4 was altered in the presence of 1 mol/L NaCl, which may enhance the affinity of the enzyme for its substrates and facilitate enzymolysis. The alginate lyase AlyPM, derived from *Pseudoalteromonas* sp. SM0524, differs from AlgM4 in that the presence of NaCl did not alter the secondary structure of AlyPM. However, NaCl enhanced the affinity of AlyPM for its substrates, thereby promoting enzymolysis [17]. In addition, AlyPM is a cold-adapted enzyme, and its thermal denaturation temperature is relatively low, with a Tm of 37 °C [17]. At 1 mol/L, NaCl not only altered the secondary structure of AlgM4 but also enhanced the ability of AlgM4 to resist thermal denaturation. As a result, the midpoint of thermal denaturation (Tm) was increased from 43.3 to 48.2 °C (Figure 5B).

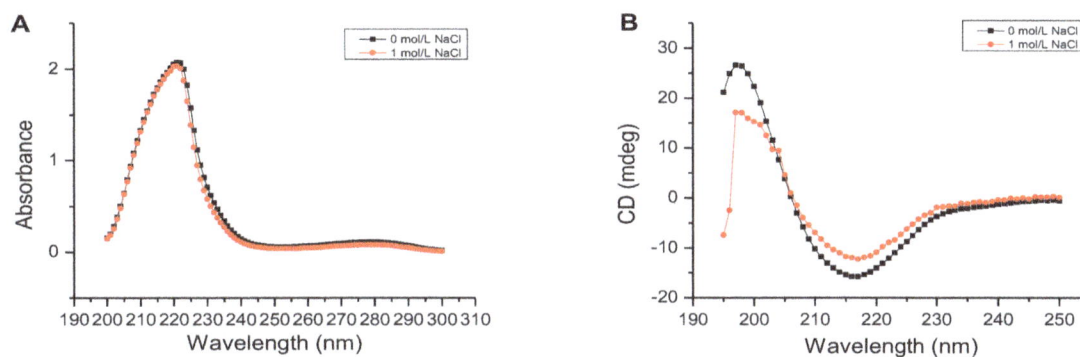

Figure 6. Determination of the secondary structure of the AlgM4. (**A**) Determination of the secondary structure by UV–Vis absorption spectra; (**B**) Determination of the secondary structure by circular dichroism.

Table 4. Secondary structure of AlgM4 as estimated by CD.

Enzyme	α-Helix (%)	β-Sheet (%)	β-Turu (%)	Random Coil (%)
AlgM4	12.4	38.2	21.3	28
AlgM4 + 1 mol/L NaCl	10.8	36.5	22.9	29.6

2.7. Analysis of the Products of AlgM4-Mediated Enzymolysis of Alginate by Ultra-Performance Liquid Chromatography (UPLC)–Quadrupole Time-of-Flight (QTOF)–Mass Spectrometry (MS)/MS

Alginate lyases with endolytic characteristics generally act on glycosidic bonds within the linear polysaccharide chain of alginate, generating unsaturated oligosaccharides that are dominated by disaccharides, trisaccharides, and tetrasaccharides [27]. Exolyases further depolymerize these oligosaccharides into mannuronic acids [18,29,30]. AlgM4 degrades both SA (Figure 7A) and polyG (Figure 7B) to produce oligosaccharides with degree of polymerization DP 2–6 [31,32]. The content of each oligosaccharide can be determined only after quantitative determination. Unlike SA and polyG, degradation of polyM produces oligosaccharides DP7, DP8, and DP9 (Figure 9). Furthermore, AlgM4 may be useful in the preparation of oligosaccharides, especially with high DP 7–9, and the study of their biological functions.

In recent years, the biological activities of oligosaccharides and the application of oligosaccharides in the medical and biotechnology fields has attracted the attention of researchers [33]. Oligosaccharides have higher degrees of polymerization and possibly better bioactivity [26]. An et al. observed that oligosaccharides (DP 6–8) derived from SA stimulated the accumulation of phytoalexin and induced phenylalanine ammonia lyase in soybean cotyledons, resulting in their acquired resistance to *Pseudomonas aeruginosa* [34]. In addition, trisaccharides, tetrasaccharides, pentasaccharides, and hexasaccharides obtained by enzymatic degradation of SA promoted the growth of lettuce seedlings [35].

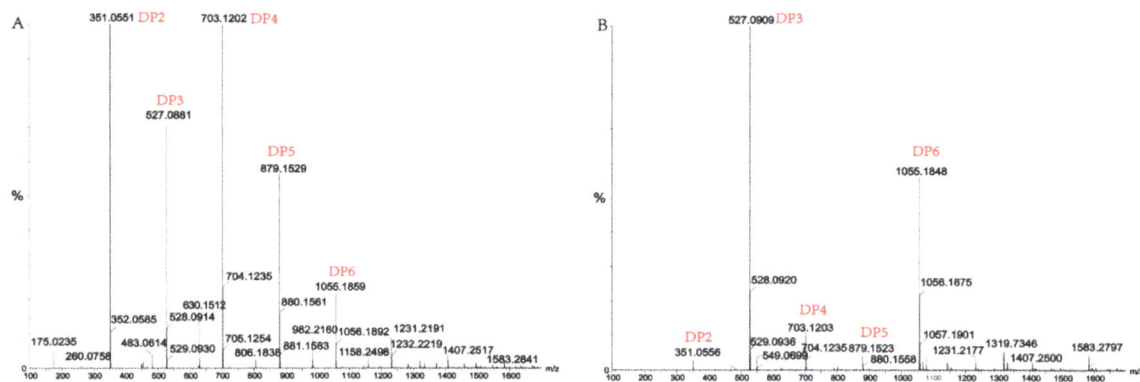

Figure 7. (**A**) Ultra-Performance Liquid Chromatography (UPLC)–Quadrupole Time-of-Flight (QTOF)–MS/MS analysis of hydrolysates of AlgM4 with sodium alginate as the substrate; (**B**) UPLC–QTOF–MS/MS analysis of hydrolysates of AlgM4 with polyG as the substrate. DP indicates the degree of polymerization of oligosaccharides from the alginate lyase hydrolysates.

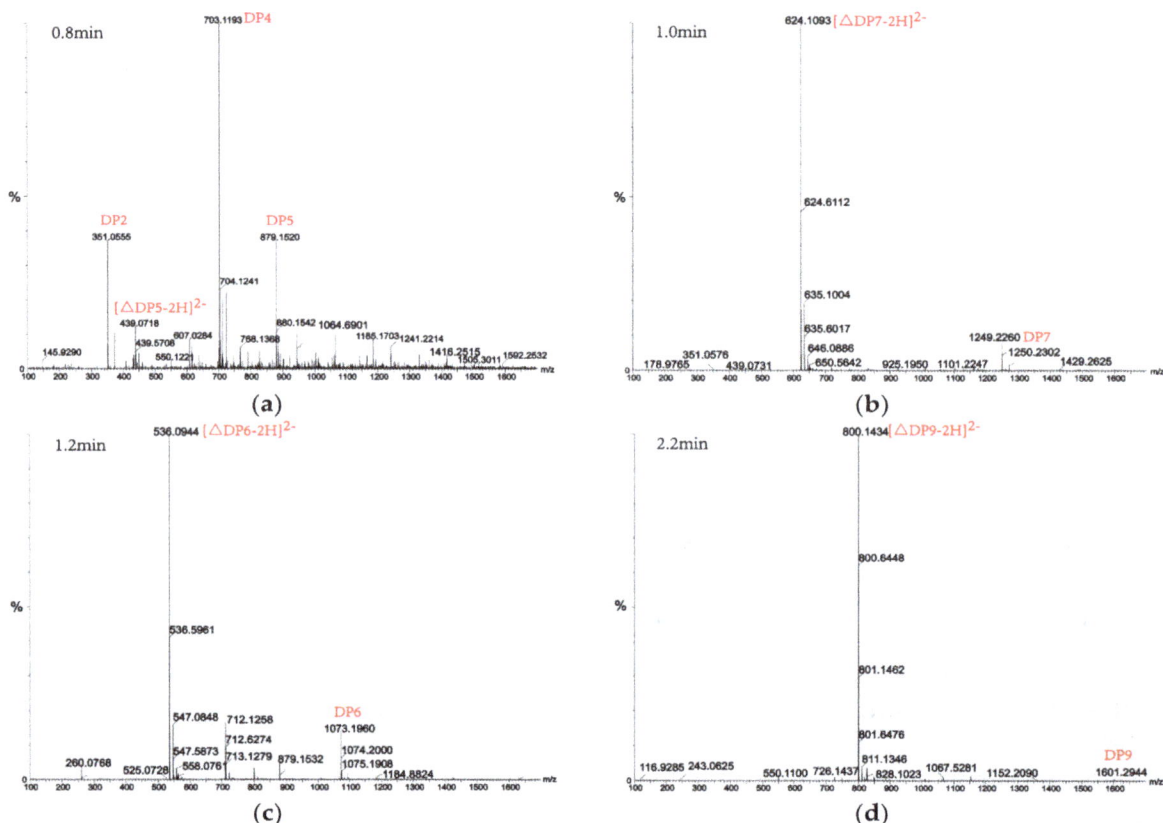

Figure 8. *Cont.*

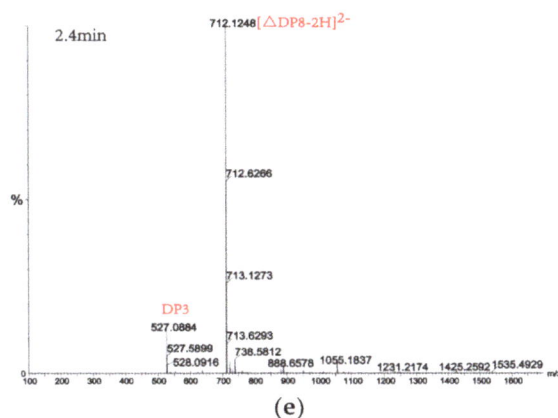

(e)

Figure 9. UPLC–QTOF–MS/MS analysis of hydrolysates of AlgM4 with polyM as the substrate. DP indicates the degree of polymerization of oligosaccharides from the alginate lyase hydrolysates. The reaction products of disaccharide, tetrasaccharide and pentasaccharide were detected at 0.8 min (**a**). The major peacks of DP7, DP6, DP9 and DP8 were detected at 1.0 min (**b**), 1.2 min (**c**), 2.2 min (**d**) and 2.4 min (**e**), respectively.

3. Materials and Methods

3.1. The Bacterium

The marine bacterium *V. weizhoudaoensis* M0101 was isolated from rotten Sargassum collected from Weizhou Island, Beihai, Guangxi Province, China.

3.2. Cloning and Expression of the algM4 Gene

The primers were designed according to the nucleic acid sequence of the *algM4* gene. The following primers were used to amplify the *algM4* gene: upstream primer, 5′-GGA ATTCCATATGCTTGCATCTTCTGTG-3′ (the *Nde*I restriction site is underlined); downstream primer, 5′-CCGCTCGAGACCTTTATAAGAACCGTG-3′ (the *Xho*I restriction site is underlined). The parameters of the polymerase chain reaction (PCR) were as follows: 94 °C for 2 min; followed by 30 cycles of 94 °C for 30 s, 58 °C for 30 s, and 72 °C for 1 min 40 s; with a final incubation at 72 °C for 10 min. The PCR product was double digested with the *Nde*I and *Xho*I restriction enzymes and then ligated into the pET30a(+) vector to construct the recombinant plasmid pET30a-*algM4*. To induce the expression of AlgM4, the recombinant plasmid was transformed into the *E. coli* strain BL21 (DE3). The positive transformants were picked, inoculated into 10 mL of LB (Luria-Bertani) medium containing 50 μg/mL kanamycin (Kan) and cultured at 37 °C with shaking (200 rpm) until the OD600 of the culture reached 0.5. The culture was then inoculated into LB medium containing 50 μg/mL Kan (inoculum volume: 1% (*v/v*)) and cultivated until the OD600 value reached 0.5. Subsequently, isopropyl β-D-1-thiogalactopyranoside (IPTG) was added at a final concentration of 0.5 mmol/L. After the addition of IPTG, the bacteria were cultured for another 12 h at 25 °C with shaking (200 rpm).

3.3. Purification of AlgM4

Protein isolation and purification was carried out at 0–4 °C. The IPTG-induced bacteria were harvested by centrifugation at 6000 rpm for 10 min. Subsequently, the bacteria were resuspended in buffer (10 mmol/L imidazole, 300 mmol/L NaCl, and 20 mmol/L Tris-HCl; pH 7.0) and lysed by ultrasonication. The lysates were centrifuged at 12,000 rpm for 30 min, and the resulting crude enzyme solution was collected. After washing off the protein impurities with buffer (50 mmol/L imidazole, 300 mmol/L NaCl, and 20 mmol/L Tris-HCl; pH 7.0), AlgM4 bound to the Ni^{2+} column was eluted with elution buffer containing 150 mmol/L imidazole, 300 mmol/L NaCl, and 20 mmol/L

Tris-HCl (pH 7.0). The purified protein was analyzed by SDS-PAGE, and the protein concentration was determined using the Bradford method.

3.4. Determination of the Enzymatic Properties of AlgM4

Enzymatic properties were determined using 0.1 mg/mL AlgM4. Nine hundred microliters of a 1.0% (w/v) SA solution was mixed with 100 μL of purified AlgM4 and incubated in a water bath at 30 °C for 10 min. The mixture was then boiled in a water bath for 5 min to terminate the reaction. After the reaction system was cooled to room temperature, the absorbance was measured at 235 nm. The unit of enzymatic activity (U) was defined as an increase in absorbance of 0.01 per minute. To determine the optimum reaction temperature, the enzymatic activity of AlgM4 was measured at 10–70 °C in 100 mmol/L Tris-HCl (pH 7.0) buffer using 0.2% (w/v) SA as a substrate. To determine the optimum reaction pH, the activity of AlgM4 was evaluated in 100 mmol/L citric acid–sodium citrate buffer (pH value: 4.5–6.5), 100 mmol/L Tris-HCl buffer (pH value: 7.0–9.0), or 100 mmol/L glycine-sodium hydroxide buffer (pH value: 9.5–12.0) at the optimum reaction temperature. To test the thermal stability of the enzyme, AlgM4 was first incubated at 25–50 °C for 30 min, and the residual AlgM4 activity was then measured under the optimum reaction conditions. To examine the effects of metal ions and surfactants on the enzymatic activity of AlgM4, metal ions (Ca^{2+}, K^+, Mg^{2+}, Mn^{2+}, Zn^{2+}, and Cu^{2+}), EDTA, or SDS were added to 50 μL of enzyme solution at a final concentration of 1 mmol/L. The activity of AlgM4 was then measured under the optimum reaction conditions using a 0.2% (w/v) SA solution as a substrate. The effect of NaCl on AlgM4 activity was also examined. NaCl was added to the enzyme solution at final concentrations of 0.1–1.4 mol/L, and the activity of AlgM4 was then compared to that observed under optimum reaction conditions.

3.5. Determination of the Kinetic Constants for AlgM4 Activity

The enzymatic activity of AlgM4 (0.1 mg/mL) was measured under the optimum reaction conditions using 3,5-dinitrosalicylic acid (DNS) assay. The concentrations of the substrate, SA, assayed were 0.5–10 mg/mL. Lineweaver-Burk (double-reciprocal) plots were generated using $1/[S]$ as the abscissa and $1/V$ as the ordinate. The V_{max} and K_m values were calculated using the Michaelis-Menten equation.

3.6. The Effects of NaCl on the Secondary Structure and Thermal Stability of AlgM4

In one group, NaCl was added to a 300 μL reaction containing 0.28 mg/mL of purified AlgM4 (final concentration of NaCl: 1 mol/L), while another group was not exposed to NaCl. The secondary structure of the purified AlgM4 protein was determined at 25 °C using a TU-1901 Dual Beam Ultraviolet Spectrophotometer (PERSEE, Beijing, China) (spectral range: 200–350 nm) and a Chirascan Circular Dichroism Spectrometer (Applied Photophysics Ltd., Surrey, UK) (spectral range: 195–250 nm; optical path length: 10 mm; bandwidth: 0.5 nm). The composition of the secondary structural elements was analyzed using CDpro software (http://sites.bmb.colostate.edu/sreeram/CDPro/CDPro.htm). The thermal denaturation temperature of AlgM4 was measured under the following conditions: spectral range, 200–260 nm; bandwidth, 0.7 nm; 25–75 °C. The Tm value was calculated using Global 3 software (Applied Photophysics Ltd., Surrey, UK).

3.7. Analysis of the Products of AlgM4-Mediated Enzymolysis Using UPLC–QTOF–MS/MS

The purified AlgM4 (0.25 mg/mL) enzyme was mixed with an equal volume of 1.0% (w/v) SA, polyM, or polyG; incubated in a water bath at 30 °C for 6 h; and then concentrated in vacuo. After high-speed centrifugation, the supernatants were collected and filtered through filter membranes with a 0.22 μm pore size. The filtrates were analyzed using liquid chromatography (LC)-MS. The equipment used in the analysis was a UPLC-QTOF-MS/MS system (Waters Corporation, Milford, MA, USA), which consisted of a UPLC I-Class instrument (Waters Corporation, Singapore) and a XEVO G2-S mass spectrometer (Waters Corporation, Milford, MA, USA). The operating conditions of LC were as follows: ACQUITY UPLC HSS T3 C18 column (2.1 mm × 100 mm, 1.8 μm, Waters Corporation, Milford, MA, USA); gradient elution with 0.1% formic acid-water (A) and formic acid-acetonitrile (B); flow rate, 0.5 mL/min; column temperature, 35 °C; analytic time, 10 min; injection volume, 1.0 μL. The MS conditions were as follows: ion scan mode, negative-ion ESI mode; scan range, 100–1700 Da; ion source temperature, 100 °C; desolvation gas temperature, 400 °C; desolvation gas flow rate, 1000 L/h; capillary voltage, 2.5 kV; cone voltage, 40 V; low collision energy, 6 V; high collision energy, 35–50 V; data acquisition software, MassLynx 4.1 SCN 884 (Waters Corporation, Milford, MA, USA); data acquisition mode, MSE.

4. Conclusions

AlgM4 is a new salt-activated and bifunctional alginate lyase of the PL7 family with endolytic activity derived from the marine bacterium *V. weizhoudaoensis* M0101. Compared with the observed AlgM4 activity in the absence of NaCl, the enzymatic activity of AlgM4 increased in the presence of various concentrations of NaCl (0.1–1.4 mol/L). The addition of 1 mol/L NaCl resulted in a more than sevenfold increase in AlgM4 activity. Therefore, AlgM4 tolerates high-salinity environments. NaCl not only altered the composition of the secondary structural elements in AlgM4 but also enhanced its thermal stability. AlgM4 hydrolyzed SA, polyM, and polyG via its endolytic activity, producing oligosaccharides (DP 2–9). The alginate lyase AlgM4 has an important application value in the preparation of bioactive oligosaccharides and in the processes of alginate saccharification and ethanol fermentation.

Acknowledgments: This study was supported by grants from National Natural Science Foundation of China (No. 31560017), Key Program of National Natural Science Foundation of Guangxi (No. 2014GXNSFDA118012), Key Research and Development Program of Guangxi (No. AB16380071), The Special Project for the Base of Guangxi Science and Technology and Talents (No. AD17129019), the Fundamental Research Funds for Guangxi Academy of Sciences (2017YJJ23020), high-level innovation teams of Guangxi colleges and universities and academic excellence program (Gui-Jiao-Ren, No. 2016/42).

Author Contributions: Shushi Huang conceived and designed the experiments; Shushi Huang and Guiyuan Huang took charge of the preparation of the manuscript. Guiyuan Huang, Qiaozhen Wang, Mingqian Lu, Chao Xu and Fei Li performed genome analysis of strain *V. weizhoudaoensis* M0101, expression and purification of recombinant enzyme, and determination of activity and enzymatic kinetics of AlgM4 lyases. Guiyuan Huang, Rongcan Zhang and Wei liao analyzed the data of UPLC-QTOF-MS/MS from alginate using AlgM4 lyases. Guiyuan Huang and Chao Xu performed the determination of the secondary structure of AlgM4 with UV-Vis and circular dichroism under NaCl stress.

References

1. Hernandez-carmona, G.; Mchugh, D.J.; Iopez-gutierrez, F. Pilot plant scale extraction of alginates from *Macrocystis pyrifera*. *Appl. Phycol.* **2000**, *11*, 493–502. [CrossRef]
2. Liu, H.; Yin, H.; Zhang, Y.H.; Zhao, X.M. Research progress on biological activities of alginate oligosaccharides. *Nat. Prod. Res.* **2012**, *S1*, 201–204.
3. Gacesa, P. Alginate. *Carbohydr. Polym.* **1988**, *8*, 161–182. [CrossRef]
4. Tako, M. Chemical characterization of acetyl fucoidan and alginate from commercially cultured cladosiphon okamuranus. *Bot. Mar.* **2000**, *43*, 393–398. [CrossRef]
5. Tako, M.; Kiyuna, S.; Uechi, S.; Hongo, F. Isolation and characterization of alginic acid from commercially cultured Nemacystus decipiens (Itomozuku). *Biosci. Biotechnol. Biochem.* **2001**, *65*, 654–657. [CrossRef] [PubMed]
6. Qian, L.; Tang, L.W.; Huang, S.S.; Chagan, I. Research progress of bioethanol from alginate fermentation. *China Biotechnol.* **2013**, *33*, 122–127.
7. Schaumann, K.; Weide, G. Enzymatic degradation of alginate by marine fungi. *Hydrobiologia* **1990**, *204*, 589–596. [CrossRef]

8. Lee, S.I.; Choi, S.H.; Lee, E.Y.; Kim, H.S. Molecular cloning, purification, and characterization of a novel polyMG-specific alginate lyase responsible for alginate MG block degradation in *Stenotrophomas maltophilia* KJ-2. *Appl. Microbiol. Biotechnol.* **2012**, *95*, 1643–1653. [CrossRef] [PubMed]

9. Rahman, M.M.; Inoue, A.; Tanaka, H.; Ojima, T. cDNA cloning of an alginate lyase from a marine gastropod *Aplysia kurodai* and assessment of catalytically important residues of this enzyme. *Biochimie* **2011**, *93*, 1720–1730. [CrossRef] [PubMed]

10. Rahman, M.M.; Wang, L.; Inoue, A.; Ojima, T. cDNA cloning and bacterial expression of a PL-14 alginate lyase from a herbivorous marine snail *Littorina brevicula*. *Carbohydr. Res.* **2012**, *360*, 69–77. [CrossRef] [PubMed]

11. Sheng, D.; Jie, Y.; Zhang, X.Y.; Mei, S.; Song, X.Y.; Chen, X.L.; Zhang, Y.Z. Cultivable Alginate Lyase-Excreting Bacteria Associated with the Arctic Brown Alga *Laminaria*. *Mar. Drugs* **2012**, *10*, 2481–2491.

12. Xiao, L.; Han, F.; Yang, Z.; Lu, X.Z.; Yu, W.G. A Novel Alginate Lyase with High Activity on Acetylated Alginate of *Pseudomonas aeruginosa*, FRD1 from *Pseudomonas* sp. QD03. *World J. Microbio. Biotechnol.* **2006**, *22*, 81–88. [CrossRef]

13. Li, S.; Yang, X.; Zhang, L.; Yu, W.; Han, F. Cloning, Expression, and Characterization of a Cold-Adapted and Surfactant-Stable Alginate Lyase from Marine Bacterium *Agarivorans* sp. L11. *J. Microbiol. Biotechnol.* **2015**, *25*, 681–686. [CrossRef] [PubMed]

14. Inoue, A.; Anraku, M.; Nakagawa, S.; Ojima, T. Discovery of a Novel Alginate Lyase from *Nitratiruptor* sp. SB155-2 Thriving at Deep-sea Hydrothermal Vents and Identification of the Residues Responsible for Its Heat Stability. *J. Biol. Chem.* **2016**, *291*, 15551–15563. [CrossRef] [PubMed]

15. Kobayashi, T.; Uchimura, K.; Miyazaki, M.; Nogi, Y.; Horikoshi, K. A new high-alkaline alginate lyase from a deep-sea bacterium *Agarivorans* sp. *Extremophiles* **2009**, *13*, 121–129. [CrossRef] [PubMed]

16. Uchimura, K.; Miyazaki, M.; Nogi, Y.; Kobayashi, T.; Horikoshi, K. Cloning and sequencing of alginate lyase genes from deep-sea strains of *Vibrio* and *Agarivorans* and characterization of a new *Vibrio* enzyme. *Mar. Biotechnol.* **2010**, *12*, 526–533. [CrossRef] [PubMed]

17. Chen, X.L.; Sheng, D.; Fei, X.; Fang, D.; Li, P.Y.; Zhang, X.Y.; Zhou, B.C.; Zhang, Y.Z.; Xie, B.B. Characterization of a New Cold-Adapted and Salt-Activated Polysaccharide Lyase Family 7 Alginate Lyase from *Pseudoalteromonas* sp. SM0524. *Front. Microbiol.* **2016**, *7*, e30105. [CrossRef] [PubMed]

18. Kim, H.T.; Chung, J.H.; Wang, D.; Lee, J.; Woo, H.C.; Choi, I.G.; Kim, K.H. Depolymerization of alginate into a monomeric sugar acid using Alg17C, an exo-oligoalginate lyase cloned from *Saccharophagus degradans* 2-40. *Appl. Microbiol. Biotechnol.* **2012**, *93*, 2233–2239. [CrossRef] [PubMed]

19. Matsubara, Y.; Iwasaki, K.; Muramatsu, T. Action of poly (alpha-L-guluronate)lyase from *Corynebacterium* sp. ALY-1 strain on saturated oligoguluronates. *J. Agric. Chem. Soc. Jpn.* **1998**, *62*, 1055–1060.

20. Thiangyian, W.; Preston, L.A.; Schiller, N.L. Alginate lyase: Review of major sources and enzyme characteristics, structure-function analysis, biological roles, and applications. *Annu. Rev. Microbiol.* **2000**, *54*, 289–340.

21. Sawabe, T.; Takahashi, H.; Ezura, Y.; Gacesa, P. Cloning, sequence analysis and expression of *Pseudoalteromonas elyakovii* IAM 14594 gene (alyPEEC) encoding the extracellular alginate lyase. *Carbohydr. Res.* **2001**, *335*, 11–21. [CrossRef]

22. Miyake, O.; Hashimoto, W.; Murata, K. An exotype alginate lyase in *Sphingomonas* sp. A1: Overexpression in *Escherichia coli*, purification, and characterization of alginate lyase IV (A1-IV). *Protein Expr. Purif.* **2003**, *29*, 33–41. [CrossRef]

23. Ochiai, A.; Yamasaki, M.; Mikami, B.; Hashimoto, W.; Murata, K. Crystal structure of exotype alginate lyase Atu3025 from *Agrobacterium tumefaciens*. *J. Biol. Chem.* **2010**, *285*, 24519–24528. [CrossRef] [PubMed]

24. Garron, M.L.; Cygler, M. Structural and mechanistic classification of uronic acid-containing polysaccharide lyases. *Glycobiology* **2010**, *20*, 1547–1573. [CrossRef] [PubMed]

25. Zhu, Y.; Wu, L.; Chen, Y.; Ni, H.; Xiao, A.; Cai, H. Characterization of an extracellular biofunctional alginate lyase from marine *microbulbifer* sp. ALW1 and antioxidant activity of enzymatic hydrolysates. *Microbiol. Res.* **2016**, *182*, 49–58. [CrossRef] [PubMed]

26. Huang, L.; Zhou, J.; Li, X.; Peng, Q.; Lu, H.; Du, Y. Characterization of a new alginate lyase from newly isolated *Flavobacterium* sp. S20. *J. Ind. Microbiol. Biotechnol.* **2013**, *40*, 113–122. [CrossRef] [PubMed]

27. Kim, H.T.; Ko, H.J.; Kim, N.; Kim, D.; Lee, D.; Choi, I.G.; Woo, H.C.; Kim, M.D.; Kim, K.H. Characterization of a recombinant endo-type alginate lyase (Alg7D) from *Saccharophagus degradans*. *Biotechnol. Lett.* **2012**, *34*, 1087–1092. [CrossRef] [PubMed]

28. Zhu, B.; Chen, M.; Yin, H.; Du, Y.; Ning, L. Enzymatic hydrolysis of alginate to produce oligosaccharides by a new purified endo-type alginate lyase. *Mar. Drugs* **2016**, *14*, 108. [CrossRef] [PubMed]

29. Jagtap, S.S.; Hehemann, J.H.; Polz, M.F.; Lee, J.K.; Zhao, H. Comparative biochemical characterization of three exolytic oligoalginate lyases from *Vibrio splendidus* reveals complementary substrate scope, temperature, and pH adaptations. *Appl. Environ. Microbio.* **2014**, *80*, 4207–4214. [CrossRef] [PubMed]

30. Park, H.H.; Kam, N.; Lee, E.Y.; Kim, H.S. Cloning and characterization of a novel oligoalginate lyase from a newly isolated bacterium *Sphingomonas* sp. MJ-3. *Mar. Biotechnol.* **2012**, *14*, 189–202. [CrossRef] [PubMed]

31. Li, S.; Yang, X.; Bao, M.; Wu, Y.; Yu, W.; Han, F. Family 13 carbohydrate-binding module of alginate lyase from *Agarivorans* sp. L11 enhances its catalytic efficiency and thermostability, and alters its substrate preference and product distribution. *FEMS Microbiol. Lett.* **2015**, *362*. [CrossRef] [PubMed]

32. Kurakake, M.; Kitagawa, Y.; Okazaki, A.; Shimizu, K. Enzymatic Properties of Alginate Lyase from *Paenibacillus* sp. S29. *Appl. Biochem. Biotechnol.* **2017**, 1–10. [CrossRef] [PubMed]

33. Tusi, S.K.; Khalaj, L.; Ashabi, G.; Kiaei, M.; Khodagholi, F. Alginate oligosaccharide protects against endoplasmic reticulum- and mitochondrial-mediated apoptotic cell death and oxidative stress. *Biomaterials* **2011**, *32*, 5438–5458. [CrossRef] [PubMed]

34. An, Q.D.; Zhang, G.L.; Wu, H.T.; Zhang, Z.C.; Zheng, G.S.; Luan, L.; Murata, Y.; Li, X. Alginate-deriving oligosaccharide production by alginase from newly isolated *Flavobacterium* sp. LXA and its potential application in protection against pathogens. *J. Appl. Microbiol.* **2009**, *106*, 161–170. [CrossRef] [PubMed]

35. Iwasaki, K.I.; Matsubara, Y. Purification of Alginate Oligosaccharides with Root Growth-promoting Activity toward Lettuce. *Biosci. Biotechnol. Biochem.* **2000**, *64*, 1067–1070. [CrossRef] [PubMed]

Identification of 2-keto-3-deoxy-D-Gluconate Kinase and 2-keto-3-deoxy-D-Phosphogluconate Aldolase in an Alginate-Assimilating Bacterium, *Flavobacterium* sp. Strain UMI-01

Ryuji Nishiyama, Akira Inoue and Takao Ojima *

Laboratory of Marine Biotechnology and Microbiology, Faculty of Fisheries Sciences, Hokkaido University, Hakodate, Hokkaido 041-8611, Japan; nsym2480rj@eis.hokudai.ac.jp (R.N.); inouea21@fish.hokudai.ac.jp (A.I.)
* Correspondence: ojima@fish.hokudai.ac.jp.

Academic Editor: Antonio Trincone

Abstract: Recently, we identified an alginate-assimilating gene cluster in the genome of *Flavobacterium* sp. strain UMI-01, a member of *Bacteroidetes*. Alginate lyase genes and a 4-deoxy-L-erythro-5-hexoseulose uronic acid (DEH) reductase gene in the cluster have already been characterized; however, 2-keto-3-deoxy-D-gluconate (KDG) kinase and 2-keto-3-deoxy-6-phosphogluconate (KDPG) aldolase genes, i.e., *flkin* and *flald*, still remained uncharacterized. The amino acid sequences deduced from *flkin* and *flald* showed low identities with those of corresponding enzymes of *Saccharophagus degradans* 2-40^T, a member of *Proteobacteria* (Kim et al., Process Biochem., 2016). This led us to consider that the DEH-assimilating enzymes of *Bacteroidetes* species are somewhat deviated from those of *Proteobacteria* species. Thus, in the present study, we first assessed the characteristics in the primary structures of KDG kinase and KDG aldolase of the strain UMI-01, and then investigated the enzymatic properties of recombinant enzymes, recFlKin and recFlAld, expressed by an *Escherichia coli* expression system. Multiple-sequence alignment among KDG kinases and KDG aldolases from several *Proteobacteria* and *Bacteroidetes* species indicated that the strain UMI-01 enzymes showed considerably low sequence identities (15%–25%) with the *Proteobacteria* enzymes, while they showed relatively high identities (47%–68%) with the *Bacteroidetes* enzymes. Phylogenetic analyses for these enzymes indicated the distant relationship between the *Proteobacteria* enzymes and the *Bacteroidetes* enzymes, i.e., they formed distinct clusters in the phylogenetic tree. recFlKin and recFlAld produced with the genes *flkin* and *flald*, respectively, were confirmed to show KDG kinase and KDPG aldolase activities. Namely, recFlKin produced 1.7 mM KDPG in a reaction mixture containing 2.5 mM KDG and 2.5 mM ATP in a 90-min reaction, while recFlAld produced 1.2 mM pyruvate in the reaction mixture containing 5 mM KDPG at the equilibrium state. An in vitro alginate-metabolizing system constructed from recFlKin, recFlAld, and previously reported alginate lyases and DEH reductase of the strain UMI-01 could convert alginate to pyruvate and glyceraldehyde-3-phosphate with an efficiency of 38%.

Keywords: alginate degradation; 4-deoxy-L-erythro-5-hexoseulose uronic acid (DEH) metabolism; *Bacteroidetes*; *Proteobacteria*; *Flavobacterium*; 2-keto-3-deoxy-D-gluconate (KDG) kinase; 2-keto-3-deoxy-6-phosphogluconate (KDPG) aldolase; alginate-derived products

1. Introduction

Alginate is an acidic heteropolysaccharide comprising two kinds of uronic acid, β-D-mannuronate and α-L-guluronate [1–3]. This polysaccharide exists as a structural material in cell-wall matrices of brown algae and biofilms of certain bacteria. Since alginate solution shows high viscosity and forms an elastic gel upon chelating Ca^{2+}, it has long been used as viscosifier and gelling agent in the fields of food and pharmaceutical industries. Alginate oligosaccharides produced by alginate lyases have also been recognized as functional materials since they exhibit various biological functions; e.g., promotion of root growth in higher plants [4,5], acceleration of growth rate of *Bifidobacterium* sp. [6], and promotion of penicillin production in *Penicillium chrysogenum* [7]. Anti-oxidant [8], anti-coagulant [9], anti-inflammation [10], and anti-infectious disease [11] are also bioactivities of alginate oligosaccharides. Recently, 4-deoxy-L-erythro-5-hexoseulose uronic acid (DEH), an end reaction product of alginate lyases, was proven to be available as a carbon source for ethanol fermentation by the genetically modified microbes [12–14]. Furthermore, 2-keto-3-deoxyaldonic acids like 2-keto-3-deoxy-D-gluconate (KDG) and 2-keto-3-deoxy-6-phosphogluconate (KDPG), which are intermediates in alginate metabolism, have been expected as leading compounds for antibiotics, antiviral agents, and other drugs and medicines [15]. Thus, such alginate-derived products are regarded as promising materials in various practical applications.

Alginate-degrading enzymes have been investigated in many organisms such as soil bacteria [16⁻21], marine bacteria [22–29], marine gastropods [30–33], and seaweeds [3,34]. Endolytic and exolytic alginate lyases split glycosyl linkages of alginate via β-elimination mechanism producing unsaturated oligosaccharides and monosaccharide, where a double bond is introduced between C4 and C5 of the newly formed non-reducing terminus [35]. Unsaturated monosaccharide, the end product of alginate lyases, is spontaneously [20] and/or enzymatically [36] converted to an open chain form, DEH, and further converted to KDG by the NAD(P)H-dependent DEH reductase. The KDG is phosphorylated to KDPG by KDG kinase and then split to pyruvate and glyceraldehyde-3-phosphate (GAP) by KDPG aldolase. The alginate-derived pyruvate and GAP are finally metabolized by Kreb's cycle. Bacterial alginate lyases have been identified in many species, e.g., *Sphingomonas* sp. [16,17], *Flavobacterium* sp. [26,27], *Saccharophagus* sp. [22,23], *Vibrio* sp. [29], and *Pseudomonas* sp. [20,21]. *Sphingomonas* sp. strain A1 possesses four kinds of alginate lyases, A1-I–IV, whose sequential action completely depolymerizes alginate to DEH [16,17]. *Flavobacterium* sp. strain UMI-01 also possesses four kinds of alginate lyases, FlAlyA, FlAlyB, FlAlyC and FlAlex, whose cooperative action efficiently degrades alginate to DEH [27]. Meanwhile, *Saccharophagus degradans* strain 2-40^T possesses two kinds of alginate lyases, Alg7D and Alg17C, which degrade alginate to unsaturated disaccharide and DEH [22,23]. The alginate-derived DEH is reduced to KDG by NAD(P)H-dependent DEH reductases as described above. Recently, this enzyme was identified in *Sphingomonas* sp. strain A1 [18,19], *Flavobacterium* sp. strain UMI-01 [28], *S. degradans* strain 2-40^T [24], *Vibrio splendidus* 12B01 [13], and marine gastropod *Haliotis discus hannai* [37]. The bacterial DEH reductases were classified under short-chain dehydrogenases/reductases (SDR) superfamily, while the gastropod enzyme was identified as a member of the aldo-keto reductase (AKR) superfamily. Information about alginate lyases and DEH reductases has been continuously accumulated; however, KDG kinase and KDPG aldolase have not been so well investigated.

Under these circumstances, DEH reductase, KDG kinase, and KDPG aldolase were recently characterized in *S. degradans* 2-40^T, a member of the phylum *Proteobacteria* [25]. The combined action of these enzymes could convert DEH to pyruvate and GAP in vitro. On the other hand, we also found the existence of alginate-assimilating gene cluster in the genome of *Flavobacterium* sp. strain UMI-01, a member of the phylum *Bacteroidetes* [27,28]. The endolytic and exolytic alginate lyase genes, *flalyA* and *flalyB*, and a DEH reductase gene, *flred*, are located in operon A, and KDG kinase-like gene *flkin* (GenBank accession number, BAQ25538) and KDPG aldolase-like gene *flald* (GenBank accession number, BAQ25539) are in operon B (Figure 1). The alginate lyases and DEH reductase of this bacterium have already been characterized [26–28]; however, KDG kinase and KDPG aldolase have not been identified yet. The amino acid sequences deduced from *flkin* and *flald* showed only 19% and 22% identities, respectively, with those of the corresponding enzymes from *S. degradans* 2-40^T [25]. These low sequence identities suggest that the properties of *Flavobacterium* (*Bacteroidetes*) enzymes may be somewhat different from those of *Saccharophagus* (*Proteobacteria*) enzymes. Therefore, in the present study, we first characterized the primary structures of KDG kinase and KDPG aldolase, FlKin and FlAld, of the strain UMI-01 compared with those of other bacterial enzymes. Then, we investigated enzymatic properties of proteins encoded by *flkin* and *flald* using recombinant enzymes, recFlKin and recFlAld. Furthermore, we constructed an in vitro alginate-metabolizing system using recFlKin and recFlAld, along with recombinant alginate lyases and DEH reductase of this bacterium to confirm that this enzyme system can produce pyruvate and GAP from alginate in vitro.

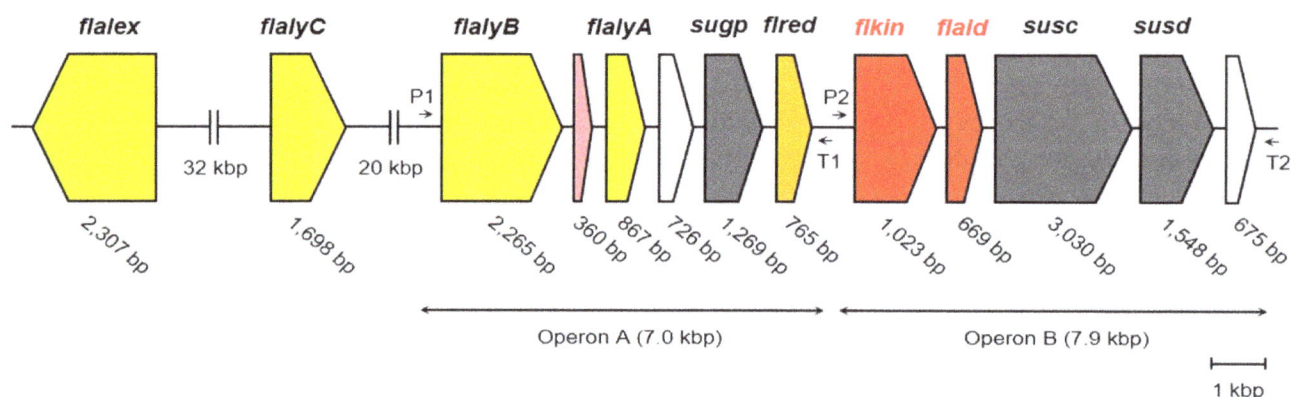

Figure 1. Alginate-assimilating enzyme genes in the genome of *Flavobacterium* sp. strain UMI-01. Yellow, alginate-lyase genes; pink, KdgF-like protein gene; white, transcriptional regulator-like protein genes; gray, membrane transporter-like genes; orange, 4-deoxy-L-erythro-5-hexoseulose uronic acid (DEH) reductase gene; red, 2-keto-3-deoxy-D-gluconate (KDG) kinase-like gene and 2-keto-3-deoxy-6-phosphogluconate (KDPG) aldolase-like gene. Arrows P1 and P2 and arrows T1 and T2 indicate predicted promoters and terminators, respectively.

2. Results

2.1. Characteristics in the Primary Structures of FlKin and FlAld

Deduced amino acid sequences of *flkin* and *flald* were compared with those of KDG kinases and KDPG aldolases from several *Proteobacteria* and *Bacteroidetes* species. Enzymes from two *Archaea* species are also included in the comparison of KDG kinases. FlKin showed considerably low amino acid identity (15%–26%) with KDG kinases from *Proteobacteria* species, i.e., *Escherichia coli* (GenBank accession number, WP_024175791) [38], *Serratia marcescens* (GenBank accession number, ABB04497) [39], and *S. degradans* 2-40^T (GenBank accession number, ABD82535) [25], and archaea, i.e., *Sulfolobus solfataricus* (GenBank accession number, WP_009991690) [40–42] and *Thermus thermophiles* (GenBank accession number, WP_011229211) [43] (Figure 2). Meanwhile, the sequence of FlKin showed relatively high identities (47%–68%) with the enzymes from *Bacteroidetes* species, i.e., *Gramella forsetii* KT0803 (GenBank accession number, CAL66135), *Dokdonia* sp. MED134 (GenBank accession number, WP_016501275), and *Lacinutrix* sp. 5H-3-7-4 (GenBank accession number, AEH01605). However, substrate-recognition residues of KDG kinase, which were identified in the *S. solfataricus* enzymes [42], i.e., Gly34, Tyr90, Tyr106, Arg108, Arg166, Asp258, and Asp294, were entirely conserved in FlKin as Gly34, Tyr89, Tyr104, Arg106, Arg169, Asp280, and Asp317, respectively. FlAld also showed low amino acid identity (22%–25%) with KDPG aldolases from *Proteobacteria* species such as *E. coli* (GenBank accession number, WP_000800517) [44,45], *Zymomonas mobilis* (GenBank accession number, S18559) [44], *Pseudomonas putida* (GenBank accession number, WP_016501275) [44,46], and *S. degradans* 2-40^T (GenBank accession number, ABD80644) [25] (Figure 3). Meanwhile, the sequence identities between FlAld and enzymes from other *Bacteroidetes* species such as *G. forsetii* KT0803 (GenBank accession number, KT0803), *Dokdonia* sp. MED134 (GenBank accession number, WP_013749799), and *Lacinutrix* sp. 5H-3-7-4 (GenBank accession number, AEH01606) were 61%–65%. Catalytic residue Lys133 and substrate-recognition residues, Glu45, Arg49, Thr73, Pro94 and Phe135 identified in the *E. coli* enzyme [45], were conserved in FlAld except for the substitution of Thr73 by Ser. Phylogenetic analyses for KDG kinases and KDPG aldolases (Figure 4A,B) suggested that the *Bacteroidetes* enzymes are somewhat phylogenetically deviated from the *Proteobacteria* (and *Archaea*) enzymes. Therefore, we decided to examine if FlKin and FlAld of the strain UMI-01 actually possess KDG kinase and KDPG aldolase activities.

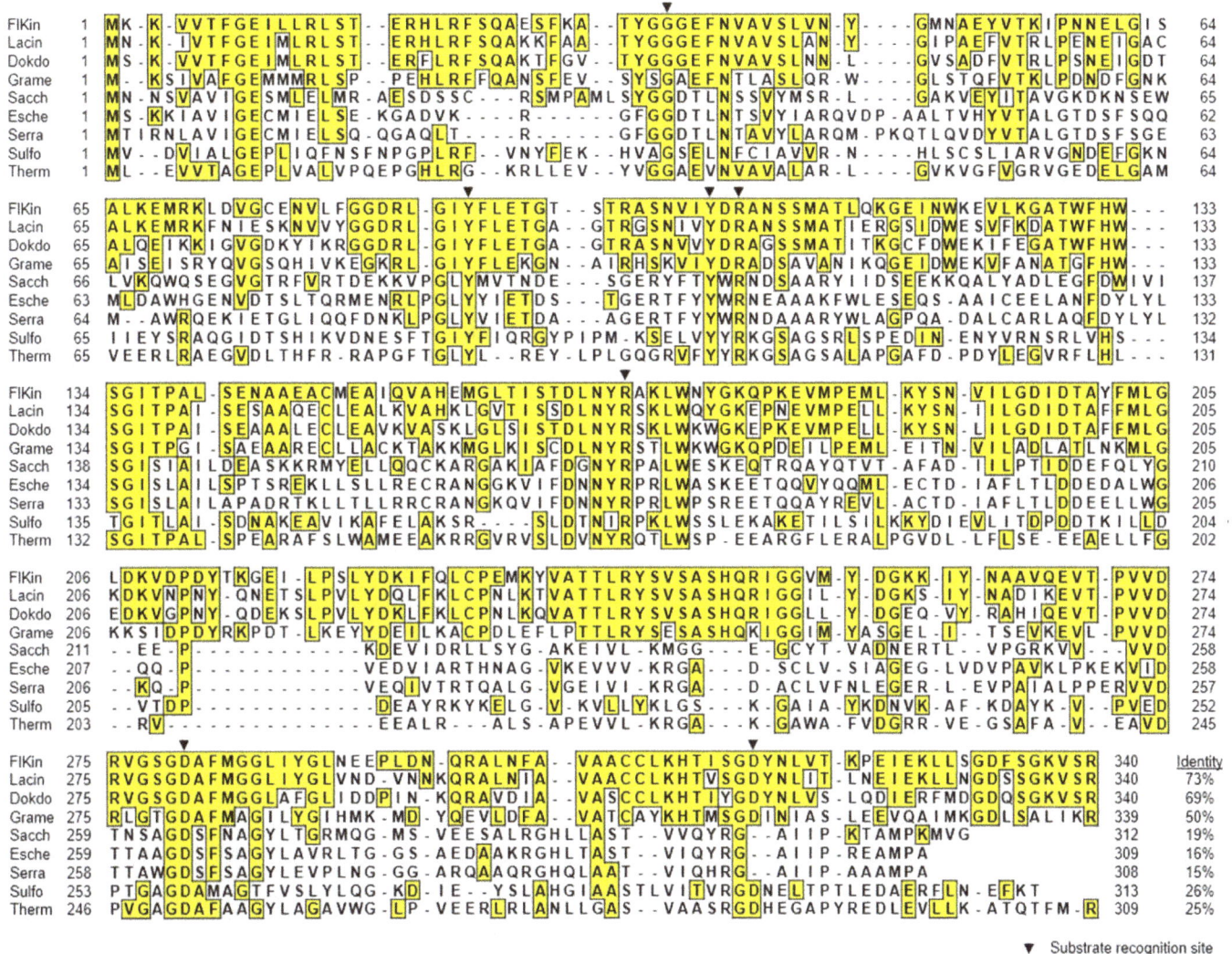

```
FlKin   1   MK.K-VVTFGEILLRLST..ERHLRFSQAESFKA.-TYGGGEFNVAVSLVN.-Y---GMNAEYVTKIPNNELG IS   64
Lacin   1   MN.K-IVTFGEIMLRLST..ERHLRFSQAKKFAA.-TYGGGEFNVAVSLAN.-Y---GIPAEFVTRLPENEIGAC    64
Dokdo   1   MS.K-VVTFGEIMLRLST..ERFLRFSQAKTFGV.-TYGGGEFNVAVSLNN-L---GVSADFVTRLPSNEIGDT    64
Grame   1   M..KSIVAFGEMMMRLSP..PEHLRFFQANSFEV.-SYSGAEFNTLASLQR-W---GLSTQFVTKLPDNDFGNK    64
Sacch   1   MN-NSVAVIGESMLELMR-AESDSSC---RSMPAMLSYGGDTLNSSVYMSR-L---GAKVEYITAVGKDKNSEW    65
Esche   1   MS.KKIAVIGECMIELSE-KGADVK.--R-----GFGGDTLNTSVYIARQVDP-AALTVHYVTALGTDSFSQQ    62
Serra   1   MTIRNLAVIGECMIELSQ-QGAQLT.---R-----GFGGDTLNTAVYLARQM-PKQTLQVDYVTALGTDSFSGE    63
Sulfo   1   MV--DVIALGEPLIQFNSFNPGPLRF.-VNYFEK-HVAGSELNFCIAVVR-N---HLSCSLIARVGNDEFGKN    64
Therm   1   ML-.EVVTAGEPLVALVPQEPGHLRG.-KRLLEV-YVGGAEVNVAVALAR-L----GVKVGFVGRVGEDELGAM    64

FlKin   65  ALKEMRKLDVGCENVLFGGDRL-GIYFLETGT.-STRASNVIYDRANSSMATLQKGEINWKEVLKGATWFHW---   133
Lacin   65  ALKEMRKFNIESKNVVYGGDRL-GIYFLETGA.-GTRGSNVIYDRANSSMATIERGSIDWESVFKDATWFHW---   133
Dokdo   65  ALQEIKKIGVGDKYIKRGGDRL-GIYFLETGA.-GTRASNVVYDRAGSSMATITKGCFDWEKIFEGATWFHW---   133
Grame   65  AISEISRYQVGSQHIVKEGKRL-GIYFLEKGN.-AIRHSKVIYDRADSAVANIKQGEIDWEKVFANATGFHW---   133
Sacch   66  LVKQWQSEGVGTRFVRTDEKKVPGLYMVTNDE.---SGERYFTYWRNDSAARYIIDSEEKKQALYADLEGFDWIVI   137
Esche   63  MLDAWHGENVDTSLTQRMENRLPGLYYIETDS.--TGERTFYYWRNEAAAKFWLESEQS-AAICEELANFDYLYL   133
Serra   64  M.-AWRQEKIETGLIQQFDNKLPGLYVIETDA.--AGERTFYYWRNDAAARYWLAGPQA-DALCARLAQFDYLYL   132
Sulfo   65  IIEYSRAQGIDTSHIKVDNESFTGIYFIQRGYPIPM.KSELVYYRKGSAGSRLSPEDIN-ENYVRNSRLVHS---   134
Therm   65  VEERLRAEGVDLTHFR-RAPGFTGLYL..-REY-LPLGQGRVFYYRKGSAGSALAPGAFD-PDYLEGVRFLHL---   131

FlKin   134 SGITPAL-SENAAEACMEAIQVAHEMGLTISTDLNYRAKLWNYGKQPKEVMPEML-KYSN-VILGDIDTAYFMLG   205
Lacin   134 SGITPAI-SESAAQECLEALKVAHKLGVTISSDLNYRSKLWQYGKEPNEVMPELL-KYSN-IILGDIDTAFFMLG   205
Dokdo   134 SGITPAI-SEAAALECLEAVKVASKSLGISISTDLNYRSKLWKWGKQPDEILPEML-KYSN-LILGDIDTAFFMLG   205
Grame   134 SGITPGI-SAEAARECLLACKTAKKMGLKISCDLNYRSTLWKWGKQPDEILPEML-EITN-VILADLATLNKMLG   205
Sacch   138 SGISIAILDEASKKRMYELLQQCKARGAKIAFDGNYRPALWESKEQTRQAYQTVT-AFAD-IILPTIDDEFQLYG   210
Esche   134 SGISLAILSPTSREKLLSLLRECRANGGKVIFDNNYRPRLWASKEETQQVYQQML-ECTD-IAFLTLDDEDALWG   206
Serra   133 SGISLALAPADRTKLLTLLRRCRANGKQVIFDNNYRPRLWPSREETQQAYREVL-ACTD-IAFLTLDDEELLWG   205
Sulfo   135 TGITLAI-SDNAKEAVIKAFELAKSR.--.-SLDTNIRPKLWSSLEKAKETILSILKKYDIEVLITDPDDTKILLD   204
Therm   132 SGITPAL-SPEARAFSLWAMEEAKRRGVRVSLDVNYRQTLWSP-EEARGFLERALPGVDL-LFLSE-EEAELLFG   202

FlKin   206 LDKVDPDYTKGEI-LPSLYDKIFQLCPEMKYVATTLRYSVSASHQRIGGVM-Y-DGKK-IY-NAAVQEVT-PVVD   274
Lacin   206 KDKVNPNY-QNETSLPVLYDQLFKLCPNLKTVATTLRYSVSASHQRIGGIL-Y-DGKS-IY-NADIKEVT-PVVD   274
Dokdo   206 EDKVGPNY-QDEKSLPVLYDKLFKLCPNLKQVATTLRYSVSASHQRIGGLL-Y-DGEQ-VY-RAHIQEVT-PVVD   274
Grame   206 KKSIDPDYRKPDT-LKEYQDEILKACPDLEFLPTTLRYSESASHQKIGGIM-YASGEL-I--TSEVKEVL-PVVD   274
Sacch   211 --EE-P-----------KDEVIDRLLSYG-AKEIVL-KMGG---E-GCYT-VADNERTL-VPGRKVV---VVD   258
Esche   207 --QQ-P-----------VEDVIARTHNAG-VKEVVV-KRGA---D-SCLV-SIAGEG-LVDVPAVKLPKEKVID   258
Serra   206 --KQ-P-----------VEQIVTRTQALG-VGEIVI-KRGA---D-ACLVFNLEGER-L-EVPAIALPPERVVD   257
Sulfo   205 --VTDP-----------DEAYRKYKELG-V-KVLLYKLGS---K-GAIA-YKDNVK-AF-KDAYK-V--PVED   252
Therm   203 --RV-------------EEALR--ALS-APEVVL-KRGA---K-GAWA-FVDGRR-VE-GSAFA-V--EAVD   245

FlKin   275 RVGSGDAFMGGLIYGLNEEPLDN-QRALNFA--VAACCLKHTISGDYNLVT-KPEIEKLLSGDFSGKVSR   340   Identity
Lacin   275 RVGSGDAFMGGLIYGLVND-VNNKQRALNIA--VAACCLKHTVSGDYNLIT-LNEIEKLLNGDSSGKVSR   340   73%
Dokdo   275 RVGSGDAFMGGLAFGLIDDPIN-KQRAVDIA--VASCCLKHTIYGDYNLVS-LQDIERFMDGDQSGKVSR   340   69%
Grame   275 RLGTGDAFMAGILYGIHMK-MD-YQEVLDFA--VATCAYKHTMSGDINIAS-LEEVQAIMKGDLSALIKR   339   50%
Sacch   259 TNSAGDSFNAGYLTGRMQG-MS-VEESALRGHLLAST--VVQYRG--AIIP-KTAMPKMVG                 312   19%
Esche   259 TTAAGDSFSAGYLAVRLTG-GS-AEDAAKRGHLTAST--VIQYRG--AIIP-REAMPA                   309   16%
Serra   258 TTAWGDSFSAGYLEVPLNG-GG-ARQAAQRGHQLAAT--VIQHRG--AIIP-AAAMPA                   308   15%
Sulfo   253 PTGAGDAMAGTFVSLYLQG-KD-IE--YSLAHGIAASTLVITVRGDNELTPTLEDAERELN-EFKT            313   26%
Therm   246 PVGAGDAFAAGYLAGAVWG-LP-VEERLRLANLLGAS--VAASRGDHEGAPYREDLEVLLK-ATQTFM-R         309   25%

                                                              ▼ Substrate recognition site
```

Figure 2. Multiple alignment for amino acid sequences of FlKin and other KDG kinases. Closed triangles indicate substrate-recognition residues of KDG kinase from *Sulfolobus solfataricus* [42]. FlKin, KDG kinase from *Flavobacterium* sp. strain UMI-01 (GenBank accession number, BAQ25538); Lacin, KDG kinase-like protein from *Lacinutrix* sp. 5H-3-7-4 (GenBank accession number, AEH01605); Dokdo, KDG kinase-like protein from *Dokdonia* sp. MED134 (GenBank accession number, WP_013749800); Grame, KDG kinase-like protein from *Gramella forsetii* KT0803 (GenBank accession number, CAL66135); Sacch, KDG kinase from *Saccharophagus degradans* 2-40^T (GenBank accession number, ABD82535) [25]; Esche, KDG kinase from *Escherichia coli* (GenBank accession number, WP_024175791) [38]; Serra, KDG kinase from *Serratia marcescens* (GenBank accession number, ABB04497) [39]; Sulfo, KDG kinase from *Sulfolobus solfataricus* (GenBank accession number, WP_009991690) [40–42]; Therm, KDG kinase from *Thermus thermophiles* (GenBank accession number, WP_011229211) [43].

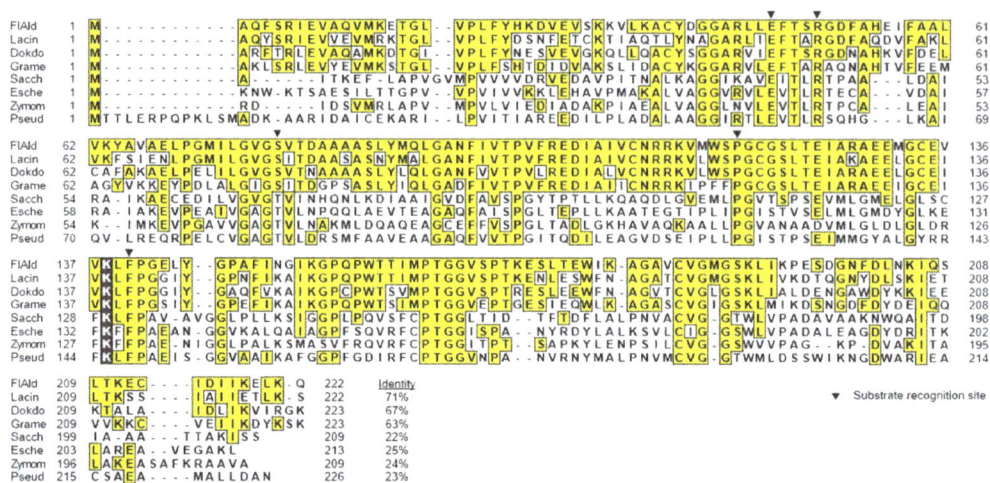

Figure 3. Multiple alignment for amino acid sequences of FlAld and other KDPG aldolases. Gray box and closed triangles indicate catalytic and substrate-recognition residues of KDPG aldolase from *E. coli* [44,45], respectively. FlAld, KDPG aldolase from *Flavobacterium* sp. strain UMI-01 (GenBank accession number, BAQ25539); Lacin, KDPG aldolase-like protein from *Lacinutrix* sp. 5H-3-7-4 (GenBank accession number, AEH01606); Dokdo, KDPG aldolase-like protein from *Dokdonia* sp. MED134 (GenBank accession number, WP_013749799); Grame, KDPG aldolase-like protein from *G. forsetii* KT0803 (GenBank accession number, CAL66136); Sacch, KDPG aldolase from *S. degradans* 2-40^T (GenBank accession number, ABD80644) [25]; Esche, KDPG aldolase from *E. coli* (GenBank accession number, WP_000800517) [44,45]; Zymom, KDPG aldolase from *Zymomonas mobilis* (GenBank accession number, S18559) [44]; Pseud, KDPG aldolase from *Pseudomonas putida* (GenBank accession number, WP_016501275) [44,46].

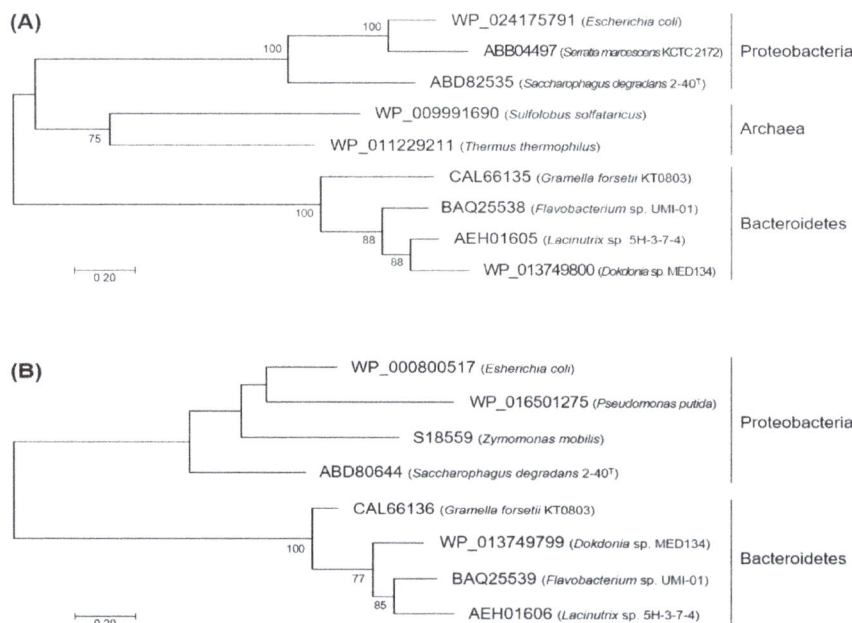

Figure 4. Phylogenetic trees for KDG kinases and KDPG aldolases. Phylogenetic analyses were carried out using amino acid sequences of KDG kinases from *Proteobacteria*, *Archaea* and *Bacteroidetes* species (**A**) and KDPG aldolases from *Proteobacteria* and *Bacteroidetes* species (**B**). Amino acid sequences of KDG kinases and KDPG aldolases were retrieved from the draft or complete genome data deposited in GenBank. Accession numbers for enzyme sequences along with the bacterial species are indicated in the right of each branch. Bootstrap values above 50% are indicated on the root of branches. Scale bar indicates 0.20 amino acid substitution.

2.2. Production of recFlKin and recFlAld, and Their Reaction Products

Coding regions of *flkin* and *flald* were amplified by PCR with specific primers listed in Table 1, cloned into pCold vector and expressed in *E. coli* BL21 (DE3). The recombinant enzymes were purified by Ni-NTA affinity chromatography. Molecular masses of recFlKin and recFlAld estimated by SDS-PAGE were 39 kDa and 26 kDa, respectively (Figure 5). These values were consistent with the calculated molecular masses of these enzymes, i.e., 39,391 Da and 25,808 Da, which include $8 \times$ Gly + $8 \times$ His-tag [26].

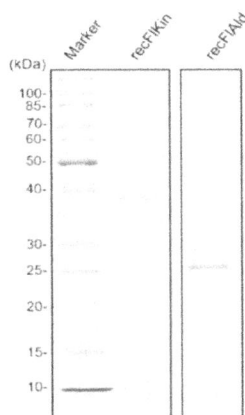

Figure 5. SDS-PAGE for recFlKin and recFlAld. Recombinant enzymes were purified Ni-NTA affinity chromatography and subjected to 0.1% SDS–10% polyacrylamide-gel electrophoresis. Proteins in the gel were stained by Coomassie Brilliant Blue R-250. Marker, molecular weight markers (Protein Ladder Broad Range, New England Biolabs, Ipswich, MA, USA).

Table 1. Primers used for amplification of *flkin* and *flald* genes.

Primer Name	Nucleotide Sequence
recFlKin-F	5′-AGGTAATACACCATGAAAAAAGTAGTCACTTTTGG-3′
recFlKin-R	5′-CACCTCCACCGGATCCTCTTGAAACTTTTCCTGAAA-3′
recFlAld-F	5′-ATGTAATACACCATGGCTCAATTTTCAAGAATAGA-3′
recFlAld-R	5′-CACCTCCACCGGATCCTTGTTTTAACTCTTTAATGA-3′

The recFlKin was allowed to react with KDG in the presence of ATP. TLC analysis suggested that the reaction product was KDPG (Figure 6A). Then, the molecular mass of the reaction product was determined by matrix-assisted laser desorption ionization-time of flight mass spectrometer (MALDI-TOF) mass spectrometry (Figure 7A,B). The 257 m/z peak was considered to be that of KDPG (MW = 258), and the 279 m/z peak was considered to be that of a sodium-salt form of KDPG. These results indicate that the reaction product of recFlKin is KDPG. Thus, we concluded that the protein encoded by *flkin* is KDG kinase. Here, it should be noted that the peak intensities of KDPG were considerably low. This was ascribable to the low ionization level of KDPG. Therefore, we attempted to improve the signal intensity of KDPG using other matrices, e.g., 2,5-dihydroxybenzoic acid and α-cyano-4-hydroxycinnamic acid. Unfortunately, signal intensity of KDPG was not improved much. We still need to investigate the suitable conditions for the detection of KDPG.

Reaction products of recFlAld were also analyzed by TLC (Figure 6B). recFlAld produced two kinds of reaction products with different mobility on TLC. According to their mobility, they were regarded as pyruvate and GAP. The staining intensity of pyruvate was significantly low compared with that of GAP. This difference was ascribable to the difference in the reactivity between pyruvate and GAP with 2,4-dinitrophenylhydrazine (DNP). Namely, GAP showed much higher reactivity with DNP than pyruvate. Then, the reaction products of recFlAld were subjected to MALDI-TOF mass spectrometry. The 87 m/z and 169 m/z peaks corresponding to pyruvate (MW = 88) and GAP

(MW = 170), respectively, were observed. The peak intensity of GAP was small (Figure 7C,D). This appeared to be due to the decomposition of GAP during the mass spectrometric analysis. Thus, we may conclude that recFlAld is the KDPG aldolase that splits KDPG to pyruvate and GAP.

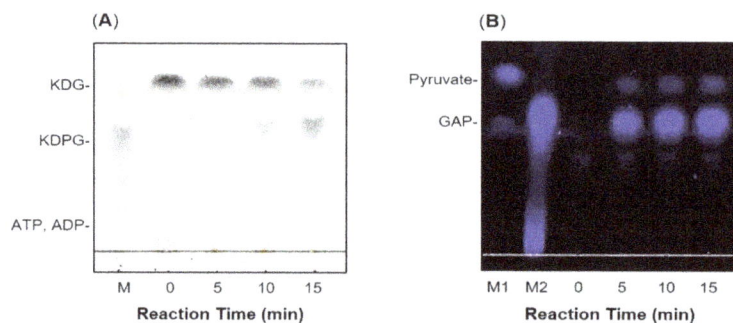

Figure 6. Thin-layer chromatography (TLC) analyses for reaction products of recFlKin and recFlAld. (**A**) Reaction products produced by recFlKin. The reaction products were visualized by spraying 10% (v/v) sulfuric acid in ethanol followed by heating at 130 °C for 10 min. M, standard KDPG; (**B**) Reaction products of recFlAld. The reaction products were visualized with 0.5% (w/v) 2,4-dinitrophenylhydrazine (DNP)–20% (v/v) sulfuric acid. The color was graphically inverted to ease the recognition of spots. M1, standard pyruvate; M2, standard glyceraldehyde-3-phosphate (GAP). Stained materials near the original position are GAP oligomers.

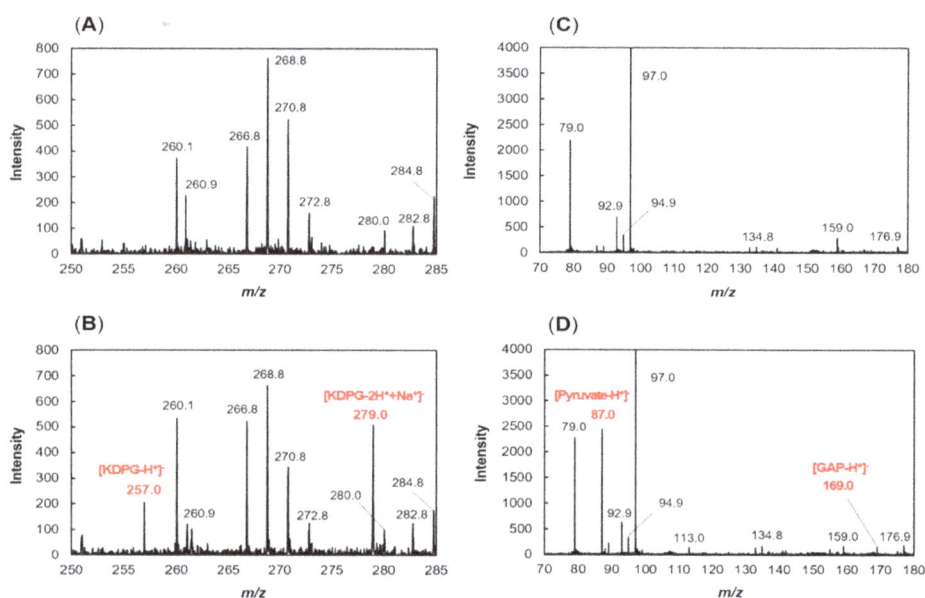

Figure 7. Mass spectrometry for reaction products of recFlKin and recFlAld. The reaction products prepared as in Section 4.10 were subjected to matrix-assisted laser desorption ionization-time of flight mass spectrometer (MALDI-TOF) mass spectrometry, and analyzed by negative-ion mode. (**A,B**) KDG before and after the recFlKin reaction, respectively; (**C,D**) KDPG before and after the recFlAld reaction, respectively. Reaction products are indicated with red letters along with molecular masses above the peaks.

2.3. Enzymatic Properties of recFlKin and recFlAld

We first investigated the kinetic parameter for recFlAld, since recFlAld was necessary for the KDG kinase assay. In the present study, the kinase activity was assayed by quantifying the pyruvate produced from KDPG by the action of recFlAld. KDPG-derived pyruvate was determined by the lactate dehydrogenase (LDH)–NADH system as described in Section 4.6. In the equilibrium state of recFlAld reaction, pyruvate concentration reached 1.2 mM. Since the KDPG concentration was originally 5 mM, that in the equilibrate state was regarded as 3.8 mM. From these values the equilibrium constant (K_{eq}) and $\Delta G°$ were calculated to be 3.8×10^{-1} M and +0.57 kcal/mol, respectively. This indicated that the equilibrium position of KDPG–aldolase reaction is slightly shifted toward the KDPG side. Next, we determined the reaction rate of recFlAld by the LDH–NADH method. By this method, the specific activity of recFlAld was estimated to be 57 U/mg at pH 7.4 and 30 °C. Coexistence of LDH–NADH in the reaction mixture could extend the aldolase reaction longer time by decreasing pyruvate concentration in the reaction equilibrium.

Next, KDG kinase activity of recFlKin was determined by using recFlAld and LDH–NADH. recFlKin was allowed to react with KDG in the presence of ATP at 30 °C and the reaction was terminated by heating at 100 °C for 3 min at the reaction times 1, 15, and 30 min. The KDPG produced in the reaction mixture was then split to pyruvate and GAP by recFlAld, and the pyruvate was quantified by the LDH–NADH system. At reaction time 90 min, recFlKin was found to produce 1.7 mM KDPG from 2.5 mM KDG at ~70% efficiency with the specific activity 0.72 U/mg. recFlKin showed an optimal temperature and pH at around 50 °C and 7.0, respectively, and was stable at 40 °C for 30 min.

2.4. Construction of In Vitro Alginate-Metabolizing System Using Recombinant Enzymes

In the present study, we identified *flkin* and *flald* in the genome of strain UMI-01 as KDG kinase and KDPG aldolase gene, respectively. Since alginate lyases and DEH reductase in this strain have already been characterized [26–28], here we examined if the sequential action of these alginate-degrading and -assimilating enzymes could convert alginate to pyruvate and GAP in vitro. Namely, recombinant alginate lyases (recFlAlyA, recFlAlyB, and recFlAlex) [26,27], DEH reductase (recFlRed) [28], KDG kinase (recFlKin), and KDPG aldolase (recFlAld) were allowed to react alginate in various combinations, and each reaction product was analyzed by TLC (Figure 8) and quantified by thiobarbituric acid (TBA) and LDH–NADH methods (Table 2). As shown in Figure 8, alginate was almost completely degraded to DEH by the simultaneous actions of recFlAlyA, recFlAlyB, and recFlAlex. The DEH was also almost completely reduced to KDG by recFlRed. Furthermore, a major part of the KDG was converted to KDPG by recFlKin, and the band of KDPG became faint by the reaction of recFlAld. This indicated the splitting of KDPG to pyruvate and GAP by the action of recFlAld. Accordingly, the sequential action of recombinant enzymes was considered to be capable of converting alginate to pyruvate and GAP in vitro. Then, the yields of intermediates in each reaction step were quantified by TBA and LDH–NADH methods (Table 2). Concentrations of the unsaturated oligo-alginates, DEH, KDG, KDPG, and pyruvate (and GAP), were determined to be 4.2 mM, 9.8 mM, 9.8 mM, 8.1 mM, and 3.8 mM, respectively. Since the initial concentration of alginate (0.2% (*w/v*)) corresponds to 10 mM monosaccharide, the yields of DEH and KDG were estimated to be ~100%, and the yields of KDPG and pyruvate were estimated to be ~80% and ~40%, respectively. These results indicated that high-value intermediates such as KDPG could be produced from alginate with fairly high efficiency by the recombinant enzymes of the strain UMI-01 in vitro.

Table 2. Quantification of reaction products produced by the recombinant enzymes.

Enzymes	Substrates/Products	Concentration (mM)	Yield (%)
None	Alginate [a]	10 [a]	-
recFlAlyA	Oligoalginates	4.2 ± 0.06	-
recFlAlyA + recFlAlyB + recFlAlex	DEH	9.8 ± 0.34	98
recFlAlyA + recFlAlyB + recFlAlex + recFlRed	KDG	9.8 ± 1.0	98
recFlAlyA + recFlAlyB + recFlAlex + recFlRed + recFlKin	KDPG	8.1 ± 0.54	81
recFlAlyA + recFlAlyB + recFlAlex + recFlRed + recFlKin + recFlAld	Pyruvate (and GAP)	3.8 ± 0.33	38

[a] 0.2% (*w/v*) sodium alginate theoretically corresponds to 10 mM monosaccharide.

DEH-
KDG-
di-
tri-
KDPG-

recFlAlyA	-	+	+	+	+	+	] Endolytic Alginate Lyase
recFlAlyB	-	-	+	+	+	+	] Exolytic Alginate Lyase
recFlAlex	-	-	+	+	+	+	
recFlRed	-	-	-	+	+	+	] DEH Reductase
recFlKin	-	-	-	-	+	+	] KDG Kinase
recFlAld	-	-	-	-	-	+	] KDPG Aldolase

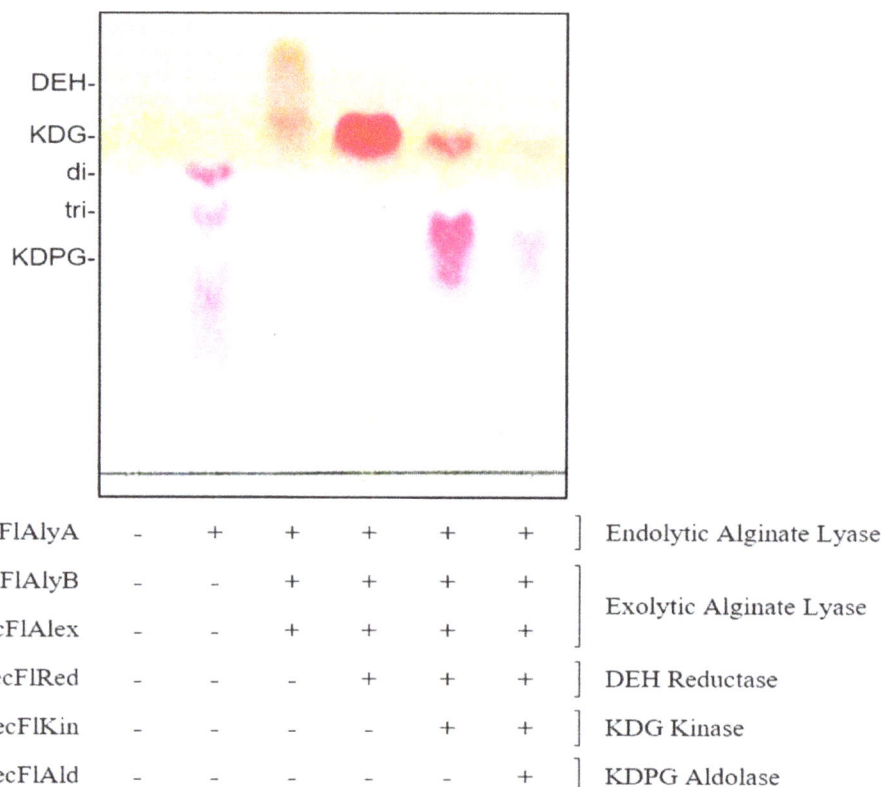

Figure 8. Construction of in vitro alginate-metabolizing system using recombinant enzymes. Alginate was allowed to react with recFlAlyA, recFlAlyB, recFlAlex, recFlRed, recFlKin, and recFlAld in various combinations at 25 °C for 12 h. The reaction products were subjected to TLC and detected by staining with 4.5% TBA. Presence and absence of each enzyme is indicated with '+' and '−', respectively. Detailed conditions are shown under Section 4.

3. Discussion

3.1. Alginate-Metabolizing Enzymes of Flavobacterium sp. Strain UMI-01

In the present study, *flkin* and *flald* in the genome of *Flavobacterium* sp. strain UMI-01 were confirmed to be the enzyme genes encoding KDG kinase and KDPG aldolase. The recombinant enzymes, recFlKin and recFlAld, showed KDG kinase and KDPG aldolase activity although low sequence identities were shown to the corresponding enzymes from other bacteria and archaea (Figures 2–4). Consequently, these genes, along with previously reported alginate lyase and DEH reductase genes were confirmed to be the genes responsible for alginate metabolism of this bacterium. The alginate-metabolizing pathway of this strain is summarized as in Figure 9. The alginate lyases degrade polymer alginate to unsaturated monomer (DEH) in the periplasmic space [24,25]. DEH reductase, KDG kinase and KDPG aldolase convert DEH to pyruvate and GAP in the cytosol. Therefore, DEH produced in the periplasmic space should be incorporated to the cytosol by certain transportation system(s). Such DEH transporters in this strain have not been identified yet; however, sugar permease-like gene *sugp* and membrane transporter-like genes *susc* and *susd* were found in the operons A and B, respectively (see Figure 1). Thus, the putative permease and transporters are also indicated in Figure 9. Another problem is how the expressions of alginate-metabolizing genes are regulated. We recently noticed that expression levels of alginate lyases were significantly low in the absence of alginate but strongly increased by the addition of alginate to the medium. This indicates that the expressions of alginate-metabolic enzymes are up-regulated by alginate. We are now searching regulatory genes for alginate-metabolizing enzyme genes in the UMI-01 strain genome.

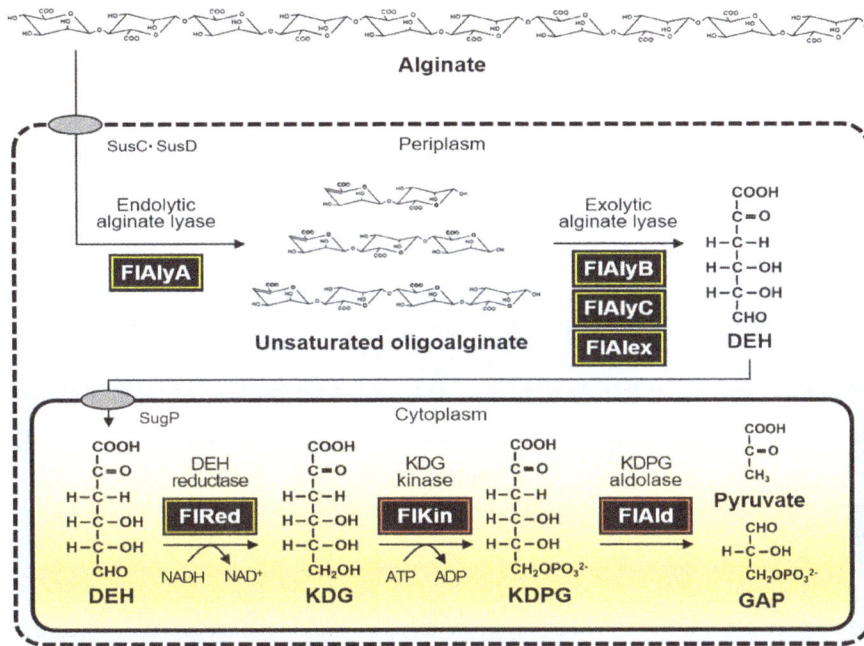

Figure 9. Alginate-metabolic system of *Flavobacterium* sp. strain UMI-01.

3.2. Properties of recFlKin and recFlAld

KDG kinase and KDPG aldolase are known to be the enzymes included in Entner–Doudoroff (ED) pathway. This pathway distributes over bacteria and archaea and play important roles in the metabolisms of glucuronate and glucose. In this pathway, KDG kinase phosphorylates KDG to KDPG, and KDPG aldolase split KDPG to pyruvate and GAP. Optimal temperature and pH of recFlKin were 50 °C and ~7.0, which were similar to those of KDG kinase from the bacteria *S. marcescens* [39]. While thermal stability of recFlKin was considerably low compared with the enzymes from archaea *S. tokodaii* [47] and *S. solfataricus* [40], e.g., these enzymes were stable up to 60–70 °C. recFlAld acts only on KDPG unlike archaea aldolases which split both KDG and KDPG [48,49]. Primary structures of bacterial aldolases showed low identity with those of archaea enzymes. The amino acid sequence of FlAld showed only 22%–25% identity with respect to *Proteobacteria* enzymes, while it showed 61%–65% identity with the *Bacteroidetes* enzymes. This suggests that somewhat deviated function between the *Proteobacteria* enzymes and *Bacteroidetes* enzymes. However, less different properties were found in recFlAld. Reverse reaction of bacterial aldolases was shown to be useful for the production of KDPG from pyruvate and GAP and also various compounds from pyruvate and aldehydes [44]. Our preliminary experiments also indicated that recFlAld could produce KDPG from pyruvate and GAP (data not shown, but see Section 2.3). Thus, recFlAld is also considered to be useful for producing novel compounds from pyruvate and various aldehydes.

3.3. Construction of In Vitro Alginate-Metabolizing System

An in vitro alginate-metabolizing system was successfully constructed from the recombinant enzymes, recFlAlyA, recFlAlyB, recFlAlex, recFlRed, recFlKin, and recFlAld. Accordingly, various kinds of intermediates could be produced by this system (Figure 8 and Table 2). Recently, alginate-assimilating enzymes of *S. degradans* 2-40^T were used for the production of KDG, KDPG, GAP and pyruvate [24,25]. However, the reaction efficiency of KDG kinase of *S. degradans* 2-40^T appeared to be lower than that of our system. Namely, the major part of KDG in the reaction mixture remained to be unphosphorylated in the *S. degradans* 2-40^T system. On the other hand, recFlKin in our system could convert KDG to KDPG with ~80% efficiency. This difference in the reaction efficiency between *S. degradans* enzyme and recFlKin may be derived from the origin of this enzyme, namely, from *Proteobacteria* species or *Bacteroidetes* species. To confirm this, we have to directly compare the KDG kinase properties between the enzymes from *Proteobacteria* and *Bacteroidetes* in future.

3.4. Production of a High-Value Product KDPG from Alginate

KDPG is a valuable leading compound for novel drugs and medicines. Synthesis of KDPG has been attempted by several methods [44,48,50]. For example, KDPG was first produced from gluconate with archaea enzymes [48]. However, this method required high-temperature reaction since the archaea enzymes are thermophilic. Reverse reaction of KDPG aldolase was also used for the production of KDPG from pyruvate and GAP [44,50]. However, this method required GAP, a significantly expensive raw material. On the other hand, we could produce KDPG from a much cheaper material, alginate, using the enzymes from the strain UMI-01. High recovery of KDPG from alginate (~80%) also indicated the practical potentiality of this enzyme. Thus, *Flavobacterium* sp. strain UMI-01 was considered to be a useful enzyme source for the production of value-added materials from alginate.

4. Experimental Section

4.1. Materials

Sodium alginate (*Macrocystis pyrifera* origin) was purchased from Sigma-Aldrich (St. Louis, MO, USA). Alginate-assimilating bacteria, *Flavobacterium* sp. strain UMI-01, was cultivated at 25 °C in a mineral salt (MS) medium including 1% (*w*/*v*) sodium alginate as described in our previous report [26]. Cell lysate (crude enzyme) of this strain was extracted from cell pellets by freeze and thaw followed by sonication as described previously [28]. DEH was prepared by the digestion of sodium alginate with the crude enzyme and purified by SuperQ-650S (Tosoh, Tokyo, Japan) anion-exchange chromatography [28]. Standard KDG, KDPG, pyruvate, and GAP were purchased from Sigma-Aldrich. pCold I expression vector was purchased from TaKaRa (Shiga, Japan) and modified to the form that can add 8 × Gly + 8 × His-tag to the C-terminus of the expressed proteins [26]. *E. coli* DH5α and BL21 (DE3) were purchased from TaKaRa. Ni-NTA resin was purchased from Qiagen (Hilden, Germany). A TLC silica gel 60 plate was purchased from Merk KGaA (Darmstadt, Germany). TSKgel DEAE-2SW (4.6 × 250 mm) and Superdex peptide 10/300 GL were purchased from Tosoh Bioscience LLC (King of Prussia, PA, USA) and GE Healthcare (Little Chalfont, Buckinghamshire, UK), respectively. Lactate dehydrogenase (LDH; porcine heart origin) and NADH were purchased from Oriental Yeast Co., LTD. (Tokyo, Japan). ATP and 9-aminoacridine were purchased from Sigma-Aldrich. Other chemicals were purchased from Wako Pure Chemical Industries Ltd. (Osaka, Japan).

4.2. Phylogenetic Analysis for KDG Kinases and KDPG Aldolases

Phylogenetic analysis was carried out using the amino acid sequences of KDG kinases or KDPG aldolases from *Proteobacteria*, *Bacteroidetes* and *Archaea* currently available. *Bacteroidetes* enzymes used are from *Gramella forsetii* KT0803, *Lacinutrix* sp. 5H-3-7-4, and *Dokdonia* sp. MED134, which were reported to be located in the alginolytic gene cluster of each species [51]. These amino acid sequences were first aligned with the sequences of FlKin or FlAld by the ClustalW program, then aligned sequences were trimmed with GBlocks. Phylogenetic trees were generated by the maximum likelihood algorithm on the basis of the LG model implemented in the Molecular Evolutionary Genetics Analysis version 6.0 (MEGA 6) software. The bootstrap values were calculated from 1000 replicates.

4.3. Cloning, Expression, and Purification of Recombinant FlKin and FlAld

Genomic DNA of strain UMI-01 was prepared with ISOHAIR DNA extraction kit (Nippon Gene, Tokyo, Japan). Coding regions of *flkin* and *flald*, 1023 bp and 669 bp, respectively, were amplified by PCR using specific primers including restriction sites, NcoI and BamHI, in the 5′-terminal regions (Table 1). Genomic PCR was performed in a medium containing 10 ng of genomic DNA, 0.2 μM each primer, and Phusion DNA polymerase (New England Biolabs, Ipswich, MA, USA). The reaction medium was preincubated at 95 °C for 2 min, and a reaction cycle of 95 °C for 10 s, 55 °C for 20 s, and 72 °C for 60 s was repeated 30 times. The PCR product was ligated to pCold I vector pre-digested by *Nco*I and *Bam*HI using In-Fusion cloning system (Clontech Laboratories, Mountain View, CA, USA). Insertion of the genes in the vector was confirmed by nucleotide sequencing with DNA sequencer 3130xl (Applied Biosystems, Foster, CA, USA). Recombinant enzymes, recFlKin and recFlAld, were expressed with the pCold I–*E. coli* BL21 (DE3) system. The transformed BL21 (DE3) was inoculated to 500 mL of 2× YT medium and cultivated at 37 °C for 16 h. Then, the temperature was lowered to 15 °C and isopropyl β-D-1-thiogalactopyranoside was added to make the final concentration of 0.1 mM. After 24-h induction, bacterial cells were harvested by centrifugation at 5000× g for 5 min and suspended in a buffer containing 10 mM imidazole-HCl (pH 8.0), 0.5 M NaCl, 1% (v/v) TritonX-100, and 0.01 mg/mL lysozyme. The suspension was sonicated at 20 kHz (30W) for a total of 4 min (30 s × 8 times with each 1 min interval) and centrifuged at 10,000× g for 10 min. The supernatant containing recombinant proteins was mixed with 1 mL of Ni-NTA resin and incubated for 30 min on ice with occasional suspension. The resin was set on a disposal plastic column (1 × 5 cm) and washed three times with 20 mL of 30 mM imidazole-HCl (pH 8.0)–0.5 M NaCl. The recombinant proteins adsorbed to the resin were eluted with 250 mM imidazole-HCl (pH 8.0)–0.5 M NaCl and collected as 1 mL fractions. The fractions containing the recombinant proteins were pooled and dialyzed against 20 mM Tris-HCl (pH 7.4)–0.1 M NaCl.

4.4. Preparation of KDG

KDG was prepared from alginate using the crude enzyme of the strain UMI-01 as follows; 0.5% (w/v) sodium alginate (50 mL) was digested at 30 °C with 1 mg/mL of the crude enzyme, which contains alginate lyases and other metabolic enzymes. NADH was added to the mixture to make the final concentration 10 mM to reduce DEH with DEH reductase contained in the crude enzyme. After 12 h, four volumes of −20 °C 2-propanol were added to terminate the reaction and the proteins and NADH precipitated were removed by centrifugation at 10,000× g for 10 min. The supernatant containing KDG was dried up in a rotary evaporator at 35 °C. The dried powder was dissolved in 50 mL of distilled water and subjected to a TOYOPEARL SuperQ-650S column (2.4 × 22 cm) equilibrated with distilled water. The absorbed KDG and trace amount of unsaturated disaccharide were separately eluted by a linear gradient of 0–0.2 M NaCl in distilled water (total 400 mL). Elution of KDG and unsaturated disaccharide was detected by TBA reaction. In this chromatography, KDG was eluted at around 80 mM NaCl, while disaccharides were eluted at around 120 mM. Approximately 90 mg of KDG was obtained from 0.25 g of sodium alginate.

4.5. Preparation of KDPG

KDPG was prepared from the KDG by using recFlKin. Namely, recFlKin was (final concentration 10 μg/mL) was added to the reaction mixture (10 mL) containing 2.5 mM KDG, 2.5 mM ATP, 5 mM MgCl$_2$, 20 mM Tris-HCl (pH 7.4), 100 mM KCl, and 1 mM dithiothreitol, and incubated at 40 °C for 3 h. The mixture was lyophilized, dissolved in 500 μL of distilled water and the supernatant was subjected to a Superdex peptide 10/300 GL column equilibrated with 0.1 M CH$_3$COONH$_4$. KDPG and KDG, which eluted together in this chromatography, were lyophilized, dissolved in 1 mL of distilled water, and subjected to HPLC (Shimadzu Prominence LC-6AD, Tokyo, Japan) equipped by TSKgel DEAE-2SW (Tosoh). KDG and KDPG were separately eluted at around 150 mM and 320 mM CH$_3$COONH$_4$ by the linear gradient of 0–0.4 M CH$_3$COONH$_4$. The amount of KDPG was quantified by the system comprising recFlAld and LDH–NADH using authentic KDPG as a standard. By the above procedure, 1.2 mg of KDPG was obtained from 4.5 mg of KDG.

4.6. Assay for KDPG Aldolase Activity

KDPG aldolase activity of recFlAld was assayed by the determination of pyruvate using a lactate dehydrogenase (LDH)–NADH coupling system [50]. Namely, the aldolase reaction was conducted at

30 °C in a reaction mixture containing 5 mM KDPG, 20 mM Tris-HCl (pH 7.4), 100 mM KCl, 1 mM DTT, and 1 µg/mL recFlAld in the presence of 0.2 mM NADH and 1 unit/mL LDH. The reaction rate was estimated from the decrease in the Abs 340 nm due to the oxidation of NADH accompanied by the reduction of pyruvate. One unit (U) of KDPG aldolase activity was defined as the amount of enzyme that produced 1 µmol of pyruvate per min.

4.7. Assay for KDG Kinase Activity

KDG kinase activity was assayed as follows. The reaction mixture containing 2.5 mM KDG, 2.5 mM ATP, 5 mM MgCl$_2$, 20 mM Tris-HCl (pH 7.4), 100 mM KCl, 1 mM DTT, and 10 µg/mL recFlKin was incubated at 30 °C. At reaction times, 1, 15, and 30 min, an aliquot (160 µL) of the reaction mixture was taken out and heated at 100 °C for 3 min to terminate the reaction. To the mixture, 240 µL of a buffer containing 84 mM Tris-HCl (pH 7.4), 167 mM KCl, 0.67 mM NADH, 2.5 µg/mL recFlAld, and 1 unit of LDH was added and the pyruvate released was determined by the LDH–NADH system. One unit (U) of KDG kinase activity was defined as the amount of enzyme that produced 1 µmol of KDPG per min. Temperature dependence of recFlKin was determined at 10–60 °C. Thermal stability of recFlKin was assessed by measuring the activity remaining after the incubation at 10–50 °C for 30 min. pH dependence of recFlKin was determined with reaction mixtures adjusted to pH 4.5–5.3 with 20 mM CH$_3$COONa buffer, pH 5.6–7.3 with 20 mM PIPES-NaOH buffer, pH 7.1–8.8 with 20 mM Tris-HCl buffer, and pH 9.1–9.7 with 20 mM glycine–NaOH buffer. The activity assay was conducted three times and the mean value was shown with standard deviation in each figure.

4.8. Construction of In Vitro Alginate-Metabolizing System from Recombinant Enzymes

An in vitro alginate-metabolizing system was constructed using recombinant alginate lyases (recFlAlyA, recFlAlyB, and recFlAlex) [26,27], recombinant DEH reductase (recFlRed) [28], and recFlKin and recFlAld prepared in the present study. Alginate-metabolizing reaction was conducted at 25 °C in a mixture containing 0.2% (w/v) sodium alginate, 10 mM NADH, 10 mM ATP, 10 mM MgCl$_2$, 20 mM sodium phosphate (pH 7.4), 1 mM DTT, and various combinations of recFlAlyA, recFlAlyB, recFlAlex, recFlRed, recFlKin, and recFlAld with each final concentration at 10 µg/mL, 10 µg/mL, 10 µg/mL, 2.5 µg/mL, 10 µg/mL, and 1 µg/mL, respectively. After 12-h reaction, unsaturated oligo-alginates, DEH, and KDG, were analyzed by TLC and TBA reaction [52]. KDPG and pyruvate concentrations were determined by the LDH–NADH reaction.

4.9. Determination of Unsaturated Sugars

Unsaturated sugars were determined by the TBA method [52]. The sample containing unsaturated sugars (150 µL) was mixed with 150 µL of 20 mM NaIO$_4$–0.125 M H$_2$SO$_4$ and allowed to react for 1 h on ice. Then, 100 µL of NaAsO$_2$–0.5 N HCl was added to the mixture and incubated for 10 min at room temperature. To the mixture, 600 µL of 0.6% (w/v) TBA was added and heated for 10 min at 100 °C. The unsaturated sugars were determined by measuring Abs 548 nm, adopting the absorption coefficient for DEH and KDG, $\varepsilon = 41 \times 10^3$ M$^{-1}\cdot$cm^{-1}, which we determined in the present study using KDG and DEH standards.

4.10. Thin-Layer Chromatography

TLC silica gel 60 plate was used for the analysis of the reaction products produced by recFlKin and recFlAld. The reaction product of recFlKin was prepared with a reaction mixture containing 2.5 mM KDG and 2.5 mM ATP and 200 µg/mL recFlKin. The reaction was carried out at 30 °C for 0–15 min and terminated by heating at 100 °C for 2 min. Four microliters of each reaction mixture was applied to a TLC plate. The reaction product was developed with 1-butanol:acetic acid:water = 2:1:1 (v:v:v) and detected by heating at 130 °C for 10 min after spraying 10% (w/v) sulfuric acid–90% (w/v) ethanol. The reaction product of recFlAld was prepared with a reaction mixture containing 5 mM KDPG and 1 µg/mL recFlAld. After the reaction at 30 °C for 0–15 min, six microliters of the reaction mixture were applied to TLC plate and developed with the same solvent as described above. The reaction product on the plate was detected with 0.5% (w/v) 2,5-dinitrophenylhydrazine (DNP)–20% (v/v) sulfuric acid–60% (v/v) ethanol. In case of unsaturated sugars, they were visualized with 4.5% (w/v) TBA after the periodic acid treatment.

4.11. Mass Spectrometry

Phosphorylation of KDG by recFlKin was detected by mass spectrometry. The KDG phosphorylated by recFlKin in the conditions described in Section 4.10 was mixed with 6.7 mg/mL 9-aminoacridine–methanol at 1:3 ($v:v$). One microliter of the mixture was applied to a sample plate and air-dried at room temperature. The sample was subjected to a matrix-assisted laser desorption ionization-time of flight mass spectrometer (MALDI-TOF-MS) (Proteomics Analyzer 4700, Applied Biosystems, Foster City, CA, USA) and analyzed in a negative-ion mode.

4.12. SDS-PAGE

SDS-PAGE was performed by the method of Porzio and Pearson [53] using 10% polyacrylamide gel. Proteins in the gel were stained with 0.1% (w/v) Coomassie Brilliant Blue R-250–50% (v/v) methanol–10% (v/v) acetic acid and the background of the gel was destained with 5% (v/v) methanol–7% (v/v) acetic acid.

4.13. Determination of Protein Concentration

Protein concentration was determined by the method of Lowry [54] using bovine serum albumin fraction V as a standard.

5. Conclusions

Enzymes responsible for the metabolism of alginate-derived DEH had not been well characterized in alginolytic bacteria. In the present study, KDG kinase-like gene *flkin* and KDPG aldolase-like gene *flald* in the genome of *Flavobacterium* sp. strain UMI-01 were investigated and the activities of the proteins encoded by these genes were assessed by using recombinant enzymes recFlKin and recFlAld. Analyses for reaction product of recFlKin and recFlAld indicated that these enzymes were KDG kinase and KDPG aldolase, respectively. Thus, the alginate metabolism of *Flavobacterium* sp. strain UMI-01 was considered to be achieved by the actions of FlKin and FlAld along with alginate lyases FlAlyA, FlAlyB and FlAlex, and DEH reductase FlRed. An in vitro alginate-metabolizing system was successfully constructed from the above enzymes. This system could convert alginate to pyruvate and GAP with 38% efficiency. This result indicates that the UMI-01 enzymes are available for the production of high-value materials like KDPG from alginate.

Acknowledgments: This study was supported in part by the Program for Constructing "Tohoku Marine Science Bases" promoted by Ministry of Education, Culture, Sports, Science and Technology, Japan.

Author Contributions: Ryuji Nishiyama took charge of designing the research and performed biochemical analysis. Akira Inoue performed cloning of *flkin* and *flald* genes and expression of recombinant enzymes. Ryuji Nishiyama, Akira Inoue and Takao Ojima took charge of preparation of manuscript.

References

1. Haug, A.; Larsen, B.; Smidsrod, O. Studies on the sequence of uronic acid residues in alginic acid. *Acta Chem. Scand.* **1967**, *21*, 691–704. [CrossRef]
2. Gacesa, P. Alginates. *Carbohydr. Polym.* **1988**, *8*, 161–182. [CrossRef]
3. Wong, T.Y.; Preston, L.A.; Schiller, N.L. Alginate lyase: Review of major sources and enzyme characteristics, structure-function analysis, biological roles, and applications. *Annu. Rev. Microbiol.* **2000**, *54*, 289–340. [CrossRef] [PubMed]

4.   Tomoda, Y.; Umemura, K.; Adachi, T. Promotion of barley root elongation under hypoxic conditions by alginate lyase-lysate (A.L.L.). *Biosci. Biotechnol. Biochem.* **1994**, *58*, 202–203. [CrossRef] [PubMed]

5.   Xu, X.; Iwamoto, Y.; Kitamura, Y.; Oda, T.; Muramatsu, T. Root growth-promoting activity of unsaturated oligomeric uronates from alginate on carrot and rice plants. *Biosci. Biotechnol. Biochem.* **2003**, *67*, 2022–2025. [CrossRef] [PubMed]

6.   Akiyama, H.; Endo, T.; Nakakita, R.; Murata, K.; Yonemoto, Y.; Okayama, K. Effect of depolymerized alginates on the growth of bifidobacteria. *Biosci. Biotechnol. Biochem.* **1992**, *56*, 355–356. [CrossRef] [PubMed]

7.   Ariyo, B.; Tamerler, C.; Bucke, C.; Keshavarz, T. Enhanced penicillin production by oligosaccharides from batch cultures of *Penicillium chrysogenum* in stirred-tank reactors. *FEMS Microbiol. Lett.* **1998**, *166*, 165–170. [CrossRef] [PubMed]

8.   Trommer, H.; Neubert, R.H.H. Screening for new antioxidative compounds for topical administration using skin lipid model systems. *J. Pharm. Pharm. Sci.* **2005**, *8*, 494–506. [PubMed]

9.   Khodagholi, F.; Eftekharzadeh, B.; Yazdanparast, R. A new artificial chaperone for protein refolding: Sequential use of detergent and alginate. *Protein J.* **2008**, *27*, 123–129. [CrossRef] [PubMed]

10. Mo, S.-J.; Son, E.-W.; Rhee, D.-K.; Pyo, S. Modulation of tnf-α-induced icam-1 expression, no and h2O2 production by alginate, allicin and ascorbic acid in human endothelial cells. *Arch. Pharm. Res.* **2003**, *26*, 244–251. [CrossRef] [PubMed]

11. An, Q.D.; Zhang, G.L.; Wu, H.T.; Zhang, Z.C.; Zheng, G.S.; Luan, L.; Murata, Y.; Li, X. Alginate-deriving oligosaccharide production by alginase from newly isolated *Flavobacterium* sp. LXA and its potential application in protection against pathogens. *J. Appl. Microbiol.* **2009**, *106*, 161–170. [CrossRef] [PubMed]

12. Enquist-Newman, M.; Faust, A.M.E.; Bravo, D.D.; Santos, C.N.S.; Raisner, R.M.; Hanel, A.; Sarvahowman, P.; Le, C.; Regitsky, D.D.; Cooper, S.R.; et al. Efficient ethanol production from brown macroalgae sugars by a synthetic yeast platform. *Nature* **2014**, *505*, 239–243. [CrossRef] [PubMed]

13. Wargacki, A.J.; Leonard, E.; Win, M.N.; Regitsky, D.D.; Santos, C.N.S.; Kim, P.B.; Cooper, S.R.; Raisner, R.M.; Herman, A.; Sivitz, A.B.; et al. An engineered microbial platform for direct biofuel production from brown macroalgae. *Science* **2012**, *335*, 308–313. [CrossRef] [PubMed]

14. Takeda, H.; Yoneyama, F.; Kawai, S.; Hashimoto, W.; Murata, K. Bioethanol production from marine biomass alginate by metabolically engineered bacteria. *Energy Environ. Sci.* **2011**, *4*, 2575. [CrossRef]

15. Miyake, H.; Yamaki, T.; Nakamura, T.; Ishibashi, H.; Fukuiri, Y.; Sakuma, A.; Komatsu, H.; Anso, T.; Togashi, K.; Umetani, H. Gluconate dehydratase. U.S. Patent 7,125,704 B2, 24 October 2006.

16. Yoon, H.J.; Hashimoto, W.; Miyake, O.; Okamoto, M.; Mikami, B.; Murata, K. Overexpression in *Escherichia coli*, purification, and characterization of *Sphingomonas* sp. A1 alginate lyases. *Protein Expr. Purif.* **2000**, *19*, 84–90. [CrossRef] [PubMed]

17. Hashimoto, W.; Miyake, O.; Momma, K.; Kawai, S.; Murata, K. Molecular identification of oligoalginate lyase of *Sphingomonas* sp. strain A1 as one of the enzymes required for complete depolymerization of alginate. *J. Bacteriol.* **2000**, *182*, 4572–4577. [CrossRef] [PubMed]

18. Takase, R.; Ochiai, A.; Mikami, B.; Hashimoto, W.; Murata, K. Molecular identification of unsaturated uronate reductase prerequisite for alginate metabolism in *Sphingomonas* sp. A1. *Biochim. Biophys. Acta Proteins Proteom.* **2010**, *1804*, 1925–1936. [CrossRef] [PubMed]

19. Takase, R.; Mikami, B.; Kawai, S.; Murata, K.; Hashimoto, W. Structure-based conversion of the coenzyme requirement of a short-chain dehydrogenase/reductase involved in bacterial alginate metabolism. *J. Biol. Chem.* **2014**, *289*, 33198–33214. [CrossRef] [PubMed]

20. Preiss, J.; Ashwell, G. Alginic acid metabolism in bacteria I. *J. Biol. Chem.* **1962**, *237*, 309–316. [PubMed]

21. Preiss, J.; Ashwell, G. Alginic acid metabolism in bacteria II. *J. Biol. Chem.* **1962**, *237*, 317–321. [PubMed]

22. Kim, H.T.; Ko, H.J.; Kim, N.; Kim, D.; Lee, D.; Choi, I.G.; Woo, H.C.; Kim, M.D.; Kim, K.H. Characterization of a recombinant endo-type alginate lyase (Alg7D) from *Saccharophagus degradans*. *Biotechnol. Lett.* **2012**, *34*, 1087–1092. [CrossRef] [PubMed]

23. Kim, H.T.; Chung, J.H.; Wang, D.; Lee, J.; Woo, H.C.; Choi, I.G.; Kim, K.H. Depolymerization of alginate into a monomeric sugar acid using Alg17C, an exo-oligoalginate lyase cloned from *Saccharophagus degradans* 2-40. *Appl. Microbiol. Biotechnol.* **2012**, *93*, 2233–2239. [CrossRef] [PubMed]

24. Takagi, T.; Morisaka, H.; Aburaya, S.; Tatsukami, Y.; Kuroda, K.; Ueda, M. Putative alginate assimilation process of the marine bacterium *Saccharophagus degradans* 2-40 based on quantitative proteomic analysis. *Mar. Biotechnol.* **2016**, *18*, 15–23. [CrossRef] [PubMed]

25. Kim, D.H.; Wang, D.; Yun, E.J.; Kim, S.; Kim, S.R.; Kim, K.H. Validation of the metabolic pathway of the alginate-derived monomer in *Saccharophagus degradans* 2-40^T by gas chromatography–mass spectrometry. *Process Biochem.* **2016**, *51*, 1374–1379. [CrossRef]

26. Inoue, A.; Takadono, K.; Nishiyama, R.; Tajima, K.; Kobayashi, T.; Ojima, T. Characterization of an alginate lyase, FlAlyA, from *Flavobacterium* sp. strain UMI-01 and its expression in *Escherichia coli*. *Mar. Drugs* **2014**, *12*, 4693–4712. [CrossRef] [PubMed]

27. Inoue, A.; Nishiyama, R.; Ojima, T. The alginate lyases FlAlyA, FlAlyB, FlAlyC, and FlAlex from *Flavobacterium* sp. UMI-01 have distinct roles in the complete degradation of alginate. *Algal Res.* **2016**, *19*, 355–362. [CrossRef]

28. Inoue, A.; Nishiyama, R.; Mochizuki, S.; Ojima, T. Identification of a 4-deoxy-l-erythro-5-hexoseulose uronic acid reductase, FlRed, in an alginolytic bacterium *Flavobacterium* sp. strain UMI-01. *Mar. Drugs* **2015**, *13*, 493–508. [CrossRef] [PubMed]

29. Badur, A.H.; Jagtap, S.S.; Yalamanchili, G.; Lee, J.-K.; Zhao, H.; Rao, C.V. Alginate lyases from alginate-degrading *Vibrio splendidus* 12B01 are endolytic. *Appl. Environ. Microbiol.* **2015**, *81*, 1865–1873. [CrossRef] [PubMed]

30. Shimizu, E.; Ojima, T.; Nishita, K. cDNA cloning of an alginate lyase from abalone, *Haliotis discus hannai*. *Carbohydr. Res.* **2003**, *338*, 2841–2852. [CrossRef] [PubMed]

31. Suzuki, H.; Suzuki, K.I.; Inoue, A.; Ojima, T. A novel oligoalginate lyase from abalone, *Haliotis discus hannai*, that releases disaccharide from alginate polymer in an exolytic manner. *Carbohydr. Res.* **2006**, *341*, 1809–1819. [CrossRef] [PubMed]

32. Rahman, M.M.; Inoue, A.; Tanaka, H.; Ojima, T. Isolation and characterization of two alginate lyase isozymes, AkAly28 and AkAly33, from the common sea hare *Aplysia kurodai*. *Comp. Biochem. Physiol. B Biochem. Mol. Biol.* **2010**, *157*, 317–325. [CrossRef] [PubMed]

33. Rahman, M.M.; Inoue, A.; Tanaka, H.; Ojima, T. cDNA cloning of an alginate lyase from a marine gastropod *Aplysia kurodai* and assessment of catalutically important residues of this enzyme. *Biochimie* **2011**, *93*, 1720–1730. [CrossRef] [PubMed]

34. Inoue, A.; Mashino, C.; Uji, T.; Saga, N.; Mikami, K.; Olima, T. Characterization of an eukaryotic PL-7 alginate lyase in the marine red alga *Pyropia yezoensis*. *Curr. Biotechnol.* **2015**, *4*, 240–248. [CrossRef]

35. Gacesa, P. Enzymic degradation of alginates. *Int. J. Biochem.* **1992**, *24*, 545–552. [CrossRef]

36. Hobbs, J.K.; Lee, S.M.; Robb, M.; Hof, F.; Bar, C.; Abe, K.T.; Hehemann, J.H.; McLean, R.; Abbott, D.W.; Boraston, A.B. KdgF, the missing link in the microbial metabolism of uronate sugars from pectin and alginate. *Proc. Natl. Acad. Sci. USA* **2016**, *113*, 6188–6193. [CrossRef] [PubMed]

37. Mochizuki, S.; Nishiyama, R.; Inoue, A.; Ojima, T. A novel aldo-keto reductase HdRed from the pacific abalone *Haliotis discus hannai*, which reduces alginate-derived 4-deoxy-L-erythro-5-hexoseulose uronic acid to 2-keto-3-deoxy-D-gluconate. *J. Biol. Chem.* **2015**, *290*, 30962–30974. [CrossRef] [PubMed]

38. Cynkin, M.A.; Ashwell, G. Uronic acid metabolism in bacteria. *J. Biol. Chem.* **1960**, *235*, 1576–1579. [PubMed]

39. Lee, Y.S.; Park, I.H.; Yoo, J.S.; Kim, H.S.; Chung, S.Y.; Chandra, M.R.G.; Choi, Y.L. Gene expression and characterization of 2-keto-3-deoxy-gluconate kinase, a key enzyme in the modified Entner-Doudoroff pathway of *Serratia marcescens* KCTC 2172. *Electron. J. Biotechnol.* **2009**, *12*. [CrossRef]

40. Kim, S.; Lee, S.B. Characterization of *Sulfolobus solfataricus* 2-keto-3-deoxy-D-gluconate kinase in the modified Entner-Doudoroff pathway. *Biosci. Biotechnol. Biochem.* **2006**, *70*, 1308–1316. [CrossRef] [PubMed]

41. Lamble, H.J.; Theodossis, A.; Milburn, C.C.; Taylor, G.L.; Bull, S.D.; Hough, D.W.; Danson, M.J. Promiscuity in the part-phosphorylative Entner-Doudoroff pathway of the archaeon *Sulfolobus solfataricus*. *FEBS Lett.* **2005**, *579*, 6865–6869. [CrossRef] [PubMed]

42. Potter, J.A.; Kerou, M.; Lamble, H.J.; Bull, S.D.; Hough, D.W.; Danson, M.J.; Taylor, G.L. The structure of *Sulfolobus solfataricus* 2-keto-3-deoxygluconate kinase. *Acta Crystallogr. Sect. D Biol. Crystallogr.* **2008**, *64*, 1283–1287. [CrossRef] [PubMed]

43. Ohshima, N.; Inagaki, E.; Yasuike, K.; Takio, K.; Tahirov, T.H. Structure of *Thermus thermophilus* 2-keto-3-deoxygluconate kinase: Evidence for recognition of an open chain substrate. *J. Mol. Biol.* **2004**, *340*, 477–489. [CrossRef] [PubMed]

44. Shelton, M.C.; Cotterill, I.C.; Novak, S.T.A.; Poonawala, R.M.; Sudarshan, S.; Toone, E.J. 2-Keto-3-deoxy-6-phosphogluconate aldolases as catalysts for stereocontrolled carbon-carbon bond formation. *J. Am. Chem. Soc.* **1996**, *118*, 2117–2125. [CrossRef]

45. Allard, J.; Grochulski, P.; Sygusch, J. Covalent intermediate trapped in 2-keto-3-deoxy-6-phosphogluconate (KDPG) aldolase structure at 1.95-A resolution. *Proc. Natl. Acad. Sci. USA* **2001**, *98*, 3679–3684. [CrossRef] [PubMed]

46. Bell, B.J.; Watanabe, L.; Lebioda, L.; Arni, R.K. Structure of 2-keto-3-deoxy-6-phosphogluconate (KDPG) aldolase from *Pseudomonas putida*. *Acta Crystallogr. D Biol. Crystallogr.* **2003**, *59*, 1454–1458. [CrossRef] [PubMed]

47. Ohshima, T.; Kawakami, R.; Kanai, Y.; Goda, S.; Sakuraba, H. Gene expression and characterization of 2-keto-3-deoxygluconate kinase, a key enzyme in the modified Entner-Doudoroff pathway of the aerobic and acidophilic hyperthermophile *Sulfolobus tokodaii*. *Protein Expr. Purif.* **2007**, *54*, 73–78. [CrossRef] [PubMed]

48. Ahmed, H.; Ettema, T.J.G.; Tjaden, B.; Geerling, A.C.M.; Oost, J.V.D.; Siebers, B. The semi-phosphorylative Entner-Doudoroff pathway in hyperthermophilic archaea: A re-evaluation. *Biochem. J.* **2005**, *390*, 529–540. [CrossRef] [PubMed]

49. Reher, M.; Fuhrer, T.; Bott, M.; Schönheit, P. The nonphosphorylative entner-doudoroff pathway in the thermoacidophilic euryarchaeon *Picrophilus torridus* involves a novel 2-Keto-3-deoxygluconate-specific aldolase. *J. Bacteriol.* **2010**, *192*, 964–974. [CrossRef] [PubMed]

50. Cotterill, I.C.; Shelton, M.C.; Machemer, D.E.W.; Henderson, D.P.; Toone, E.J. Effect of phosphorylation on the reaction rate of unnatural electrophiles with 2-keto-3-deoxy-6-phosphogluconate aldolase. *J. Chem. Soc. Trans.* **1998**, *1*, 1335–1341. [CrossRef]

51. Kabisch, A.; Otto, A.; Konig, S.; Becher, D.; Albrecht, D.; Schuler, M.; Teeling, H.; Amann, R.I.; Scheweder, T. Functional characterization of polysaccharide utilization loci in the marine *Bacteroidetes* 'Gramella forsetii' KT0803. *ISME J.* **2014**, *8*, 1492–1502. [CrossRef] [PubMed]

52. Weissbach, A.; Hurwitz, J. The Formation of 2-Keto-3-deoxyheptonie Acid in Extracts of *Escherichia coli* B. I. Identification. *J. Biol. Chem.* **1959**, *234*, 705–710. [PubMed]

53. Porzio, M.A.; Pearson, A.M. Improved resolution of myofibrillar proteins with sodium dodecyl sulfate-polyacrylamide gel electrophoresis. *Biochim. Biophys. Acta (BBA) Protein Struct.* **1977**, *490*, 27–34. [CrossRef]

54. Lowry, O.H.; Rosebrough, N.J.; Farr, A.L.; Randall, R.J. Protein measurement with the dolin phenol reagent. *J. Biol. Chem.* **1951**, *193*, 265–275. [PubMed]

6

Structure-Based Design and Synthesis of a New Phenylboronic-Modified Affinity Medium for Metalloprotease Purification

Shangyong Li [1,†], Linna Wang [1,†], Ximing Xu [2,†], Shengxiang Lin [3,*], Yuejun Wang [1], Jianhua Hao [1,4] and Mi Sun [1,4,*]

[1] Key Laboratory of Sustainable Development of Marine Fisheries, Ministry of Agriculture, Yellow Sea Fisheries Research Institute, Chinese Academy of Fishery Sciences, 106 Nanjing Road, Qingdao 266071, China; lshywln@163.com (S.L.); wlnwfllsy@163.com (L.W.); wangyj@ysfri.ac.cn (Y.W.); haojh@ysfri.ac.cn (J.H.)

[2] Institute of Bioinformatics and Medical Engineering, School of Electrical and Information Engineering, Jiangsu University of Technology, Changzhou 213000, China; ximing.xu@jsut.edu.cn

[3] Laboratory of Oncology and Molecular Endocrinology, CHUL Research Center (CHUQ) and Laval University, 2705 Boulevard Laurier, Ste-Foy, Ville de Québec, QC G1V 4G2, Canada

[4] Laboratory for Marine Drugs and Bioproducts, Qingdao National Laboratory for Marine Science and Technology, Qingdao 266237, China

* Correspondence: sheng-xiang.lin@crchul.ulaval.ca (S.L.); sunmi0532@yahoo.com (M.S.).

† These authors contributed equally to this paper.

Academic Editors: Vassilios Roussis and Antonio Trincone

Abstract: Metalloproteases are emerging as useful agents in the treatment of many diseases including arthritis, cancer, cardiovascular diseases, and fibrosis. Studies that could shed light on the metalloprotease pharmaceutical applications require the pure enzyme. Here, we reported the structure-based design and synthesis of the affinity medium for the efficient purification of metalloprotease using the 4-aminophenylboronic acid (4-APBA) as affinity ligand, which was coupled with Sepharose 6B via cyanuric chloride as spacer. The molecular docking analysis showed that the boron atom was interacting with the hydroxyl group of Ser176 residue, whereas the hydroxyl group of the boronic moiety is oriented toward Leu175 and His177 residues. In addition to the covalent bond between the boron atom and hydroxyl group of Ser176, the spacer between boronic acid derivatives and medium beads contributes to the formation of an enzyme-medium complex. With this synthesized medium, we developed and optimized a one-step purification procedure and applied it for the affinity purification of metalloproteases from three commercial enzyme products. The native metalloproteases were purified to high homogeneity with more than 95% purity. The novel purification method developed in this work provides new opportunities for scientific, industrial and pharmaceutical projects.

Keywords: metalloprotease; adsorption analysis; molecular docking; affinity purification; aminophenylboronic acid

1. Introduction

Proteases are enzymes that catalyze the hydrolysis of peptide bonds. Based on the mechanism of catalysis, proteases can be classified into six classes, including metallo, serine, aspartic, cysteine, glutamic, and threonine proteases [1]. Proteases are the most important industrial enzymes, accounting for more than 60% of the total enzyme market [2,3]. They have broad applications in the pharmaceutical, leather, food, and detergent industries [1–4]. Proteases play critical roles in normal biological processes; their unusual activities have been implicated in the development and progression of many diseases, e.g., fibrosis, arthritis, cancer, cardiovascular diseases, nephritis, and central nervous system disorders [5–7]. Among all of the six classes of proteases, only untagged serine proteases can be purified in one step using *p*-aminobenzamidine-modified affinity medium [8,9]. This simple procedure of affinity purification significantly accelerated the pharmaceutical application of many serine proteases [10–15].

Currently, there is no straightforward and efficient protocol for the purification of metalloproteases [14–16]. The traditional protocol that has multiple steps is expensive and results in low recovery [17–20]. Although some reports refer to the high-yield purification of metalloproteases (more than 90% purity) in one-step procedure, these protocols were based on immobilized metal affinity chromatography (IMAC) that has its disadvantages [21,22]. The first one is the use of high concentrations of imidazole and salt in the elution buffer of the IMAC procedure, which necessitates additional dialysis or a desalting step [23,24]. Also, it is well known that purification of a metalloprotein via a metal ion chelated by the resin in a similar manner results in the exchange of metal ion from resin with a metal ion from metalloprotein. This metal transfer causes a decrease in the stability of purified metalloprotein [21]. In addition, the use of chelating agents during purification has to be avoided as these compounds can remove the metal ion from the enzyme active site [23–25]. Therefore, design, synthesis and application of a new specific and efficient medium for the purification of metalloproteases are important tasks.

The structure-based design of the affinity ligand that serves as a specific inhibitor or substrate analogue is an efficient and commonly used approach in the affinity purification of enzymes [26–28]. An alkaline metalloprotease, MP (accession no. ACY25898) from marine bacterium *Flavobacterium* sp. YS-80-122, has been previously isolated in our laboratory [3]. This enzyme is a typical Zn-containing metalloprotease with antioxidant activity, and it has been commercially used as a detergent additive. The analysis of its crystal structure (PDB: 3U1R) [29] allowed suggestion of a novel affinity ligand that could reversibly bind to the active site and could be used for the affinity purification of the enzyme. Our preliminary virtual screening and experimental verification indicated that boronic acid derivatives (BADs) could reversibly inhibit the activity of MP [30]. Phenylboronate group, which can form a temporary covalent bond with any molecule that contains a 1,2-*cis*-diol group, is widely used in the affinity purification of 1,2-*cis*-diol-containing biomolecules such as glycoproteins, glycopeptides, nucleosides, and nucleic acids [31–35]. However, application of the resins modified by phenylboronate in the purification of metalloproteases has never been reported.

Here, the phenylboronate-modified resin was synthesized through the coupling of 4-aminophenylboronic acid (4-APBA) with epoxy-activated Sepharose 6B via cyanuric chloride spacer. The binding site and structure-activity relationship between 4-APBA-modified medium and MP were analyzed using molecular docking and adsorption determination, correspondingly. The synthesized medium was used for development of one-step affinity purification of metalloproteases. Three commercially available metalloproteases were efficiently purified with a high purity (more than 95%) using the protocol developed. Our research provides new opportunities for the development of industrial methods of metalloprotease purification.

2. Results and Discussion

2.1. Design and Synthesis of Affinity Medium for Metalloprotease Purification

Our initial virtual screening showed that some BADs could inhibit MP catalytic activity [30]. To confirm that BADs could inhibit MP, ten BADs were purchased or synthesized, and then their inhibitory effect was tested on MP. Surprisingly, three compounds were strong MP inhibitors with apparent K_i value of 0.8–1.2 μM. Thus, we focused our efforts on the design of BADs-based affinity medium for metalloprotease purification. Immobilisation of a ligand onto the epoxy-activated resin should be achieved via a nucleophilic group present in the ligand, often a primary amine [27].

Aminephenylboronic acid was chosen as an affinity ligand for our study because it was commercially available and had a favorable configuration for synthesis affinity medium.

The nature of the immobilized complex or, in other words, the choice of affinity ligand and spacer arm, has a major influence on the outcome of a biomimetic affinity of purification procedure [27–29,36]. To obtain an optimal affinity medium, two types of APBA-based ligands, 4-APBA, and 3-APBA, were tested. The affinity ligands were coupled with activated Sepharose 6B via cyanuric chloride spacer. To estimate the effect of the presence of a boron atom in the affinity ligand, another type of the affinity ligand lacking of boron atom (aniline ligand) was synthesized. The scheme for the synthesis of 4-APBA-modified Sepharose 6B is shown in Figure 1. To confirm the ligand structure, the medium was hydrolyzed with 6 M HCl, and then the resultant with molecular formula of $C_{12}H_{15}BClN_5O_4$ and molecular mass of 339.5. Because the chlorine on the triazine ring was unstable in acidic condition, the hydrolysis with 6 M HCl would replace the chlorine on the ligand with a hydroxyl group [37], thus the theoretical structure of the purified ligand should be with a molecular formula of $C_{12}H_{16}BN_5O_5$ and molecular mass of 321.1. The ligand may be broken into fragments as $C_9H_{10}BN_5O_3$ at cone voltage of 170 V and molecular mass of 247.01. As shown in Figure S1, the main peak, 247.02, showed good agreement with $[M-C_3H_6O_2-H]^+$. The possible structures of chemicals in principal peaks are also shown in Figure S1. These results showed that the synthesized ligands had a good reliability.

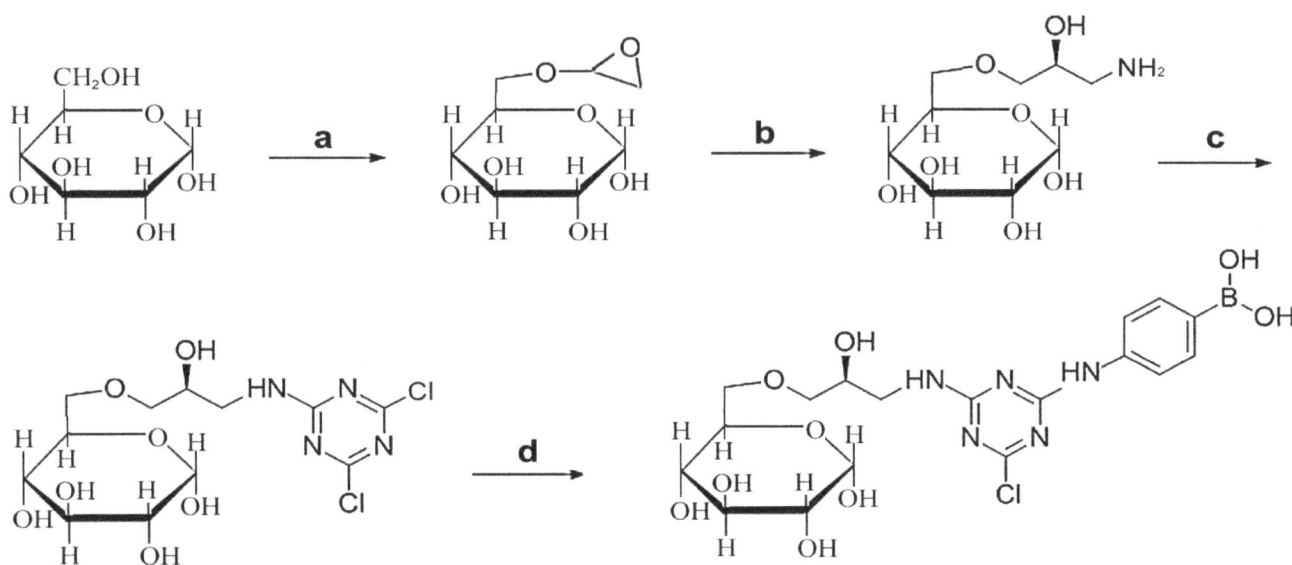

Figure 1. Synthesis protocol and scheme of the 4-APBA ligand coupled with actived Sepharose 6B via cyanuric chloride spacer. Reagents and conditions: (**a**) epichlorohydrin, DMSO, NaOH aqueous solution, 2.5 h; (**b**) 35% saturated ammonia, overnight; (**c**) cyanuric chloride, 50% acetone, pH 7–8; (**d**) 4-APBA, sodium carbonate, 24 h.

The 3-APBA-modified medium and aniline-modified medium were synthesized using the same concentration of 3-APBA or aniline as for 4-APBA (Figure 2A,B). The density of the free amino groups was determined by the ninhydrin test before the adding of the APBA ligands, giving equal ligand densities (tab:marinedrugs-15-00005-t001). Equilibrium adsorption studies were performed to characterize the affinity value of MP and these three affinity media (Figure 3A). Desorption constant for the 4-APBA-modified medium was 14.9 µg/mL which was significantly lower than that for the 3-APBA medium (21.5 µg/mL) and aniline medium (67.2 µg/mL). Meanwhile, the theoretical maximum absorption (Q_{max}) for the 4-APBA medium (29.6 mg/g) was significantly higher than it was for the other two media (24.9 mg/g and 10.6 mg/g, respectively) (tab:marinedrugs-15-00005-t001), indicating the high affinity of 4-APBA-modified Sepharose 6B towards MP. Therefore, 4-APBA was chosen as the affinity ligand for the further design and synthesis of affinity medium.

Table 1. Ligand densities, desorption constant (K_d) and theoretical maximum absorption (Q_{max}) analysis of the affinity media.

Ligands	Spacer Arms	Ligand Density (μmol/mL)	K_d (μg/mL)	Q_{max} (mg/g)
Aniline	Cyanuric chloride	20.9	67.2	10.6
3-APBA [a]	Cyanuric chloride	20.9	21.5	24.9
4-APBA	Cyanuric chloride	20.9	14.9	29.6
4-APBA	10-atom spacer	41.8	24.4	24.6
4-APBA	5-atom spacer	27.8	46.3	22.3

[a] APBA represents aminophenylboronic acid.

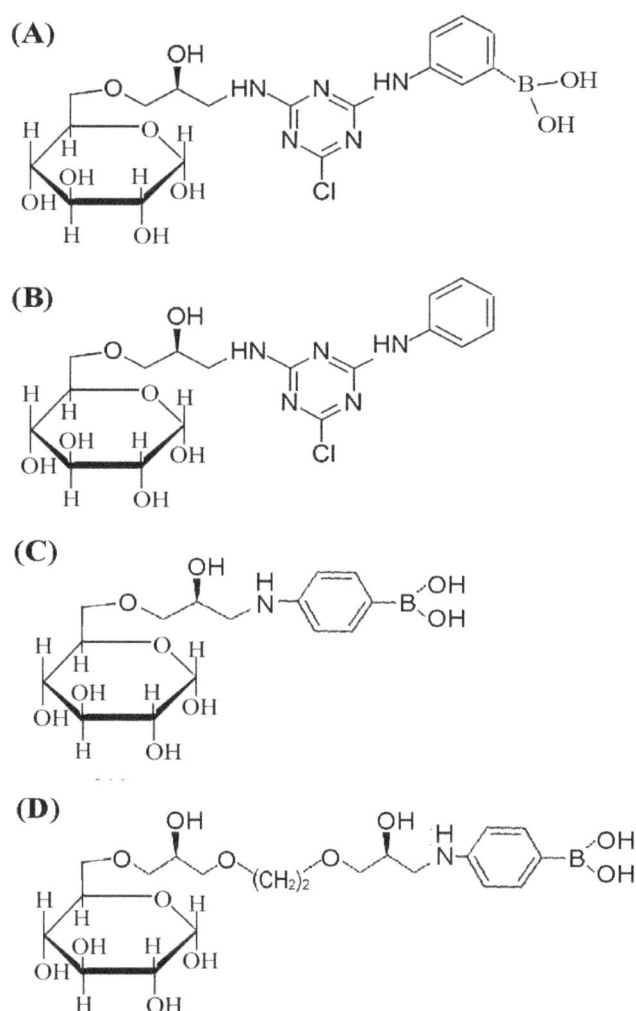

Figure 2. The scheme of four different affinity media. (**A**) 3-APBA ligand coupled with activated Sepharose 6B via cyanuric chloride spacer; (**B**) Aniline ligand coupled with activated Sepharose 6B via cyanuric chloride spacer; (**C**) 4-APBA ligand coupled with activated Sepharose 6B via 5-atom spacer arm; (**D**) 4-APBA ligand coupled with activated Sepharose 6B via 10-atom spacer arm.

To find the optimal spacer arm, two different lengths of linear arms (5-atom spacer and 10-atom spacer) and a cyclic arm (cyanuric chloride) were tested. Cyanuric chloride is a typical cyclic compound containing the s-triazine (C_3N_3) ring that could supply a higher mechanical strength for the ligand stabilization and was widely used in the affinity medium synthesis [38–41]. The scheme for the synthesis of media with the 5-atom spacer and the 10-atom spacer are shown in Figure 2C,D, correspondingly. In the adsorption analysis (Figure 3B), 4-APBA ligand with cyclic spacer arm exhibited the highest adsorption value, even though its epoxy content (20.9 μmol/mL) was lower than the content of 5-atom linear spacer (41.8 μmol/mL) and the 10-atom linear spacer (27.8 μmol/mL) (tab:marinedrugs-15-00005-t001). Thus, cyanuric chloride was chosen as the compound for generation of optimal spacer arm.

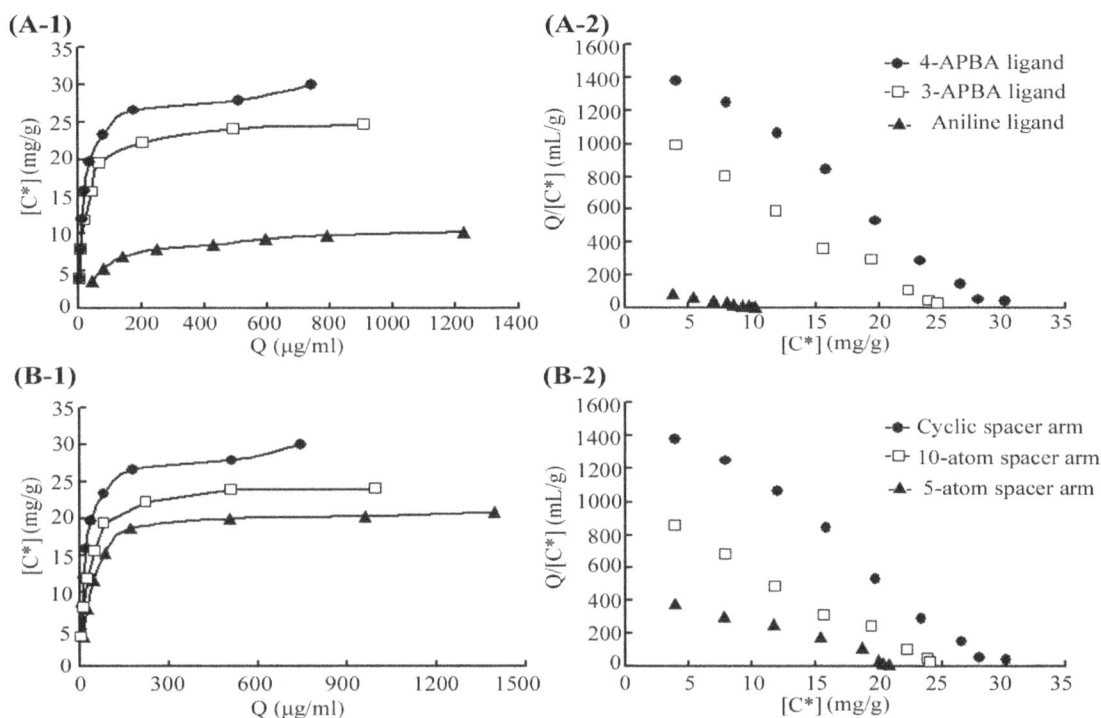

Figure 3. Adsorption analyses of different affinity media. (**A**) Adsorption analysis of affinity media with three different ligands via the same spacer arm (cyanuric chloride); (**B**) Adsorption analysis of affinity media with the same ligand (4-APBA) via three different spacer arms. (**1**) Equilibrium adsorption of metalloprotease (MP) on the affinity medium in a batch system (50 mM Gly-NaOH buffer, pH 8.6, 25 °C), (**2**) Plot describing the equilibrium of the absorption on the medium and the enzyme concentration in the liquid phase.

2.2. Binding Analysis for 4-APBA-Modified Medium and MP

Quite a few of studies show that boron-containing small molecules interacted with proteins through a covalent bond between the boron atom and the oxygen atom in the hydroxyl group of a serine [42]. In this study, the molecular docking analysis also indicated that the boron atom interacted with the hydroxyl group of Ser176 residue through covalent bonding, whereas the hydroxyl group of the boronic moiety is oriented toward Leu175 and His177 residues (Figure 4). We found that several secondary interactions could contribute to the stabilization of MP interaction with 4-APBA-modified medium. For example, the benzene ring of the 4-APBA ligand formed a π-π interaction with His171 residue of MP. In addition, the hydrogen bond between the s-triazine ring of the spacer and the molecule of water was observed, as well as the hydrogen bond between the hydroxyl group of an atom of the Ala128 residue. The aniline ligand bound with Sepharose 6B via cyanuric chloride also exhibited a low affinity (K_d, 67.2 μg/mL; Q_{max}, 10.6 mg/g) toward MP, implying that several secondary interactions can occur in addition to the interaction with the boronate ion.

Boronate affinity materials have gained increasing attention in recent years [31–33]. The mechanism involved is similar to other conventional boronate affinity chromatography. Moreover, other possible binding mechanisms were also exhibited in the molecular docking performance. One performance showed that it could be possible for Ser176 and His177 to interact with the hydroxyl groups of the boronic acid (not the boron atom) through hydrogen binding [31]. This binding mechanism relied on the hydrogen binding, which exhibited much lower affinity than the conventional binding. In the adsorption analysis, the aniline ligand exhibited a much lower affinity than APBA ligand with boronic acid, implying that the boronic acid was very important in the binding mechanism. The other possible performance is for the boron atom to coordinate with the water molecule through intermolecular B-N coordination [34]. The Ser176 residue was located in the bottom of the active-site pocket that had enough space for binding with a molecule larger than the molecule of water. Also, the 4-APBA ligand with 10-atom linear spacer showed a similar adsorption value with that for the 5-atom spacer, even though its epoxy content (27.8 μmol/mL) was smaller than the 5-atom spacer (41.8 μmol/mL) (tab:marinedrugs-15-00005-t001). This probably occurred because the longer spacer arm provided the larger spatial distance and thus provided a better accessibility of the Ser176 residue in the cavity of active site. Summarizing, the boron atom bound to MP by trapping the Ser176 hydroxyl group in the active site pocket.

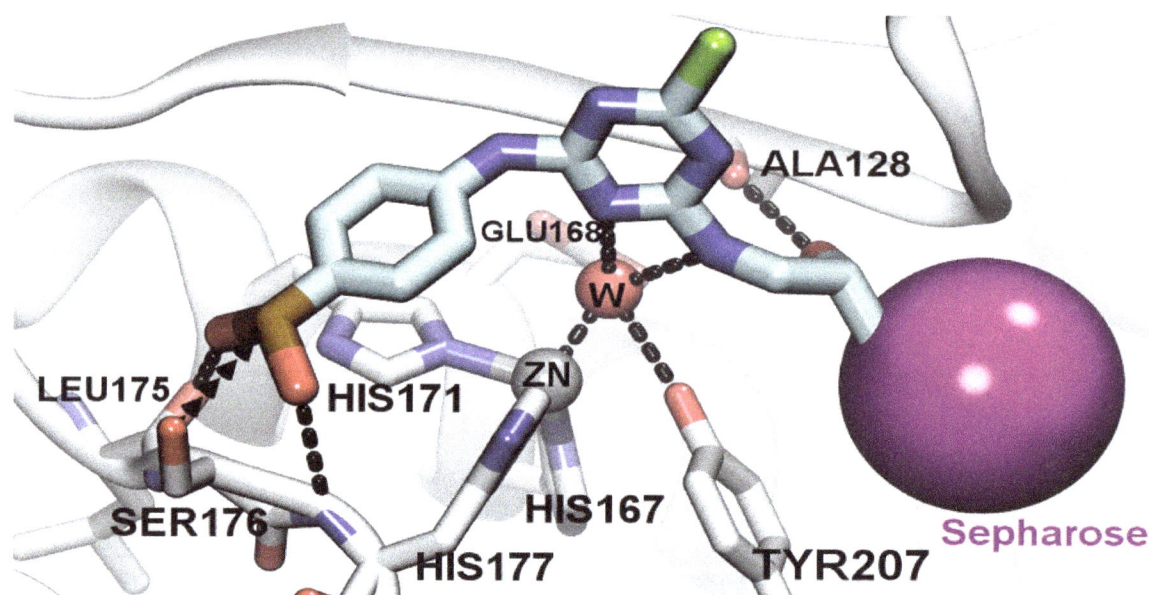

Figure 4. The binding mode of MP and the 4-APBA-modified medium. The atom force field maps were generated using Autogrid4 software for AutoDock4 (Zn); binding conformation was analyzed by Lamarckian Genetic Algorithm-Local Search combined algorithm with default searching parameter.

2.3. One-Step Affinity Purification of Commercial Metalloprotease Products

Three commercially available products (MP, DENIE-B LPS-P and ViscozymeL) containing metalloproteases were dissolved in the loading buffer to a final concentration of 10 mg/mL each. Then, the enzymes were purified by a one-step purification protocol using the 4-APBA-modified Sepharose 6B medium. We tested different loading and elution conditions to optimize the yield of metalloproteases. Almost all of the metalloproteases contained seven or eight calcium ions stabilizing their three-dimensional structure [29]. Thus, to obtain properly folded enzymes, both of the loading and elution buffers contained 1 mM $CaCl_2$. The 0.1 M Gly-NaOH buffer, pH 8.6, was chosen as the loading buffer because of the highest affinity of MP to the beads and stability of all three enzymes at this pH. Different acetic acid buffers (pH ranging from 4.0 to 6.0) were tested to select an optimal pH for MP elution, as low acidity favored the disruption of the H-bond interactions between MP and the medium. The highest protein yield was obtained at pH 5.4. Thus, 0.1 M acetic acid (pH 5.4) was chosen as the elution buffer. The SDS-PAGE analysis of the crude and purified metalloproteases is shown in Figure S2. The activity and purity of purified enzymes are shown in tab:marinedrugs-15-00005-t002.

Table 2. Comparison of affinity and traditional purification methods for three available metalloprotease products.

Enzymes	Purification Method	Activity Recovery (%)	Protein Purity (%)	Specific Activity (U/mg)	Time Requirement
MP	Affinity protocol [a]	64.1	98.8	95.6	~1 h
	Traditional protocol [b]	8.9	97.6	96.2	>48 h
DENIE-B LPS-P	Affinity protocol [a]	45.2	95.9	64.6	~1 h
	Traditional protocol [c]	10.7	91.4	62.3	>48 h
ViscozymeL	Affinity protocol [a]	37.8	97.1	51.4	~1 h
	Traditional protocol [d]	3.4	58.3	27.8	>96 h

[a] In the affinity protocol, enzymes were purified by 4-APBA-modified medium; [b] The traditional purification protocol of MP was composed of five steps, including ultrafiltration, ammonium sulfate precipitation, desalting, anion-exchange and gel-filtration chromatography; [c] The traditional purification protocol of DENIE-B LPS-P was composed of three steps, including ammonium sulfate precipitation, desalting and anion-exchange chromatography; [d] The traditional purification protocol of ViscozymeL was composed of six steps, including ammonium sulfate precipitation, hydrophobic chromatography, desalting, anion-exchange chromatography, and two steps of gel-filtration chromatography.

In our previous work, the five-step purification protocol for the purification of MP was developed. It included ammonium sulfate precipitation, desalting, anion-exchange and gel-filtration chromatography and took more than 48 h of work [3,21]. Here, we report a simple and efficient one-step MP purification procedure that takes less than one hour. Our protocol is based on the 4-APBA-modified Sepharose 6B medium that efficiently bound native MP from natural sources. Here we compared this one-step protocol with the previous reported methods, along with all the different purification steps, activity yields, specific activities and time requirement. According to the measurements of MP activity in the initial sample and purified protein, almost 64.1% of initial MP was purified, whereas only 8.9% of initial MP was recovered using the traditional protocol. The specific activity of MP purified by the APBA-modified protocol (95.6 U/mg) is similar with the value obtained through the traditional purification protocol (96.2 U/mg) and the IMAC protocol (94.8 U/mg). However, the purity of MP is different, being higher with the APBA-modified affinity purification (98.8%) with respect to the traditional and IMAC methods (92.5% and 94.7%, respectively). Even if the IMAC protocol results an activity recovery higher than the 4-APBA protocol, it is longer. Moreover, the APBA-modified affinity protocol avoids the use of toxic imidazole and the loss of metallic ions in the MP active pocket that reduce the enzyme stability. Based on all the positive features of this affinity protocol, such as one-step of chromatography, shorter times, and higher purity, it is clear there is potential in this approach for the industrial production of high-purity MP.

To determine whether our medium has an affinity value to metalloproteases from other sources, two other commercial metalloprotease products, DENIE-B LPS-P and ViscozymeL, were used for enzyme purification. DENIE-B LPS-P was an enzyme concentrate produced from *Bacillus subtilis* that was widely used in leather softening [43]. Based on the activity measurement in our study, only 10.7% of metalloprotease was purified from this commercial product using the traditional three-step purification protocol, including ammonium sulfate precipitation, desalting and anion-exchange chromatography on a Q Sepharose column. ViscozymeL was a cell wall degrading enzyme complex from *Aspergillus* sp., containing a wide range of carbohydrases and metalloprotease [44]. Traditional purification of the metalloprotease from ViscozymeL resulted in only less than 60% pure enzyme, which required six steps, including ammonium sulfate precipitation, hydrophobic chromatography, desalting, anion-exchange chromatography, and two steps of gel-filtration chromatography. Meanwhile, IMAC (Cu-IDA ligand) purification of these two protein products resulted in less than 60% purity of metalloproteases (data not shown). However, the 4-APBA-modified medium could efficiently purify metalloproteases from those two products (tab:marinedrugs-15-00005-t002). The activity recoveries of DENIE-B LPS-P and ViscozymeL were 45.2% and 37.8%, respectively. Meanwhile, when the purified enzymes were analyzed by HPLC with a TSK3000SW gel filtration column, both of them were more than 95% pure

(Figure 5). To sum, our novel methodology5 had multiple advantages in comparison with all known techniques of metalloprotease purification.

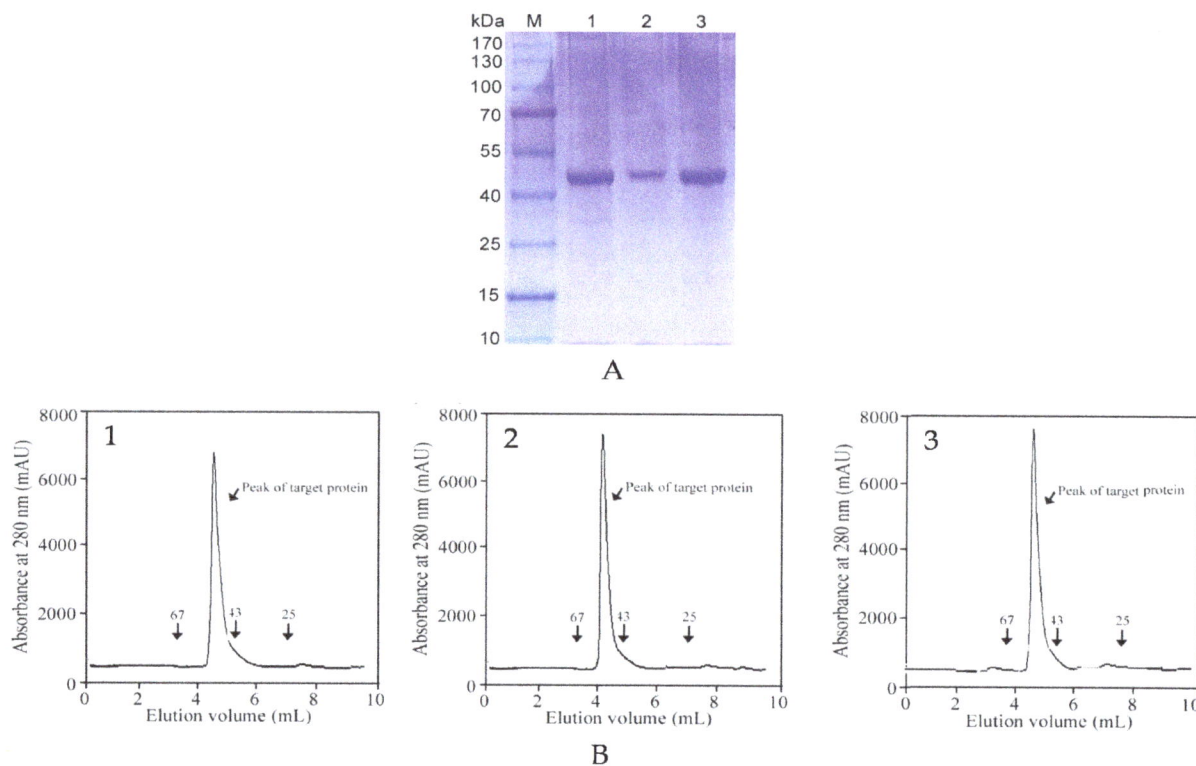

Figure 5. Purity analysis of three purified enzyme products. (**A**) SDS-PAGE (10.0%) analysis showed that the enzymes were purified to an apparent homogeneous population with a molecular mass of 48 kDa and the purity was more than 95%. *Lane M*, molecular mass standard protein marker; *Lane 1*, the purified MP; *Lane 2*, the purified DENIE-B LPS-P; *Lane 3*, the purified ViscozymeL; (**B**) HPLC analysis using the size exclusion by gel filtration of the purified MP (**1**), DENIE-B LPS-P (**2**) and ViscozymeL (**3**) on a TSK 3000SW column.

3. Materials and Methods

3.1. Materials

The dried powders of crude metalloprotease, MP, were yielded from marine bacterium *Flavobacterium* sp. YS-80-122. A commercial metalloprotease concentrate produced from *Bacillus subtilis*, DENIE-B LPS-P, was purchased from Denykem Ltd. (Shanghai, China). Cell wall degrading enzyme complex from *Aspergillus* sp., ViscozymeL, containing a wide range of carbohydrases and metalloprotease was obtained from Novozymes, Denmark. The 4-aminophenylboronic acid, 3-aminophenylboronic acid (3-APBA), aniline and cyanuric chloride (2,4,6-trichloro-1,3,5-triazine) were purchased from Sigma-Aldrich, St. Louis, MO, USA. Activated Sepharose 6B with two different spacer arm lengths (5-atoms, 10-atoms) were from Beijing Weishibohui Chromatography Technology Co., Beijing, China. All remaining reagents were of analytical grade (Sinopharm Chemical Reagent, Shanghai, China).

3.2. Synthesis of Affinity Medium

The affinity media were prepared according to the methods developed previously [45,46]. The scheme of the synthesis procedure is shown in Figure 1. Initially, Sepharose 6B was modified by epichlorohydrin to form activated amino-sepharose. Briefly, Sepharose 6B (100 g) was thoroughly washed with deionized water at a 1:10 ratio until the pH value of the eluate reached 7.0 and the beads were dried. To activate Sepharose 6B, the beads were resuspended in 50 mL of activating solution (1 M NaOH, 2.5 g DMSO, and 10 mL epichlorohydrin) followed by incubation at 40 °C for 2.5 h with shaking (Figure 1a). Then, 35% saturated ammonia (150 mL) was added to the activated Sepharose 6B resuspended in 350 mL distilled water. The beads were incubated overnight at 30 °C on a rotary 39 shaker to form aminated Sepharose 6B (Figure 1b). To attach cyanuric chloride to the amino groups of aminated Sepharose 6B, the beads were resuspended in 350 mL 50% (v/v) acetone in an ice-salt bath, and then 8 g of cyanuric chloride dissolved in 70 mL acetone was added with a flow rate of 0.5 mL/min in the shaking station. The neutral pH was maintained by simultaneous addition of 1 M NaOH. The beads were washed with 50% (v/v) acetone to remove the free cyanuric chloride (Figure 1c). The density of the free amino group was determined by the ninhydrin test in the following procedure: a small aliquot of beads was smeared on filter paper, sprayed with ninhydrin solution (0.2% (w/v) in acetone), and heated briefly with a hair dryer. The appearance of purple color indicated the presence of free amino groups, whereas the color disappearance indicated that cyanuric chloride had been linked to the amino groups [27]. Then, a twofold excess of 4-APBA dissolved in 2 M sodium carbonate was added to the dichlorotriazinylated Sepharose 6B beads. After 24 h of stirring at room temperature, the beads were filtered, washed well with water and stored in 0.02% (w/v) sodium azide (Figure 1d). To confirm the conformation of the 4-APBA ligand on the medium, 100 mg dried medium was incubated with 6 M HCl at boiling condition for 24 h, and HCl was removed by vacuum evaporation. The hydrolyzed chemical was purified and analyzed with ESI-MS (HP1100LC MSD, Agilent, San Francisco, CA, USA) according to the methods reported [37].

To generate control beads with 3-APBA and/or aniline, the affinity medium with 3-APBA or aniline instead of 4-APBA was synthesized according to the described method above (Figure 2A,B). To generate control beads with two different spacer arms, the 4-APBA-modified Sepharose 6B beads with 5-atom or 10-atom spacer arms were synthesized according to the published method [21,28,36]. The schemes for the generation of these beads are shown in Figure 2C,D. Briefly, 5 g of 4-APBA dissolved in 80 mL of 2 M sodium carbonate was added to the previously activated Sepharose 6B. After 24 h of incubation at room temperature with stirring, the beads were filtered, washed well with water and stored in 0.02% (w/v) sodium azide [13,14].

3.3. Adsorption Value Analysis

To characterize the interaction of MP with five different types of affinity media, an equilibrium adsorption study was performed. The constant of desorption (K_d) and the theoretical maximum adsorption capacity (Q_{max}) of these affinity media were analyzed according to the Scatchard analysis model [24,47]. Briefly, one milliliter of increasing concentrations of purified metalloprotease (0.1–0.9 mg/mL in 20 mM Gly-NaOH buffer, pH 8.6) was mixed with 0.5 g of each affinity medium and shaken for 2 h at 4 °C until the solution reached adsorption equilibrium. Then, the mixtures were centrifuged at 1500 g for 5 min. The protease activity and protein concentration were measured in the supernatants.

The analysis of equilibrium adsorption provided a relationship between the concentration of metalloprotease in the solution and the amount of enzyme absorbed on the affinity medium. The data obtained were analyzed using the Scatchard plot according to the following equation:

$$Q = \frac{Q_{max}[C^*]}{K_d + [C^*]}$$

Therein, Q is the adsorption amount of enzyme to the medium (mg/g), Q_{max} is the theoretical maximum of metalloprotease absorption to the affinity medium (mg/g), $[C^*]$ is the concentration of metalloprotease in solution (mg/mL), and K_d is the desorption constant.

3.4. Molecular Docking Analysis

The MP protein structure (PDB ID 3U1R) [26] was prepared by AutoDockTools (The Scripps Research Institute, San Diego, CA, USA). Briefly, hydrogens and gasteiger charge were added and waters were removed, except the water molecules bound to the zinc ion, which was treated as hydrogen acceptor. Ca^{2+} and Zn^{2+} were kept in the structure. The protein structure was then prepared according to the reference [48], using an improved zinc force field for AutoDock4 (Zn) (The Scripps Research Institute, San Diego, CA, USA). For the ligand, the sepharose part was not considered in molecular docking, as it was an inert polymeric support, and frequently used for coupling the "active" affinity ligands to the matrix. The ligand structure was built and minimized with Maestro (Schrödinger LLC., Cambridge, MA, USA). The type of boron atom was set to be sp^3 hybridization to mimic its binding with the hydroxyl group in Ser/Thr amino acids. Finally, the ligand was converted to the pdbqt format by AutoDockTools. The atom force field maps were generated using Autogrid4 software for AutoDock4 (Zn); binding conformation was searched by Lamarckian Genetic Algorithm-Local Search combined algorithm with default searching parameter. Fifty conformations were generated for further analysis. The representation was visualized with VMD 1.9.2 software (The Scripps Research Institute, San Diego, CA, USA) [49].

3.5. Traditional and Affinity Purification of Three Commercial Metalloproteases

Two grams of dried powder of three commercially available products containing three different metalloproteases, MP, DENIE-B LPS-P, and ViscozymeL, were dissolved in 50 mL sample loading buffer (0.1 M Gly-NaOH buffer, pH 8.6) each. The traditional purification protocol of MP was composed of five steps, including ultrafiltration, ammonium sulfate precipitation on 60% saturation, desalting, anion-exchange on a Q-sepharose column and gel-filtration chromatography on Sephacryl S-200 HR [3]. Meanwhile, the other two commercial metalloproteases were purified used traditional column purification protocol in this study. The traditional purification protocol of DENIE-B LPS-P was composed of three steps, including ammonium sulfate precipitation on 60% saturation, desalting and anion-exchange chromatography on a Q-sepharose column. The traditional purification protocol of ViscozymeL was composed of six steps, including ammonium sulfate precipitation on 40% saturation, hydrophobic chromatography on a phenyl column, desalting, anion-exchange chromatography on a diethylaminoethanol(DEAE)-sepharose column, and two step of gel-filtration chromatography on Sephacryl S-200 HR (GE Healthcare, Madison, WI, USA).

In the affinity purification protocol, the supernatant was loaded onto 10 mL pre-equilibrated column and washed with washing buffer (0.1 M Gly-NaOH buffer, pH 8.6) until the eluate exhibited no detectable absorbance at 280 nm. The target protein was eluted with elution buffer (0.1 M acetic acid buffer, pH 5.4). The flow rate of the mobile phase was 3.0 mL/min. The concentrations of each elution peak were measured by the Bradford method, using bovine serum albumin (BSA) as a standard. The purified enzyme was further characterized by 10% SDS-PAGE and high performance liquid chromatography (HPLC) analysis. The purification process was repeated more than five times.

3.6. Enzymatic Activity Assay

One hundred microliters of enzyme solution were mixed with 4.9 mL of casein solution (0.6% (w/v) in 25 mM borate buffer, pH 10.0) and incubated at 25 °C for 10 min. The relative enzyme activity was measured using Folin-Ciocalteu's method [3,4]. One unit was defined as the amount of enzyme causing the release of 1 μg tyrosine per minute under the above conditions.

3.7. Protein Purity Analysis

SDS-PAGE analysis was carried out on a Mini-protean II system from Bio-Rad (Hercules, CA, USA). The purity of the purified proteases was calculated by a gel imaging analysis system (Gelpro Analyzer 3.2 (Thermo Fisher Scientific, Waltham, MA, USA) according to the integration of the lane darkness. HPLC (Agilent 1260, San Francisco, CA, USA) analysis was performed with a TSK3000SW gel filtration column (Tosoh Co., Tokyo, Japan) monitored at 280 nm [27]. The solvent phase was 0.1 M PBS, 0.1 M Na_2SO_4, 0.05% NaN_3, pH 6.7. The flow rate was 0.6 mL/min.

4. Conclusions

In this study, an affinity medium more efficient for metalloprotease purification than other currently available techniques was designed, synthesized and experimentally characterized. Testing the adsorption properties of five designed resins, the Sepharose 6B media coupled with 4-APBA via cyanuric chloride spacer was selected for the purification of native metalloproteases from three commercial products. Metalloproteases from these sources were purified in one step with high efficiency and purity (more than 95%). Compared with the previously reported methods, this protocol resulted in several positive features, such as fewer steps, better activity recoveries, and higher purity. Coupled with efficacy, time-saving procedure and accessible reagents, this novel affinity purification protocol represents a potential important tool for industrial application.

Supplementary Materials: Figure S1: The ESI-MS analysis of the affinity ligand. The possible structures of the chemicals in principal peaks are shown. The ESI-MS cone voltage (170 V) was selected. Scanning was performed from m/z 100 to 1000 in 10 s, and several scans were summed to obtain the final spectrum, Figure S2: SDS-PAGE analysis of three purified and crude commercial metalloproteases. (A) Analysis of crude (Line 1) and purified (Line 2) MP;(B) Analysis of crude (Line 2) and purified (Line 1) DENIE-B LPS-P; (C) Analysis of crude (Line 2) and purified (Line 1) ViscozymeL.

Acknowledgments: This project was funded by National Science Foundation-Shandong province Joint Fund (U1406402-5); International Science and Technology Cooperation and Exchanges (2014DFG30890); Postdoctoral Science Foundation of China (2016M590673 and 2015M582170); Postdoctoral Science Foundation of Shandong; Postdoctoral Researcher Applied Research Project of Qingdao (Q51201601 and Q51201613); National Natural Science Foundation of China (41376175); The Scientific and Technological Innovation Project Financially Supported by Qingdao National Laboratory for Marine Science and Technology (2015ASKJ02); National Hi-tech R&D Program (2014AA093516); Science and Technique Plan of Qingdao (14-2-4-11-jch). International collaboration (PSR-SIIRI) supported by the Ministry of Science and Technology of China and the Ministry of Economy, Science and Innovation of Quebec, Canada (MESI).

Author Contributions: S.L., S.L. and M.S. conceived and designed the experiments. S.L., L.W., J.L., Y.W. and J.H. performed the experiments. S.L., L.W., Y.W. and X.X. analyzed the data. S.L., X.X. and M.S. wrote the main manuscript text. All authors reviewed the manuscript.

References

1. Li, Q.; Yi, L.; Marek, P.; Iverson, B.L. Commercial proteases: Present and future. *FEBS Lett.* **2014**, *587*, 1155–1163. [CrossRef] [PubMed]
2. Gupta, R.; Beg, Q.; Khan, S.; Chauhan, B. An overview on fermentation, downstream processing and properties of microbial alkaline proteases. *Appl. Microbiol. Biotechnol.* **2002**, *60*, 381–395. [PubMed]
3. Wang, F.; Hao, J.H.; Yang, C.Y.; Sun, M. Cloning, expression, and identification of a novel extracellular cold-adapted alkaline protease gene of the marine bacterium strain YS-80-122. *Appl. Biochem. Biotechnol.* **2010**, *162*, 1497–1505. [CrossRef] [PubMed]
4. Hao, J.H.; Sun, M. Purification and characterization of a cold alkaline protease from a psychrophilic *Pseudomonas aeruginosa* HY1215. *Appl. Biochem. Biotechnol.* **2015**, *175*, 715–722. [CrossRef] [PubMed]
5. Kunamneni, A.; Durvasula, R. Streptokinase—A drug for thrombolytic therapy: A patent review. *Recent Adv. Cardiovasc. Drug Discov.* **2014**, *9*, 106–121. [CrossRef] [PubMed]
6. Craik, C.S.; Page, M.J.; Madison, E.L. Proteases as therapeutics. *Biochem. J.* **2011**, *435*, 1–16. [CrossRef] [PubMed]
7. Wang, W.J.; Yu, X.H.; Wang, C.; Yang, W.; He, W.S.; Zhang, S.J.; Yan, Y.G.; Zhang, J. MMPs and ADAMTSs in intervertebral disc degeneration. *Clin. Chim. Acta* **2015**, *448*, 238–246. [CrossRef] [PubMed]
8. Erban, T. Purification of tropomyosin, paramyosin, actin, tubulin, troponin and kinases for chemiproteomics and its application to different scientific fields. *PLoS ONE* **2011**, *6*, e22860. [CrossRef] [PubMed]
9. Nakamura, K.; Suzuki, T.; Hasegawa, M.; Kato, Y.; Sasaki, H.; Inouye, K. Characterization of *p*-aminobenzamidine-based sorbent and its use for high-performance affinity chromatography of trypsin-like proteases. *J. Chromatogr. A* **2003**, *1009*, 133–139. [CrossRef]

10. Braganza, V.J.; Simmons, W.H. Tryptase from rat skin: Purification and properties. *Biochemistry* **1991**, *30*, 4997–5007. [CrossRef] [PubMed]

11. Burchacka, E.; Witkowska, D. The role of serine proteases in the pathogenesis of bacterial infections. *Postep. Hig. Med. Doświadczalnej* **2016**, *70*, 678–694. [CrossRef] [PubMed]

12. Jean, M.; Raghavan, A.; Charles, M.L.; Robbins, M.S.; Wagner, E.; Rivard, G.É.; Charest-Morin, X.; Marceau, F. The isolated human umbilical vein as a bioassay for kinin-generating proteases: An in vitro model for therapeutic angioedema agents. *Life Sci.* **2016**, *155*, 180–188. [CrossRef] [PubMed]

13. McMurray, J.J.; Dickstein, K.; Kober, L.V. Aliskiren, enalapril, or aliskiren and enalapril in heart failure. *N. Engl. J. Med.* **2016**, *374*, 1521–1532. [CrossRef] [PubMed]

14. Häse, C.C.; Finkelstein, R.A. Bacterial extracellular zinc-containing metalloproteases. *Microbiol. Rev.* **1993**, *12*, 823–837.

15. Adekoya, O.A.; Sylte, I. The thermolysin family (M4) of enzymes: Therapeutic and biotechnological potential. *Chem. Biol. Drug Des.* **2009**, *73*, 7–16. [CrossRef] [PubMed]

16. Gowda, C.D.; Shivaprasad, H.V.; Kumar, R.V.; Rajesh, R.; Saikumari, Y.K.; Frey, B.M.; Frey, F.J.; Sharath, B.K.; Vishwanath, B.S. Characterization of major zinc containing myonecrotic and procoagulant metalloprotease 'malabarin' from non lethal trimeresurus malabaricus snake venom with thrombin like activity: Its neutralization by chelating agents. *Curr. Top. Med. Chem.* **2011**, *11*, 2578–2588. [CrossRef] [PubMed]

17. Lei, F.F.; Cui, C.; Zhao, H.F.; Tang, X.L.; Zhao, M.M. Purification and characterization of a new neutral metalloprotease from marine *Exiguobacterium* sp. SWJS2. *Biotechnol. Appl. Biochem.* **2016**, *63*, 238–248. [CrossRef] [PubMed]

18. Majumder, R.; Banik, S.P.; Khowala, S. Purification and characterisation of κ-casein specific milk-clotting metalloprotease from *Termitomyces clypeatus* MTCC 5091. *Food Chem.* **2015**, *173*, 441–448. [CrossRef] [PubMed]

19. Zhao, H.L.; Yang, J.; Chen, X.L.; Su, H.N.; Zhang, X.Y.; Huang, F.; Zhou, B.C.; Xie, B.B. Optimization of fermentation conditions for the production of the M23 protease Pseudoalterin by deep-sea *Pseudoalteromonas* sp. CF6-2 with artery powder as an inducer. *Molecules* **2014**, *19*, 4779–4790. [CrossRef] [PubMed]

20. Shao, X.; Ran, L.Y.; Liu, C.; Chen, X.L.; Zhang, X.Y.; Qin, Q.L.; Zhou, B.C.; Zhang, Y.Z. Culture condition optimization and pilot scale production of the M12 metalloprotease myroilysin produced by the deep-sea bacterium *Myroides profundi* D25. *Molecules* **2015**, *20*, 11891–11901. [CrossRef] [PubMed]

21. Li, S.Y.; Wang, L.N.; Yang, J.; Liu, J.Z.; Lin, S.X.; Hao, J.H.; Sun, M. Affinity purification of metalloprotease from marine bacterium using immobilized metal affinity chromatography. *J. Sep. Sci.* **2016**, *39*, 2050–2056. [CrossRef] [PubMed]

22. Chessa, J.P.; Petrescu, I.; Bentahir, M.; Van Beeumen, J.; Gerday, C. Purification, physico-chemical characterization and sequence of a heat labile alkaline metalloprotease isolated from a psychrophilic *Pseudomonas* species. *Biochim. Biophys. Acta* **2000**, *1479*, 265–274. [CrossRef]

23. Martínez, C.A.; Seidel-Morgenstern, A. Purification of single-chain antibody fragments exploiting pH-gradients in simulated moving bed chromatography. *J. Chromatogr. A* **2016**, *1434*, 29–38. [CrossRef] [PubMed]

24. Chen, B.; Li, R.; Li, S.Y.; Chen, X.L.; Yang, K.D.; Chen, G.L.; Ma, X.X. Evaluation and optimization of the metal-binding properties of a complex ligand for immobilized metal affinity chromatography. *J. Sep. Sci.* **2016**, *39*, 518–524. [CrossRef] [PubMed]

25. Cheung, R.C.; Wong, J.H.; Ng, T.B. Immobilized metal ion affinity chromatography: A review on its applications. *Appl. Microbiol. Biotechnol.* **2012**, *96*, 1411–1420. [CrossRef] [PubMed]

26. Labrou, N.E. Design and selection of ligands for affinity chromatography. *J. Chromatogr. B* **2003**, *790*, 67–78. [CrossRef]

27. Ye, L.; Xu, A.Z.; Cheng, C.; Zhang, L.; Huo, C.X.; Huang, F.Y.; Xu, H.; Li, R.Y. Design and synthesis of affinity ligands and relation of their structure with adsorption of proteins. *J. Sep. Sci.* **2011**, *34*, 3145–3150. [CrossRef] [PubMed]

28. Havlicek, V.; Lemr, K.; Schug, K.A. Current trends in microbial diagnostics based on mass spectrometry. *Anal. Chem.* **2013**, *85*, 790–797. [CrossRef] [PubMed]

29. Zhang, S.C.; Sun, M.; Li, T.; Wang, Q.H.; Hao, J.H.; Han, Y.; Hu, X.J.; Zhou, M.; Lin, S.X. Structure analysis of a new psychrophilic marine protease. *PLoS ONE* **2011**, *6*, e26939. [CrossRef] [PubMed]

30. Ji, X.F.; Zheng, Y.; Wang, W.; Sheng, J.; Hao, J.H.; Sun, M. Virtual screening of novel reversible inhibitors for marine alkaline protease MP. *J. Mol. Graph. Model.* **2013**, *46*, 125–131. [CrossRef] [PubMed]

31. Li, D.J.; Chen, Y.; Liu, Z. Boronate affinity materials for separation and molecular recognition: Structure, properties and applications. *Chem. Soc. Rev.* **2015**, *44*, 8097–8123. [CrossRef] [PubMed]

32. Wang, S.; Ye, J.; Li, X.; Liu, Z. Boronate affinity fluorescent nanoparticles for förster resonance energy transfer inhibition assay of *cis-diol* biomolecules. *Anal. Chem.* **2016**, *88*, 5088–5096. [CrossRef] [PubMed]

33. Xue, Y.; Shi, W.; Zhu, B.; Gu, X.; Wang, Y.; Yan, C. Polyethyleneimine-grafted boronate affinity materials for selective enrichment of *cis-diol*-containing compounds. *Talanta* **2015**, *140*, 1–9. [CrossRef] [PubMed]

34. Toprak, A.; Görgün, C.; Kuru, C.İ.; Türkcan, C.; Uygun, M.; Akgöl, S. Boronate affinity nanoparticles for RNA isolation. *Mater. Sci. Eng. C Mater. Biol. Appl.* **2015**, *50*, 251–256. [CrossRef] [PubMed]

35. Jiang, H.P.; Qi, C.B.; Chu, J.M.; Yuan, B.F.; Feng, Y.Q. Profiling of cis-diol-containing nucleosides and ribosylated metabolites by boronate-affinity organic-silica hybrid monolithic capillary liquid chromatography/mass spectrometry. *Sci. Rep.* **2015**, *5*, 7785. [CrossRef] [PubMed]

36. Fasoli, E.; Reyes, Y.R.; Guzman, O.M.; Rosado, A.; Cruz, V.R.; Borges, A.; Martinez, E.; Vibha Bansal, V. Para-aminobenzamidine linked regenerated cellulose membranes for plasminogen activator purification: Effect of spacer arm length and ligand density. *J. Chromatogr. B* **2013**, *930*, 13–21. [CrossRef] [PubMed]

37. Xin, X.; Dong, D.X.; Wang, T.; Li, R.X. Affinity purification of serine proteinase from *Deinagkistrodon acutus* venom. *J. Chromatogr. B* **2007**, *859*, 111–118. [CrossRef] [PubMed]

38. Van Ness, J.; Kalbfleisch, S.; Petrie, C.R.; Reed, M.W.; Tabone, J.C.; Vermeulen, N.M. A versatile solid support system for oligodeoxynucleotide probe-based hybridization assays. *Nucleic Acids. Res.* **1991**, *19*, 3345–3350. [CrossRef] [PubMed]

39. Batra, S.; Bhushan, R. Amino acids as chiral auxiliaries in cyanuric chloride-based chiral derivatizing agents for enantioseparation by liquid chromatography. *Biomed. Chromatogr.* **2014**, *28*, 1532–1546. [CrossRef] [PubMed]

40. Chien, T.E.; Li, K.L.; Lin, P.Y.; Lin, J.L. Infrared spectroscopic study of the adsorption forms of cyanuric acid and cyanuric chloride on TiO_2. *Langmuir* **2016**, *32*, 5306–5313. [CrossRef] [PubMed]

41. Mountford, S.J.; Daly, R.; Robinson, A.J.; Hearn, M.T. Design, synthesis and evaluation of pyridine-based chromatographic adsorbents for antibody purification. *J. Chromatogr. A* **2014**, *1355*, 15–25. [CrossRef] [PubMed]

42. Smoum, R.; Rubinstein, A.; Dembitsky, V.M.; Srebnik, M. Boron containing compounds as protease inhibitors. *Chem. Rev.* **2012**, *112*, 4156–4220. [CrossRef] [PubMed]

43. Huang, S.H.; Pan, S.H.; Chen, G.G.; Huang, S.; Zhang, Z.F.; Li, Y.; Liang, Z.Q. Biochemical characteristics of a fibrinolytic enzyme purified from a marine bacterium, *Bacillus subtilis* HQS-3. *Int. J. Biol. Macromol.* **2013**, *62*, 124–130. [CrossRef] [PubMed]

44. Huang, W.Q.; Zhong, L.F.; Meng, Z.Z.; You, Z.J.; Li, J.Z.; Luo, X.C. The structure and enzyme characteristics of a recombinant leucine aminopeptidase rlap1 from *Aspergillus sojae* and its application in debittering. *Appl. Biochem. Biotechnol.* **2015**, *177*, 190–206. [CrossRef] [PubMed]

45. Guo, W.; Ruckenstein, E. A new matrix for membrane affinity chromatography and its application to the purification of concanavalin A. *J. Membr. Sci.* **2001**, *182*, 227–234. [CrossRef]

46. Dong, D.X.; Liu, H.R.; Xiao, Q.S.; Li, R.X. Affinity purification of egg yolk immunoglobulins (IgY) with a stable synthetic ligand. *J. Chromatogr. B* **2008**, *870*, 51–54. [CrossRef] [PubMed]

47. Xin, Y.; Yang, H.L.; Xiao, X.L.; Zhang, L.; Zhang, Y.R.; Tong, Y.J.; Chen, Y.; Wang, W. Affinity purification of urinary trypsin inhibitor from human urine. *J. Sep. Sci.* **2012**, *35*, 1–6. [CrossRef] [PubMed]

48. Santos-Martins, D.; Forli, S.; Ramos, M.J.; Olson, A.J. AutoDock4(Zn): An improved AutoDock force field for small-molecule docking to zinc metalloproteins. *J. Chem. Inf. Model.* **2014**, *54*, 2371–2379. [CrossRef] [PubMed]

49. Humphrey, W.; Dalke, A.; Schulten, K. VMD: Visual molecular dynamics. *J. Mol. Graph. Model.* **1996**, *14*, 33–38. [CrossRef]

7

A Novel Cold-Adapted Leucine Dehydrogenase from Antarctic Sea-Ice Bacterium *Pseudoalteromonas* sp. ANT178

Yatong Wang, Yanhua Hou, Yifan Wang, Lu Zheng, Xianlei Xu, Kang Pan, Rongqi Li and Quanfu Wang *

School of Marine Science and Technology, Harbin Institute of Technology, Weihai 264209, China; wangyatong199311@163.com (Y.W.); marry7718@163.com (Y.H.); daid01@126.com (Y.W.); zhenglu0206@126.com (L.Z.); 17863108956@163.com (X.X.); m15662319228@163.com (K.P.); l1263769417@163.com (R.L.)

* Correspondence: wangquanfuhit@hit.edu.cn.

Abstract: L-*tert*-leucine and its derivatives are useful as pharmaceutical active ingredients, in which leucine dehydrogenase (LeuDH) is the key enzyme in their enzymatic conversions. In the present study, a novel cold-adapted LeuDH, *psleudh*, was cloned from psychrotrophic bacteria *Pseudoalteromonas* sp. ANT178, which was isolated from Antarctic sea-ice. Bioinformatics analysis of the gene *psleudh* showed that the gene was 1209 bp in length and coded for a 42.6 kDa protein containing 402 amino acids. PsLeuDH had conserved Phe binding site and NAD⁺ binding site, and belonged to a member of the Glu/Leu/Phe/Val dehydrogenase family. Homology modeling analysis results suggested that PsLeuDH exhibited more glycine residues, reduced proline residues, and arginine residues, which might be responsible for its catalytic efficiency at low temperature. The recombinant PsLeuDH (rPsLeuDH) was purified a major band with the high specific activity of 275.13 U/mg using a Ni-NTA affinity chromatography. The optimum temperature and pH for rPsLeuDH activity were 30 °C and pH 9.0, respectively. Importantly, rPsLeuDH retained at least 40% of its maximum activity even at 0 °C. Moreover, the activity of rPsLeuDH was the highest in the presence of 2.0 M NaCl. Substrate specificity and kinetic studies of rPsLeuDH demonstrated that L-leucine was the most suitable substrate, and the catalytic activity at low temperatures was ensured by maintaining a high k_{cat} value. The results of the current study would provide insight into Antarctic sea-ice bacterium LeuDH, and the unique properties of rPsLeuDH make it a promising candidate as a biocatalyst in medical and pharmaceutical industries.

Keywords: leucine dehydrogenase; cold-adapted; Antarctic bacterium; sea-ice; homology modeling

1. Introduction

Leucine dehydrogenase (LeuDH; EC 1.4.1.9), a NAD⁺ dependent oxidoreductase, which catalyzes reversible L-leucine and other branched chain L-amino acids deamination reaction to the formation of the corresponding α-keto acid [1].

The enzyme was first identified in *Bacillus cereus* [2], and then was found in some microorganisms *Bacillus licheniformis* [3], *Bacillus sphaericus* [4], *Citrobacter freundii* [5], and *Laceyella sacchari* [6]. Moreover, crystal structures of the LeuDH from *Sporosarcina psychrophila* [7] and *Bacillus sphaericus* have been described [8].

LeuDH is used as a biocatalyst to format amino acids for using in the pharmaceutical industry by catalyzing the corresponding α-keto acids [9]. However, some of α-keto acids are unstable and degraded during prolonged incubation at moderate temperatures, such as 37 °C [10].

Importantly, cold-adapted enzymes that exhibit high levels of activity at room temperature (20–25 °C) should be useful for converting such unstable α-keto acids. What is more, cold-adapted enzymes have better conversion rates, the specificity of substrate and product, fewer by-products, which are required in the modern industry [11].

Although many LeuDHs have already been characterized, only a few cold-adapted LeuDH have been reported, such as LeuDH from *Alcanivorax dieselolei* [12] and *Sporosarcina psychrophila* [7].

Antarctic sea-ice, due to its specific geographical location and climate, is considered as an extreme environment on the earth. To develop the ability to withstand the extreme environment, sea-ice microorganisms have evolved several adaptive strategies and would be the new and promising microbial sources of cold-adapted enzymes.

In our previous studies, some cold-adapted enzymes were isolated from Antarctic sea-ice bacteria and had become interesting for industrial applications [13,14]. It is well-known that L-*tert*-leucine and its derivatives are useful as pharmaceutical active ingredients and chiral auxiliaries, while LeuDH is a key enzyme for the enzymatic production of L-*tert*-leucine.

Here, we briefly describe the homology modeling, expression, and characterization of cold-adapted LeuDH from Antarctic sea-ice bacterium. This LeuDH had unique properties make it good candidate for future medical and pharmaceutical industry applications.

2. Results and Discussion

2.1. Gene Cloning and Sequence Analysis

The *psleudh* gene was amplified from genomic DNA of the strain ANT178. It consisted of an ORF of 1209 bp, encoded a protein of 402 amino acid resides with a theoretical p*I* of 5.08. Furthermore, the DNA sequence of *psleudh* was submitted to the GenBank database with the accession number of MH322031.

Based on sequences alignment, PsLeuDH showed the highest sequence similarity (88.0%) with LeuDH from *Pseudoalteromonas nigrifaciens* (ASM53600), followed by a sequence similarity of 65.0% with LeuDH from *Colwellia piezophila* (WP_019029130).

More importantly, PsLeuDH had a conserved Phe binding site (I344) and NAD$^+$ binding sites (G233, G235, T236, V237, D256, I257, A261, C290, A291, C312, and N314). The coenzyme binding domain of NAD$^+$ in LeuDH was capable of catalyzing the reversible oxidative deamination of L-leucine and several other branched chain amino acids to form the corresponding 2-oxo acid derivatives.

This domain could be classified as a member of the Rossmann fold superfamily, comprising a plurality of different dehydrogenases, wherein the amino acid dehydrogenase family comprises a common feature: a beta-sheet-alpha helix-beta sheet conformation [15]. PsLeuDH had this structural feature from Figure 1, further demonstrating that PsLeuDH was a member of the Glu/Leu/Phe/Val dehydrogenase family.

```
PsLeuDH                                                            α1          β1              β2
                                                               QQQQQQ     ————————>      ————————>
            1        10        20        30        40        50        60        70        80        90
PsLeuDH   MEFLCALTIKAGYLIQCCFALGFILSWYSEALSEVIKSRQFHACIPFRVNYFGVFKVAVFNQPEFDDHEQVVFCSDKASGLKAIAVHST
PnLeuDH   MEFLCALTIKAGYLIQCCFALGFILSWYSEALSEVIKSRQFHACIPFRVNYFGVFKVVVFNQVEFDNHEQVVFCTDKESGLKAIIAVHNT
CpLeuDH   .........................................................MALFDLPDFDDHEQVVYCSDDASGLKAIIAVHST
RsLeuDH   .........................................................MAVFNHSEFDNHEQVVYCSDAETGLKAIIAVHST
SbLeuDH   ...................MLWFCNKYNKQSMG...THLVGFFIVYSGEIHVAVFNHVSFDEHEQVVFCHDKESGLKAIVAIHNT
BsLeuDH   ................................................MEIFKYMEKYDYEQLVFCQDEASGLKAVIAIHDT
BtLeuDH   ................................................MALFEYLEKYDYEQVVFCQDKESGLKAIIAIHDT
LsLeuDH   ................................................MKIFEYMGKYDYEQIVLCHDEQSGLKAIICIHDT


PsLeuDH   TT      β3            α2              β4            α3          β5      TT    α4
                 ———————>   QQQQQQQQQQQQQQ   ———————>     QQQQQQQQ    ———>        QQQQQQ
                   100        110        120        130        140        150        160        170        180
PsLeuDH   NLGAAVGGTRLWDYASDEDAVNDALRLSKGMTYKNAVAGLPLGGGKAVIIGDAKELDSEQLFRAFGRQLNGLGGSYFSAEDVGINCGDVA
PnLeuDH   NLGPAVGGCRMWDYANDEDAVYDVLRLSKGMTYKNAVARLPFGGGKSVIIGNAKEIKSEQLFRAFGRQLERLNGSYYSAEDVNINCDDVA
CpLeuDH   KLGAAVGGCRMWDYASDEEALIDVLRLSKGMTYKNAMAGLKMGGGKSVIIGDAKTLKSEALFKAFGEALNGLNGRYFSAEDVNITTSDIA
RsLeuDH   ALGPAVGGCRLWDYASDEAALNDVLR.SRGMTYKNAMAGLPLGGGKAVILGDAKKIKSEQLFRAFGRMVHRLSGTYYSAEDVNITTSDIM
SbLeuDH   NLGPAVGGCRMWNYQSDDEALTDVLRLSRGMTYKNAMAGLTMGGGKSVIIADPKRQDREALFRAFGCRFINSLGGRYYSAEDVGTTTADIM
BsLeuDH   TLGPALGGARMWTYNAEEEAIEDALRLARGMTYKNAAAGLNGGGKTVIIGDPFADKNEDMFRALGRFIQGLNGRYITAEDVGTTVDDMD
BtLeuDH   TLGPALGGTRMWTYDSEEAAIEDALRLAKGMTYKNAAAGLNLGGAKTVIIGNPRKDKSEAMFRALGRYIQGLNGRYITAEDVGTTVDDMD
LsLeuDH   TLGPALGGTRMWTYDSEDAAIEDALRLARGMTYKNAAAGLNLGGGKTVVIGDPKKDKSEALFRALGRYIQGLNGRYITAEDVGTTVEDMD


PsLeuDH   QQQ         TTT  QQQQQQQQQQQQQQQ       α5      TT    β6       α6          β7         α7
                   190          200        210        220          230        240        250        260
PsLeuDH   MMNKVTSFVLGLS...GKSGDPAPFTALGTFLCIKAALNHQRGHDNFAGIKVAVQGLGTVGFGLCKHLSEAGAELFVTDISQAAIDRAVN
PnLeuDH   MMNKETSYVLGLE...GKSGNPSPFTALGYTFLGIKAALNHQRGHQNFAGIKVAVQGLGTVAYLCKHLSEAGAELYVTDINQAAIDRVVN
CpLeuDH   IANTVTPFVTGTE...GKSGNPAPFTALGTFLGIKASVKHKFNRDDLTGLKVAVQGLGSVGYLLCEHLHNAGVELVITDINQSTLDNAAN
RsLeuDH   QVHQETPYVAGLE...GKSGNPAPFTALGTYRGVKAAAKHYFGSDDLSGKTIAVQGLGCVGFYLCEHLHKEGAKLIVTDINQEAVQRAVN
SbLeuDH   IAHEETPYMAGLE...GKSGDPSPFTALGTYLGIKAAVKHRLGLGSLKLGIKAVQGVGNLGYLLCKHLHNEGAELIVTDIHQASLDRVAT
BsLeuDH   LIHQETDYVTGISPAFGSSGNPSPVTAYGVYRGMKAAAKEAFGSDSLEGLAVSVQGLGNVAKALCKKLNTEGAKLVVTDVNKAAVSAAVA
BtLeuDH   IIHEETDFVTGISPSFGSSGNPSPVTAYGVYRGMKAAAKEAFGTDNLEGKVIAVQGVGNVAYHLCKHLHAEGAKLIVTDINKEAVQRAVE
LsLeuDH   IIREETKYVTGVSPAFGSSGNPSPVTAYGVYKGMKAASKVAFGEDSLKGKVVAVQGVGNVAYNLCKHLHAEGAKLIVTDINQANGDRAVQ
                                                                        ●  ●●●                ●●    ●


PsLeuDH   QQ  β8  η1      β9         α8     β10       α9      β11  α10   α11    η2    TT
              ———>  QQQ      ———>        QQQQ    ———>    QQQQQQQQQ  ———>  QQQ   QQQQQQQQQ
                   270        280        290        300        310        320        330        340
PsLeuDH   EFGATAVGIDEIFDLDADVFAPCALGATINDATIPQLKATIIAGCANNQLAQSKHGEIIREKGILYAPDYVINAGGIINVAFETLPEGYS
PnLeuDH   EFGATAVGIDEIYDLDVDVYAPCALGATINDATIPRLKATIIAGCANNQLAQSRHGEIIREKGILYAPDYVINAGGIINVYYETLPEGYS
CpLeuDH   EFNAKVVGLDEIYDQDVDIYAPCALGATINDQTLPRIKASIIAGCANNQLAEQRHDQALLQRGILYAPDYVINAGGIINVSFENN....YD
RsLeuDH   QFGATAVGLDEIYSVEADIYSPCALGATLNDDTIPLLKAKVIAGCANNQLKEKRHGEVLRQQGILYTPDYVINAGGIINVAFEMRTEGYN
SbLeuDH   DFGAIVVAPQELYAQDVDVYAPCALGATLNDATLPQLKAKIVAGCANNQLAEVRHGEQLKEMGILYAPDYVINAGGIINVSFEKD...YD
BsLeuDH   EEGADAVAPNAIYGVTCDIFAPCALGATVLNDFTIPQLKAKVIAGSADNQLKDPRHGKYLHELGIVYAPDYVINAGGVINVADELYG..YN
BtLeuDH   EFGATAVEPNEIYGVECDIYAPCALGATVNDETIPQLKAKVIAGSANNQLKEDRHGDIIHEMGIVYAPDYVINAGGVINVADELYG..YN
LsLeuDH   EFGAEAVSPDKIYDVDCDIFSPCALGAIINDETIERLTCKVVAGAANNQLKEEKHGEMLEQKGIIYAPDYVINAGGVINVADELYG..YN
                   ●●                          ●  ●                              ▲


PsLeuDH   α12           α13                α14
          QQQQQQQQ    QQQQQQQQQQQQQQ     QQQQQQQQQQQQQQ
                   360        370        380        390        400
PsLeuDH   AAASNKHVNGIFDSLTEIFARSDKENKSTHLIADELAQEILKNGL..........
PnLeuDH   AAASNKHVEGIFDTLTEIFARSEKEQKSTHLIADELAQEILKNGL..........
CpLeuDH   MEKSRQKVSNIYDTLLDIYVKADIQNRPTGIIADEMAREIIKNGGK.........
RsLeuDH   ADESTAKVNEIYDTLLNLFQRADEQALPTSTVADLMAQEIISQGRKI........
SbLeuDH   AAKSTAKVEEIYNTLLKIFSQADAQNRTTGAVADEMAKAIIEAAKK.........
BsLeuDH   RTRAMKRVDGIYDSIEKIFAISKRDGVPSYVAADRMAEERIAKVAKARSQFLQDQRNILNGR..
BtLeuDH   RERALKRVESIYDTIAKVIEISKRDGIATYVAADRLAEERIASLKNSRSTYLRNGHDIISRR..
LsLeuDH   RDRAMKRVETIYDNMLKVFEIAKRDGIPSYKAADRMAEERIAAMRKTRSTFLVNGQSILSHRLE
```

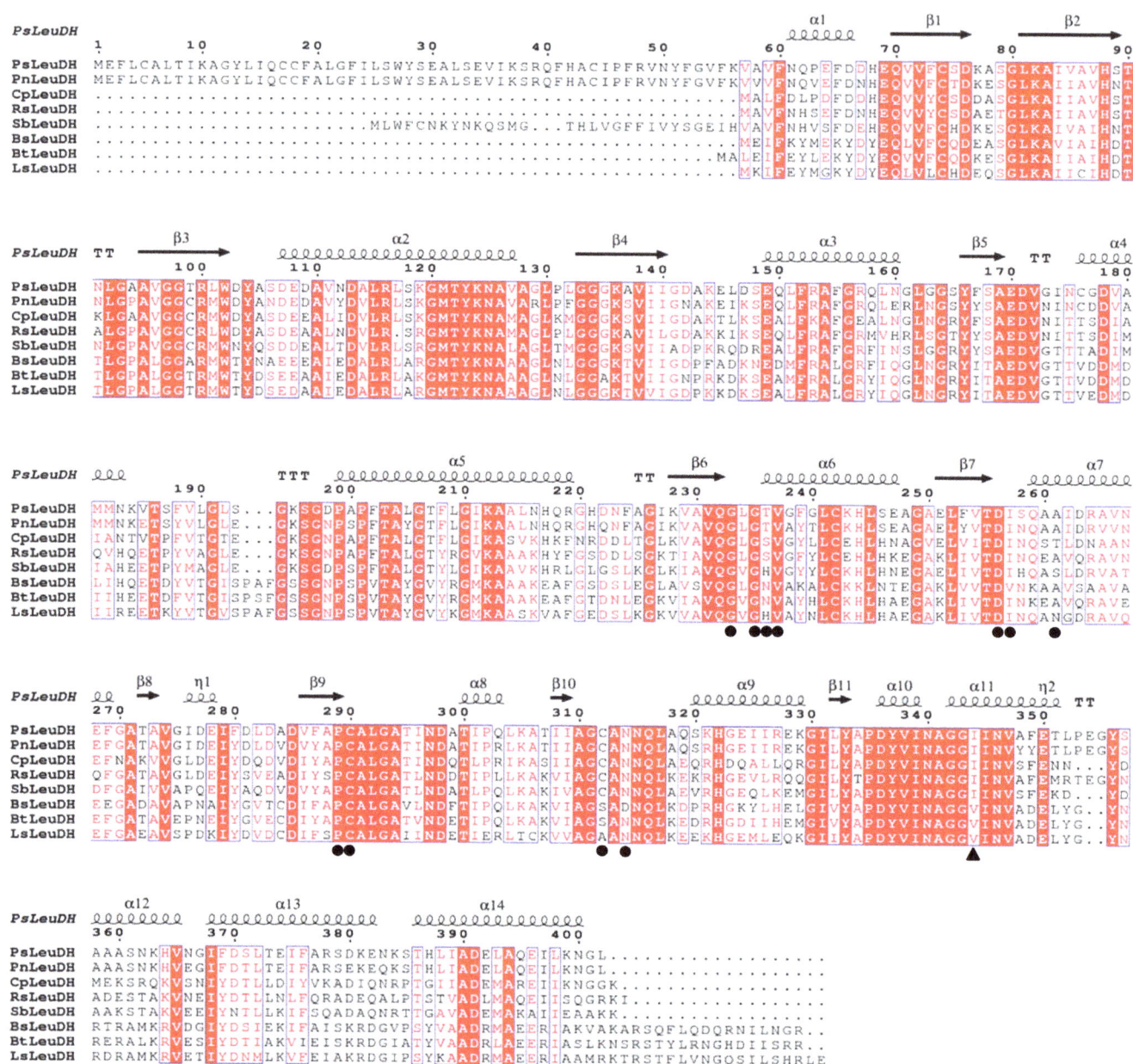

Figure 1. Amino acid sequence alignment of PsLeuDH and related LeuDH. PsLeuDH, *Pseudoalteromonas* sp. ANT178 LeuDH (MH322031); PnLeuDH, *Pseudoalteromonas nigrifaciens* (ASM53600); *Cp*LeuDH, *Colwellia piezophila* (WP_019029130); RsLeuDH, *Rheinheimera salexigens* (WP_070050751); SbLeuDH, *Shewanella baltica* BA175 (AEG11165); BsLeuDH, *Bacillus sphaericus* ATCC4525 (PDB ID:1LEH); BtLeuDH, *Bacillus thuringiensis* (WP_001162678); and LsLeuDH, *Laceyella sacchari* (KR065697). ●, NAD binding site; ▲, Phe binding site.

2.2. Homology Modeling and Analysis of PsLeuDH

BsLeuDH (PDB ID:1LEH), encoded 364 amino acids, was isolated from mesophilic bacteria *Bacillus sphaericus* ATCC4525 [16], which exhibited the highest sequence identity (51%) to PsLeuDH using DALI server. The comparative analysis of the 3D structure of PsLeuDH and the mesophilic enzyme Bs-LeuDH was shown in Figure 2. It could be seen that two LeuDHs had a similar NAD$^+$ binding site and Phe binding site.

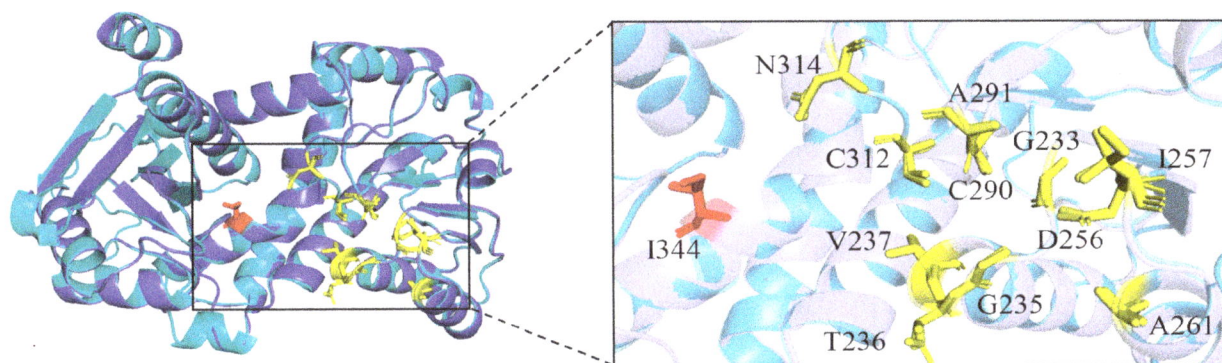

Figure 2. Three-dimensional structure comparison of PsLeuDH and BsLeuDH model. PsLeuDH, tv-blue; BsLeuDH, cyan; NAD$^+$ binding site, yellow ball stick model; Phe binding site, red ball stick model.

Comparison of structural adaptation characteristics and amino acid substitutions between PsLeuDH and BsLeuDH was shown in Table 1. It can be seen that PsLeuDH exhibited several cold-adapted features. Firstly, the number of electrostatic interactions of PsLeuDH was less than BsLeuDH, which might make the structure of PsLeuDH more flexible [17]. PsLeuDH also had less hydrophobic interactions compared to BsLeuDH, it might make PsLeuDH less rigid and contributed to decrease in structural stability [18]. Secondly, PsLeuDH revealed higher glycine residues and fewer proline and arginine residues that could affect the cold-adapted proteins properties which might offer higher flexibility to proteins [19]. Several amino acid residues in BsLeuDH were replaced by glycine residues in PsLeuDH. The glycine residues might improve the flexibility of the active site, and regulate the entropy of protein unfolding [10], thus probably improving the catalytic efficiency of the enzyme at low temperature. Additionally, proline might reduce the configuration entropy of the unfolding of protein molecules [20] and reduce the stability of enzyme molecules. Additionally, the stability of enzyme was also a significant factor to determine its catalytic characteristics. Some arginine residues in PsLeuDH were replaced by other residues at the same position in BsLeuDH. One of the stability factors in protein structure referred to salt bridges formed by arginine residues [19], arginine might make protein molecules more stable through ionic interaction. Compared with mesophilic enzyme BsLeuDH, PsLeuDH had higher flexibility and lower thermal stability, resulting in higher catalytic efficiency at low temperature [21].

Table 1. Comparison of structural adaption features and amino acid substitutions between PsLeuDH and its homolog (BsLeuDH).

Parameters	PsLeuDH	Bs-LeuDH	Expected Effect on PsLeuDH
Electrostatic interactions			
Salt Bridge (2.5 to 4.0)	17	22	
Hydrogen Bonds ($\leq$3.3 Å)	368	403	Protein stability
Cation-pi interactions	3	11	
Aromatic interactions	6	8	
Hydrophobic interactions	227	318	Thermolability
Glycine residues	42	36	
Proline residues	9	11	
Arginine residues	10	17	Flexibility
Glycine substitution (PsLeuDH → BsLeuDH)	G163 → N107, G177 → D121, G238 → A185, G240 → A187, G275 → A222, G401 →V346		
Proline substitution (PsLeuDH → BsLeuDH)	A94 → P38, A143 →P87, S320 → P267, S385 → P330		
Proline substitution (PsLeuDH → BsLeuDH)	P131 → N75, P63 → M7		Stability
Arginine substitution (PsLeuDH → BsLeuDH)	R219 → F166, R264 → A211, R327 → H274, R378 → I323		

2.3. Expression and Purification of the rPsLeuDH

The gene coding for the PsLeuDH was cloned into the pET-28a (+) vector and expressed in *E. coli* BL21 (DE3) under IPTG induction (Figure 3, Lane 3). rPsLeuDH was purified in a single step using His-tag affinity chromatography. A major band was observed on SDS-PAGE with about the molecular weight 44.4 kDa (Figure 3, Lane 4, 5). It is noteworthy that the last purified rPsLeuDH exhibited the highest specific activity of 275.13 U/mg.

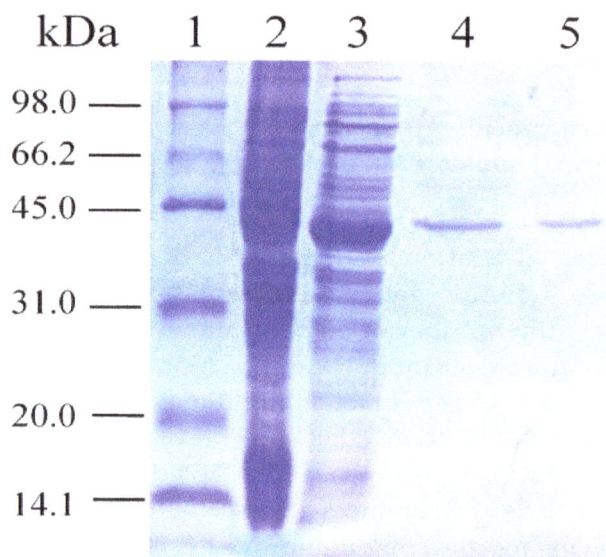

Figure 3. Expression and purification analysis of PsLeuDH. Lane 1: molecular weight standard marker; Lane 2: crude extract from the BL21/pET-28a (+); Lane 3: crude extract from the BL21/pET-28a (+)-PsLeuDH with IPTG induction; Lane 4: rPsLeuDH eluted with 50 mM imidazole; Lane 5: rPsLeuDH eluted with 100 mM imidazole.

2.4. Effects of Temperature and pH on Activity and Stability of rPsLeuDH

The temperature characteristic of rPsLeuDH was shown in Figure 4a. It exhibited the highest activity at 30 °C, and that of a cold-adapted LeuDH was 30 °C [12], whereas thermophilic LeuDH was approximately 40–65 °C [6,22], or (60–75 °C) [5]. It is worth pointing out that rPsLeuDH retained 40% of the highest activity at 0 °C, suggested that the enzyme is a cold-adapted enzyme [23]. Furthermore, the thermostability of rPsLeuDH was assessed in Figure 4b. It was stable and retained 85% of its initial activity after incubating at 30 °C after 120 min. While, after incubating at 50 °C for 20 min, it was only 30% of its activity lower than other cold-adapted LeuDHs from *Alcanivorax dieselolei* [12] and *Sporosarcina psychrophila* [7]. However, thermostable LeuDH could retain full activity after incubation at 65 °C for 10 min [24]. The above results indicated that rPsLeuDH had thermal instability, which was another significant feature of cold-adapted enzyme [25].

The effect of pH on rPsLeuDH activity was shown in Figure 4c. The activity of rPsLeuDH was higher under alkaline conditions (pH 7.0–10.0), with the highest activity at pH 9.0. Similar results were described in other LeuDHs such as *Sporosarcina psychrophile* (pH 8.5–11.0) [7], *Laceyella sacchari* (pH 9.5–11) [6] and *Citrobacter freundii* (pH 9.0 to 11.0) [5]. After 30 min of exposure to pH 6.0–10.0, the stability of rPsLeuDH showed a similar pattern with that of the activity response to pH (Figure 4d). This broad range of pH dependence for the activity and stability made the rPsLeuDH probably useful for medical industrial applications.

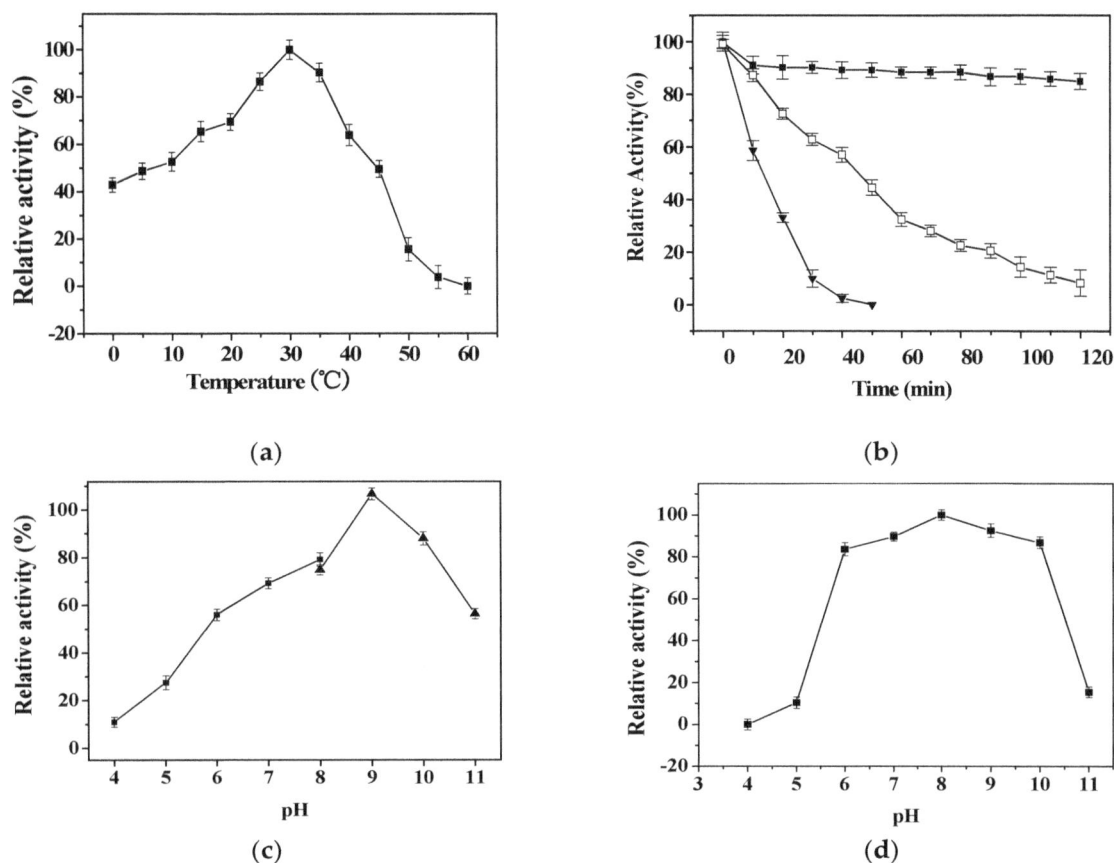

(a)

(b)

(c)

(d)

Figure 4. Effects of temperature and pH on the activity and stability of rPsLeuDH. (**a**) Effect of temperature on the activity of rPsLeuDH. (**b**) Effect of temperature on the stability of rPsLeuDH. (■) 30 °C, (□) 40 °C, (▼) 50 °C. (**c**) Effect of pH on the activity of rPsLeuDH. (**d**) Effect of pH on the stability of rPsLeuDH. Data are presented as mean ± SD ($n = 3$).

2.5. Effects of NaCl Concentration and Different Reagents on the Activity of PsLeuDH

The effect of NaCl concentration on the rPsLeuDH activity was shown in Figure 5. It could be seen that rPsLeuDH was stable at 0–3.0 M NaCl, with the highest activity at 2.0 M NaCl, which may be related to high salinity in the Antarctic sea ice environment. The similar result was also found in LeuDH from *Bacillus licheniformis* [3] and *Thermoactinomyces intermedius* [24] after high salt concentration treatment. The effect of various reagents on the rPsLeuDH activity was listed in Table 2. rPsLeuDH was completely inhibited by 1 mM $Pb(NO_3)_2$ and $BaCl_2$. Inhibitions by 1 mM $CrCl_2$ and $CdCl_2$ were 86.7% and 92.4%, respectively, while only partially inhibited by other metals salt in some extent. In addition, rPsLeuDH was sensitive to Thiourea and ethanol, but Triton X-100 kept the enzyme activity.

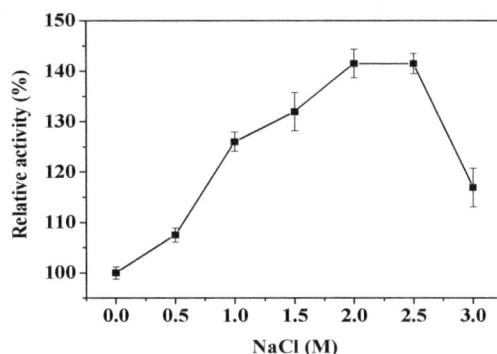

Figure 5. Effect of salt concentration on the activity of rPsLeuDH.

Table 2. Effects of different reagents on the activity of rPsLeuDH.

Reagent	Concentration	Relative Activity (%)	Reagent	Concentration	Relative Activity (%)
None		100 ± 0.0			
KCl	1 mM	99.7 ± 1.6	KCl	5 mM	40.0 ± 1.9
$CoCl_2$	1 mM	90.1 ± 1.7	$CoCl_2$	5 mM	70.0 ± 2.0
$MgCl_2$	1 mM	87.9 ± 0.8	$MgCl_2$	5 mM	65.8 ± 1.2
$CaCl_2$	1 mM	87.9 ± 0.4	$CaCl_2$	5 mM	68.1 ± 0.9
$ZnCl_2$	1 mM	80.0 ± 2.5	$ZnCl_2$	5 mM	72.2 ± 2.0
$FeCl_2$	1 mM	75.1 ± 2.2	$FeCl_2$	5 mM	62.4 ± 1.7
$CuCl_2$	1 mM	61.0 ± 2.2	$CuCl_2$	5 mM	41.0 ± 1.5
$HgCl_2$	1 mM	29.2 ± 0.3	$HgCl_2$	5 mM	12.3 ± 1.9
$CrCl_2$	1 mM	13.3 ± 0.3	$CrCl_2$	5 mM	5.8 ± 2.9
$CdCl_2$	1 mM	7.6 ± 0.5	$CdCl_2$	5 mM	0.0 ± 0.0
$Pb(NO_3)_2$	1 mM	0.0 ± 0.0	$Pb(NO_3)_2$	5 mM	0.0 ± 0.0
$BaCl_2$	1 mM	0.0 ± 0.0	$BaCl_2$	5 mM	0.0 ± 0.0
EDTA	1 mM	91.8 ± 2.7	EDTA	5 mM	84.2 ± 2.1
Thiourea	1 mM	51.5 ± 4.0	Thiourea	5 mM	34.3 ± 2.6
Triton X-100	0.2%	102.7 ± 1.4	Ethanol	25%	67.5 ± 1.4

2.6. The Substrate Specificity Analysis and Kinetic Parameters of rPsLeuDH

The substrate specificity analysis of rPsLeuDH was listed in Table 3. It could catalyze and utilize five substrates, indicating that rPsLeuDH possessed a broad spectrum of substrates in catalytic oxidation reaction. L-leucine was the most suitable substrate for rPsLeuDH, which was the similar with other microbial LeuDH [6,22]. The kinetic parameters of rPsLeuDH were determined. K_m and V_m of L-leucine were calculated as 0.33 mM and 15.24 μmol/min·mg, respectively. Besides, the k_{cat} value of L-leucine was 30.13/s, demonstrating that rPsLeuDH had a high affinity to substrates and was conducive to improving catalytic efficiency at low temperature.

Table 3. Substrate specificity analysis of rPsLeuDH.

Substrate	V_m (μmol/min·mg)	K_m (mM)	k_{cat} (1/s)	k_{cat}/K_m (mM^{-1} s^{-1})
L-lecine	15.24	0.33	30.13	91.30
L-tyrosine	13.35	0.48	26.39	54.98
L-proline	10.52	0.64	20.80	32.50
DL-methionine	8.38	0.75	16.57	22.09
L-arginine	7.13	0.84	14.09	16.77

2.7. The Thermodynamic Parameters of rPsLeuDH

Thermodynamic parameters such as ΔH, ΔS and ΔG at different temperature (0–30 °C) were calculated and listed in Table 4. At 0, 10, 20, and 30 °C, the k_{cat} value of rPsLeuDH were 12.25, 14.96, 20.20 and 30.13/s, respectively, indicating that the k_{cat} value increased with increasing temperature, which was similar to the k_{cat} change trend of cold-adapted β-D-galactosidase at different temperatures [26]. rPsLeuDH also exhibited lower ΔH, ΔS and ΔG and higher k_{cat} at low temperature, as compared to mesophilic enzyme, which may be mainly related to the conformation of cold adapted protein [27]. On the other hand, it may also be related to increasing the efficiency of binding of the substrate to the catalytic site [28].

Table 4. Thermodynamic parameter of the rPsLeuDH.

Temperature (°C)	ΔH (KJ/mol)	ΔS (J/mol K)	ΔG (KJ/mol)	k_{cat} (1/s)
0	18.27	−156.45	61.01	12.25
10	18.19	−157.75	62.90	14.96
20	18.11	−158.02	64.43	20.20
30	18.02	−157.28	65.70	30.13

3. Materials and Methods

3.1. Microorganisms and Growth Conditions

The strain *Pseudoalteromonas* sp. ANT178, isolated from sea ice in Antarctica (68°30′ E, 65°00′ S), was used as a source of *psleudh* gene. The strain ANT178 was cultivated in the 2216E sea water medium (initial pH 7.5, 5 g/L peptone, and 1 g/L yeast extract) for 96 h at 12 °C. *E. coli* BL21 (DE3) was used as the plasmid host.

3.2. Sequence Analysis of LeuDH Gene

The open reading frame and amino acid sequences of *psleudh* gene were computed (https://www.ncbi.nlm.nih.gov/orffinder/). The theoretical molecular weight and pI were also analyzed using the ExPASy Compute pI/Mw tool (http://web.expasy.org/computepi). Multiple sequence alignment of the amino acids of PsLeuDH was performed using Bioedit 7.2 and ESPript 3.0 [29].

3.3. Protein Homology Modeling

A homology model of LeuDH was built with SWISS-MODEL. LeuDH from mesophilic bacteria *Bacillus sphaericus* ATCC4525 (PDB ID:1LEH) [16] was selected as the template. The structure figures were created with PyMOL software (DeLano Scientific LLC, San Carlos, CA, USA). Salt bridges were carried out using VMD 1.9.3. (University of lllinois Urbana-Champaign, Champaign, IL, USA). For the hydrogen bonds, a cut-off distance of 3.3 Å was set. Cation-pi interactions, aromatic interactions, ionic interactions, and hydrogen bonds were predicted by the Protein Interactions Calculator program (http://pic.mbu.iisc.ernet.in).

3.4. Molecular Cloning, Expression and Purification of rPsLeuDH

The genome of *Pseudoalteromonas* sp. ANT178 was sequenced and annotated using high-throughput technologies (data not shown). The full-length gene of *psleudh* was amplified by PCR using the primers 5′-GATGGATCCATGGAATTT TTATGTG-3′ (*Bam*HI site underlined) and 5′-CAGAAGCTTGAAGACCGTTTT TAAG-3′ (*Hin*dIII site underlined) according to its genome sequence. PCR was performed with Taq DNA polymerase (TaKaRa Bio, Dalian, China). The product was then directly cloned into the corresponding sites of the pET-28a (+) vector and transformed into *E. coli* BL21. The transformants with the *psleudh* gene were grown in Luria-Bertani (LB) medium supplemented with 100 mg/L kanamycin and cultured by shaking at 37 °C until the OD_{600} reached 0.6–0.8. Then, 1.0 mM sopropyl-β-D-thiogalactopyranoside (IPTG) was added for induction. The bacterial cells were cultured at 37 °C for 2–3 h, and then the culture temperature was shifted to 28 °C to induce the protein expression for 6 h. The induced cells centrifuged at 4 °C and 7500× g for 15 min and subjected to ultrasonic disruption with 150 W (JY96-IIN, Shanghai, China). The insoluble debris was removed by centrifuged at 4 °C and 7500× g for 15 min, and the supernatant was harvested as crude protein (21.99 mg). Purification of rPsLeuDH with the His-tagged was purified using Ni-NTA affinity chromatography. The purified protein (1.11 mg) was eluted with 10, 50, 100 and 250 mM imidazole buffer (20 mM Tris-HCl, 500 mM NaCl, pH 8.0) at a flow rate of 1.0 mL/min. The purity and the molecular mass of the rPsLeuDH were determined by SDS-PAGE, using 12.0% polyacrylamide gels.

3.5. Assay of rPsLeuDH Activity

The standard enzyme assay were based on traditional method and modified on basis [1,30]. The oxidation reaction activity assay was determined by 200 µL reaction system. It contained 0.1 M Glycine-NaOH (pH 10.4) buffer, 10 mM L-leucine and 10 µL purified enzyme (0.62 µg), which incubated at 30 °C for 2 min. After adding 1 mM NAD^+, the changes of absorbance at 340 nm within 1 min were detected. Futhermore, the reductive amination reaction system containing (200 µL) 0.2 M NH_4Cl-NH_4OH buffer (pH 9.0), 5 mM TMP and 10 µL purified enzyme at 30 °C for 2 min, after adding 0.2 mM NADH, changes in absorbance at 340 nm within 1 min were measured. One unit of LeuDH activity was defined as the amount of enzyme catalyzed the formation or reduction of 1 μmoL NADH/min at 30 °C.

3.6. Characterization of the Purified rPsLeuDH

The optimal temperature of the purified rPsLeuDH was determined with the standard assay at temperatures from 0 °C to 60 °C. To evaluate the thermostability, the purified enzyme was incubated at three different temperatures (30, 40, and 50 °C) for 120 min, and the residual activity was measured by the standard enzyme assays. The optimal pH of the purified enzyme was determined at 30 °C using Citric acid/Na$_2$HPO$_4$ buffer (0.2 M) and NH$_4$Cl-NH$_4$OH buffer (0.2 M) for pH ranges 4.0–8.0 and 8.0–10.0, respectively. To assess pH stability, the rPsLeuDH was pretreated at pH 4.0–11.0 in the absence of substrate at 30 °C for 30 min, and the residual activity was measured by the standard enzyme assays. The purified rPsLeuDH was incubated at 0–3.0 M NaCl at 30 °C for 30 min, and remaining activity was assayed with the standard enzyme assays. The effects of different reagents on the rPsLeuDH activity were assayed with the standard enzyme assay after pre-incubating enzyme in different metal ions at 30 °C for 30 min. Enzyme activity assayed without any reagent was defined as control (100%).

3.7. Kinetic Parameter of the rPsLeuDH

To assess the kinetics parameters, the Lineweaver-Burk plot method was used to calculate the K_m and V_m of rPsLeuDH [31]. The kinetic constants of NADH (0.025 mM–0.4 mM), L-leucine (0.05 mM–2 mM), L-tyrosine (0.05 mM–2 mM), L-proline (0.05 mM–2 mM), DL-methionine (0.05 mM–2 mM), L-arginine (0.05 mM–2 mM), TMP (0.05 mM–2 mM), and NAD$^+$ (0.025 mM–0.4 mM) were determined by the above method in rPsLeuDH.

3.8. Thermodynamic Parameter of the rPsLeuDH

The k_{cat} parameter is the reaction rate constant for the enzymatic-substrate complex chemical conversion into the enzyme and the product. k_{cat} was calculated based on kinetics experiments, and the thermodynamic related parameters were assayed by the modification method of Feller [27] as follows:

$$k_{cat} = Ae^{\frac{-E_a}{RT}} \tag{1}$$

$$\Delta H = E_a - RT \tag{2}$$

$$\Delta S = R\left(Ink_{cat} - 24.76 - InT + \frac{E_a}{RT} \right) \tag{3}$$

$$\Delta G = \Delta H - T\Delta S \tag{4}$$

where A is the constant, E_a is the activation energy of the reaction, R is the gas constant (8.314 J mol^{-1} K^{-1}), ΔH is the enthalpy of activation, ΔS is the entropy of activation, and ΔG is the free energy of activation.

4. Conclusions

A novel cold-adapted leucine dehydrogenase gene (*psleudh*) was cloned from Antarctic sea-ice bacterium and expressed in *E. coli* (DE3). Through homology modeling and comparison with its homologous enzyme (BsLeuDH), it was suggested that more glycine residues, reduced proline residues and arginine residues might be responsible for its catalytic efficiency at low temperature. rPsLeuDH was purified and characterized with higher activity at 30 °C, high salt (3.0 M), remarkable pH stability (pH 6.0–10.0), and higher specific activity (275.13 U/mg). These unique properties of rPsLeuDH make it a promising candidate as a biocatalyst in the enzymatic production of L-*tert*-leucine at room temperature.

Author Contributions: Y.W., Y.H. and Q.W. took charge of the research and designed the experiments; Y.W., L.Z., X.X., K.P., R.L., Y.W. and Y.H. performed the experiments and analyzed the data; Y.W. and Q.W. wrote the paper.

References

1. Ohshima, T.; Wandrey, C.; Sugiura, M.; Soda, K. Screening of thermostable leucine and alanine dehydrogenases in thermophilic *Bacillus* strains. *Biotechnol. Lett.* **1985**, *7*, 871–876. [CrossRef]

2. Sanwal, B.D.; Zink, M.W. L-leucine dehydrogenase of *Bacillus cereus*. *Arch. Biochem. Biophys.* **1961**, *94*, 430–435. [CrossRef]

3. Nagata, S.; Bakthavatsalam, S.; Galkin, A.G.; Asada, H.; Sakai, S.; Esaki, N.; Soda, K.; Ohshima, T.; Nagasaki, S.; Misono, H. Gene cloning, purification, and characterization of thermostable and halophilic leucine dehydrogenase from a halophilic thermophile, *Bacillus licheniformis* TSN9. *Appl. Microbiol. Biotechnol.* **1995**, *44*, 432–438. [CrossRef] [PubMed]

4. Katoh, R.; Nagata, S.; Misono, H. Cloning and sequencing of the leucine dehydrogenase gene from *Bacillus sphaericus*, IFO 3525 and importance of the C-terminal region for the enzyme activity. *J. Mol. Catal. B Enzym.* **2003**, *23*, 239–247. [CrossRef]

5. Mahdizadehdehosta, R.; Kianmehr, A.; Khalili, A. Isolation and characterization of leucine dehydrogenase from a thermophilic *Citrobacter freundii* JK-91 strain isolated from Jask Port. *Iran. J. Microbiol.* **2013**, *5*, 278–284. [PubMed]

6. Zhu, W.J.; Li, Y.; Jia, H.H.; Wei, P.; Zhou, H.; Jiang, M. Expression, purification and characterization of a thermostable leucine dehydrogenase from the halophilic thermophile *Laceyella sacchari*. *Biotechnol. Lett.* **2016**, *38*, 855–861. [CrossRef] [PubMed]

7. Zhao, Y.; Wakamatsu, T.; Doi, K.; Sakuraba, H.; Ohshima, T. A psychrophilic leucine dehydrogenase from *Sporosarcina psychrophila*: Purification, characterization, gene sequencing and crystal structure analysis. *J. Mol. Catal. B Enzym.* **2012**, *83*, 65–72. [CrossRef]

8. Turnbull, A.P.; Ashford, S.R.; Baker, P.J.; Rice, D.W.; Rodgers, F.H.; Stillman, T.J.; Hanson, R.L. Crystallization and quaternary structure analysis of the NAD(+)-dependent leucine dehydrogenase from *Bacillus sphaericus*. *J. Mol. Biol.* **1994**, *236*, 663–665. [CrossRef] [PubMed]

9. Zhu, L.; Wu, Z.; Jin, J.M.; Tang, S.Y. Directed evolution of leucine dehydrogenase for improved efficiency of L-*tert*-leucine synthesis. *Appl. Microbiol. Biotechnol.* **2016**, *100*, 5805–5813. [CrossRef] [PubMed]

10. Galkin, A.; Kulakova, L.; Ashida, H.; Sawa, Y.; Esaki, N. Cold-adapted alanine dehydrogenases from two Antarctic bacterial strains: Gene cloning, protein characterization, and comparison with mesophilic and thermophilic counterparts. *Appl. Environ. Microb.* **1999**, *65*, 4014–4020.

11. Gerday, C.; Aittaleb, M.; Bentahir, M.; Chessa, J.P.; Claverie, P.; Collins, T.; D'Amico, S.; Dumont, J.; Garsoux, G.; Georlette, D.; et al. Cold-adapted enzymes: From fundamentals to biotechnology. *Trend Biotechnol.* **2000**, *18*, 103–107. [CrossRef]

12. Jiang, W.; Sun, D.F.; Lu, J.X.; Wang, Y.L.; Wang, S.Z.; Zhang, Y.H.; Fang, B.S. A cold-adapted leucine dehydrogenase from marine bacterium *Alcanivorax dieselolei*: Characterization and L-*tert*-leucine production. *Eng. Life Sci.* **2016**, *16*, 283–289. [CrossRef]

13. Shi, Y.L.; Wang, Q.F.; Hou, Y.H.; Hong, Y.Y.; Han, X.; Yi, J.L.; Qu, J.J.; Lu, Y. Molecular cloning, expression and enzymatic characterization of glutathione s-transferase from Antarctic sea-ice bacteria *Pseudoalteromonas* sp. ANT506. *Microbiol. Res.* **2014**, *169*, 179–184. [CrossRef] [PubMed]

14. Wang, Y.T.; Han, H.; Cui, B.Q.; Hou, Y.H.; Wang, Y.F.; Wang, Q.F. A glutathione peroxidase from Antarctic psychrotrophic bacterium *Pseudoalteromonas* sp. ANT506: Cloning and heterologous expression of the gene and characterization of recombinant enzyme. *Bioengineered* **2017**, *8*, 742–749. [CrossRef] [PubMed]

15. Kuroda, S.I.; Tanizawa, K.; Sakamoto, Y.; Tanaka, H.; Soda, K. Alanine dehydrogenases from two Bacillus species with distinct thermostabilities: Molecular cloning, DNA and protein sequence determination, and structural comparison with other NAD(P)(+)-dependent dehydrogenases. *Biochemistry* **1990**, *29*, 1009–1015. [CrossRef] [PubMed]

16. Baker, P.J.; Turnbull, A.P.; Sedelnikova, S.E.; Stillman, T.J.; Rice, D.W. A role for quaternary structure in the substrate specificity of leucine dehydrogenase. *Structure* **1995**, *3*, 693–705. [CrossRef]

17. Paredes, D.I.; Watters, K.; Pitman, D.J.; Bystroff, C.; Dordick, J.S. Comparative void-volume analysis of

psychrophilic and mesophilic enzymes: Structural bioinformatics of psychrophilic enzymes reveals sources of core flexibility. *BMC Struct. Biol.* **2011**, *11*, 42–50. [CrossRef] [PubMed]

18. Li, F.L.; Shi, Y.; Zhang, J.X.; Gao, J.; Zhang, Y.W. Cloning, expression, characterization and homology modeling of a novel water-forming NADH oxidase from *Streptococcus* mutans ATCC 25175. *Int. J. Biol. Macromol.* **2018**, *113*, 1073–1079. [CrossRef] [PubMed]

19. Mohammadi, S.; Parvizpour, S.; Razmara, J.; Abu Bakar, F.D.; Illias, R.M.; Mahadi, N.M.; Murad, A.M. Structure prediction of a novel Exo-β-1,3-Glucanase: Insights into the cold adaptation of psychrophilic yeast *Glaciozyma antarctica* PI12. *Interdiscip. Sci. Comput. Life Sci.* **2016**, *10*, 157–168. [CrossRef] [PubMed]

20. Herning, T.; Yutani, K.; Inaka, K.; Kuroki, R.; Matsushima, M.; Kikuchi, M. Role of proline residues in human lysozyme stability: A scanning calorimetric study combined with X-ray structure analysis of proline mutants. *Biochemistry* **1992**, *31*, 7077–7085. [CrossRef] [PubMed]

21. Siglioccolo, A.; Gerace, R.; Pascarella, S. "Cold spots" in protein cold adaptation: Insights from normalized atomic displacement parameters (B′-factors). *Biophys. Chem.* **2010**, *153*, 104–114. [CrossRef] [PubMed]

22. Li, J.; Pan, J.; Zhang, J.; Xu, J.H. Stereoselective synthesis of L-*tert*-leucine by a newly cloned leucine dehydrogenase from *Exiguobacterium sibiricum*. *J. Mol. Catal. B Enzym.* **2014**, *105*, 11–17. [CrossRef]

23. Feller, G.; Narinx, E.; Arpigny, J.L.; Aittaleb, M.; Baise, E.; Genicot, S.; Gerday, C. Enzymes from psychrophilic organisms. *FEMS Microbiol. Rev.* **1996**, *18*, 189–202. [CrossRef]

24. Ohshima, T.; Nishida, N.; Bakthavatsalam, S.; Kataoka, K.; Takada, H.; Yoshimura, T.; Soda, K.; Esaki, N. The purification, characterization, cloning and sequencing of the gene for a halostable and thermostable leucine dehydrogenase from *Thermoactinomyces intermedius*. *Eur. J. Biochem.* **1994**, *222*, 305–312. [CrossRef] [PubMed]

25. Michetti, D.; Brandsdal, B.O.; Bon, D.; Isaksen, G.V.; Tiberti, M.; Papaleo, E. A comparative study of cold- and warm-adapted endonucleases a using sequence analyses and molecular dynamics simulations. *PLoS ONE* **2017**, *12*, e0169586. [CrossRef] [PubMed]

26. Pawlak-Szukalska, A.; Wanarska, M.; Popinigis, A.T.; Kur, J. A novel cold-active β-D-galactosidase with transglycosylation activity from the Antarctic *Arthrobacter* sp. 32cB-Gene cloning, purification and characterization. *Process Biochem.* **2014**, *49*, 2122–2133. [CrossRef]

27. Lonhienne, T.; Gerday, C.; Feller, G. Psychrophilic enzymes: Revisiting the thermodynamic parameters of activation may explain local flexibility. *Biochim. Biophys. Acta* **2000**, *1543*, 1–10. [CrossRef]

28. Khrapunov, S.; Chang, E.; Callender, R.H. Thermodynamic and structural adaptation differences between the mesophilic and psychrophilic lactate dehydrogenases. *Biochemistry* **2017**, *56*, 3587–3595. [CrossRef] [PubMed]

29. Robert, X.; Gouet, P. Deciphering key features in protein structures with the new ENDscript server. *Nucleic Acids Res.* **2014**, *42*, W320–W324. [CrossRef] [PubMed]

30. Ohshima, T.; Nagata, S.; Soda, K. Purification and characterization of thermostable leucine dehydrogenase from *Bacillus* stearothermophilus. *Arch. Microbiol.* **1985**, *141*, 407–411. [CrossRef]

31. Shang, Z.C.; Zhang, L.L.; Wu, Z.J.; Gong, P.; Li, D.P.; Zhu, P.; Gao, H.J. The activity and kinetic parameters of oxidoreductases in phaeozem in response to long-term fertiliser management. *J. Soil Sci. Plant Nutr.* **2012**, *12*, 597–607. [CrossRef]

8

Microalgal Enzymes with Biotechnological Applications

Giorgio Maria Vingiani [1], Pasquale De Luca [2], Adrianna Ianora [1], Alan D.W. Dobson [3,4] and Chiara Lauritano [1,*]

[1] Marine Biotechnology Department, Stazione Zoologica Anton Dohrn, CAP80121 (NA) Villa Comunale, Italy
[2] Research Infrastructure for Marine Biological Resources Department, Stazione Zoologica Anton Dohrn, CAP80121 (NA) Villa Comunale, Italy
[3] School of Microbiology, University College Cork, College Road, T12 YN60 Cork, Ireland
[4] Environmental Research Institute, University College Cork, Lee Road, T23XE10 Cork, Ireland
* Correspondence: chiara.lauritano@szn.it.

Abstract: Enzymes are essential components of biological reactions and play important roles in the scaling and optimization of many industrial processes. Due to the growing commercial demand for new and more efficient enzymes to help further optimize these processes, many studies are now focusing their attention on more renewable and environmentally sustainable sources for the production of these enzymes. Microalgae are very promising from this perspective since they can be cultivated in photobioreactors, allowing the production of high biomass levels in a cost-efficient manner. This is reflected in the increased number of publications in this area, especially in the use of microalgae as a source of novel enzymes. In particular, various microalgal enzymes with different industrial applications (e.g., lipids and biofuel production, healthcare, and bioremediation) have been studied to date, and the modification of enzymatic sequences involved in lipid and carotenoid production has resulted in promising results. However, the entire biosynthetic pathways/systems leading to synthesis of potentially important bioactive compounds have in many cases yet to be fully characterized (e.g., for the synthesis of polyketides). Nonetheless, with recent advances in microalgal genomics and transcriptomic approaches, it is becoming easier to identify sequences encoding targeted enzymes, increasing the likelihood of the identification, heterologous expression, and characterization of these enzymes of interest. This review provides an overview of the state of the art in marine and freshwater microalgal enzymes with potential biotechnological applications and provides future perspectives for this field.

Keywords: microalgae; enzymes; marine biotechnology; -omics technologies; heterologous expression; homologous expression

1. Introduction

Water covers around 71% of the Earth's surface, with salt water responsible for 96.5% of this percentage [1]. Due to its molecular structure and chemical properties, water includes (and often participates in) every chemical reaction that is biologically relevant [2]. In such reactions, enzymes cover a fundamental role: They are organic macromolecules that catalyze biological reactions (so-called "biocatalysts" [3]). Due to their substrate-specificity, enzymes are commonly used in several sectors (such as food processing, detergent, pharmaceuticals, biofuel, and paper production) to improve, scale, and optimize industrial production. For example, hydrolases, which are enzymes that catalyze the hydrolysis of chemical bonds, have applications in several fields. Examples of industrially relevant hydrolases are cellulases for biofuel production [4], amylases for syrup production [5], papain, phytases and galactosidases for food processing [6], and other hydrolases which have various pharmaceutical applications [7]. The demand for new enzymes is growing every year, and many financial reports expect the global enzyme market value to surpass the $10 billion mark by 2024 (Allied Market Research, 2018, https://www.alliedmarketresearch.com/enzymes-market;ResearchandMarket.com, 2018, https://www.researchandmarkets.com/research/6zpvw9/industrial?w=4), of which $7 billion alone will be for industrial applications (BCC Research, 2018, https://www.bccresearch.com/market-research/biotechnology/global-markets-for-enzymes-in-industrial-applications.html).

Microalgae are photosynthetic unicellular organisms that can be massively cultivated under controlled conditions in photobioreactors with relatively small quantities of micro- and macro-nutrients [8], and can thus fit perfectly into this market sector. Microalgae continue to be used in a number of biotechnological applications. Searching the available literature in the PubMed database, this trend is clearly visible (search filters used were the word "microalgae" in the Title/Abstract field and the word "biotechnolog*" in the Text Word field, using the asterisk wildcard to expand the term selection; Figure 1). Considering the full 20-year interval between "1999–2018", it is clear that as of 2012, there has been a rapid increase in the number of publications involving both "microalgae" and "biotechnology", reaching a peak in the years 2015–2016.

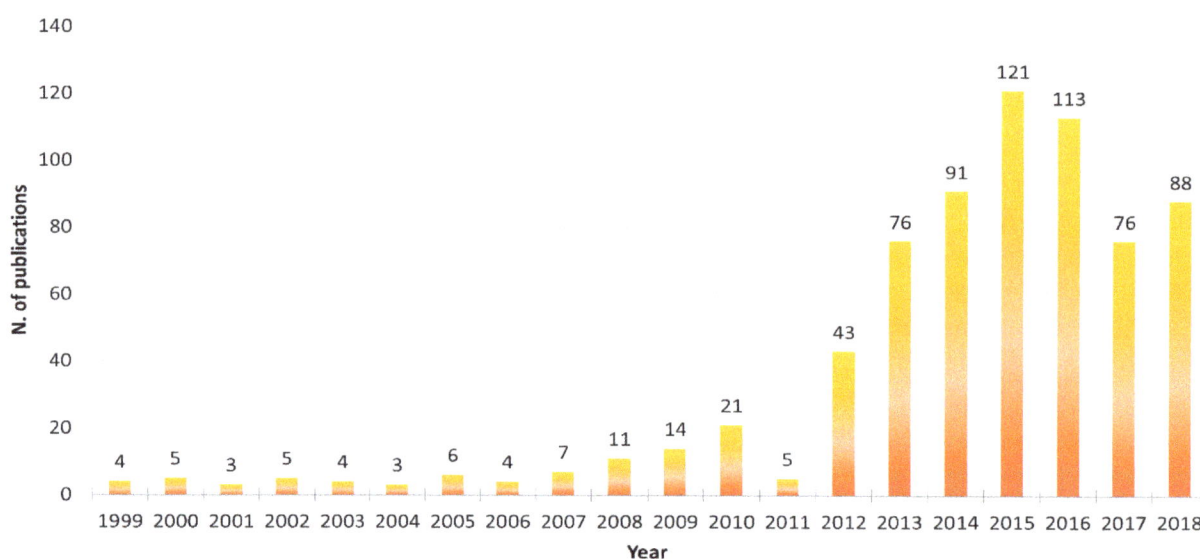

Figure 1. Microalgae Biotechnology PubMed Search Results 1999–2018. Using PubMed database search in the 20-years interval 1999–2018, the following search filters were set: The word "microalgae" in the [Title/Abstract] field and the word "biotechnolog*" in the [Text Word] field, using the asterisk (*) wildcard to expand the term selection (such as biotechnology, biotechnological, and biotechnologies).

The literature regarding the biotechnological applications of microalgae is dominated by four main research sectors: (1) Direct use of microalgal cells, for bioremediation applications and as food supplements [9]; (2) Extraction of bioactives for different applications (e.g., cosmeceutical, nutraceutical, and pharmaceutical applications, and for biofuel production [10,11]); (3) Use of microalgae as platforms for heterologous expression or endogenous gene editing and overexpression [12]; (4) Use of microalgae as sources of enzymes for industrial applications [13]. The latter field appears to be less well-studied compared to the others, due to the high costs currently involved in enzyme extraction and characterization, as well as the scarcity of annotated microalgal genomes.

Recent projects, such as those funded under the European Union Seventh Framework 2007–2017 (EU FP-7), e.g., BIOFAT (https://cordis.europa.eu/project/rcn/100477/factsheet/en) and GIAVAP (https://cordis.europa.eu/project/rcn/97420/factsheet/en), together with Horizon 2020 programs, e.g., ALGAE4A-B (http://www.algae4ab.eu/project.html) and VALUEMAG (https://www.valuemag.eu/), have resulted in an increase in –omics data (i.e., genomics, transcriptomics, proteomics and metabolomics data) available for microalgae, improving the possibility of finding new enzymes from both marine and freshwater species [14]. Mogharabi and Faramarzi recently reported the isolation of some enzymes from algae and highlighted their potential as cell factories [15]. This review aims to provide a summary of the current literature on microalgal enzymes with potential biotechnological applications with a particular focus on enzymes involved in the production of high-value added lipids and biodiesel, healthcare applications, and bioremediation.

2. Enzymes from Microalgae

2.1. Enzymes for High-Value Added Lipids and Biodiesel Production

Microalgae are known to accumulate large amounts of lipids [16], with triglycerides (TAGs) and poly-unsaturated fatty acids (PUFA) being the most studied from a biotechnological application standpoint, particularly for the production of biodiesel and nutraceuticals [9,16–18]. TAGs, esters derived from glycerol and three chained fatty acids (FA) which are usually stored in cytosol-located lipid droplets [19], can be used to produce biodiesel following acid- or base-catalyzed transesterification reactions [20]. PUFAs, for their part, have well-proven beneficial health effects [21,22], especially Ω-3 fatty acids such as docosahexaenoic acid (DHA) and eicosapentaenoic acid (EPA) (Figure 2).

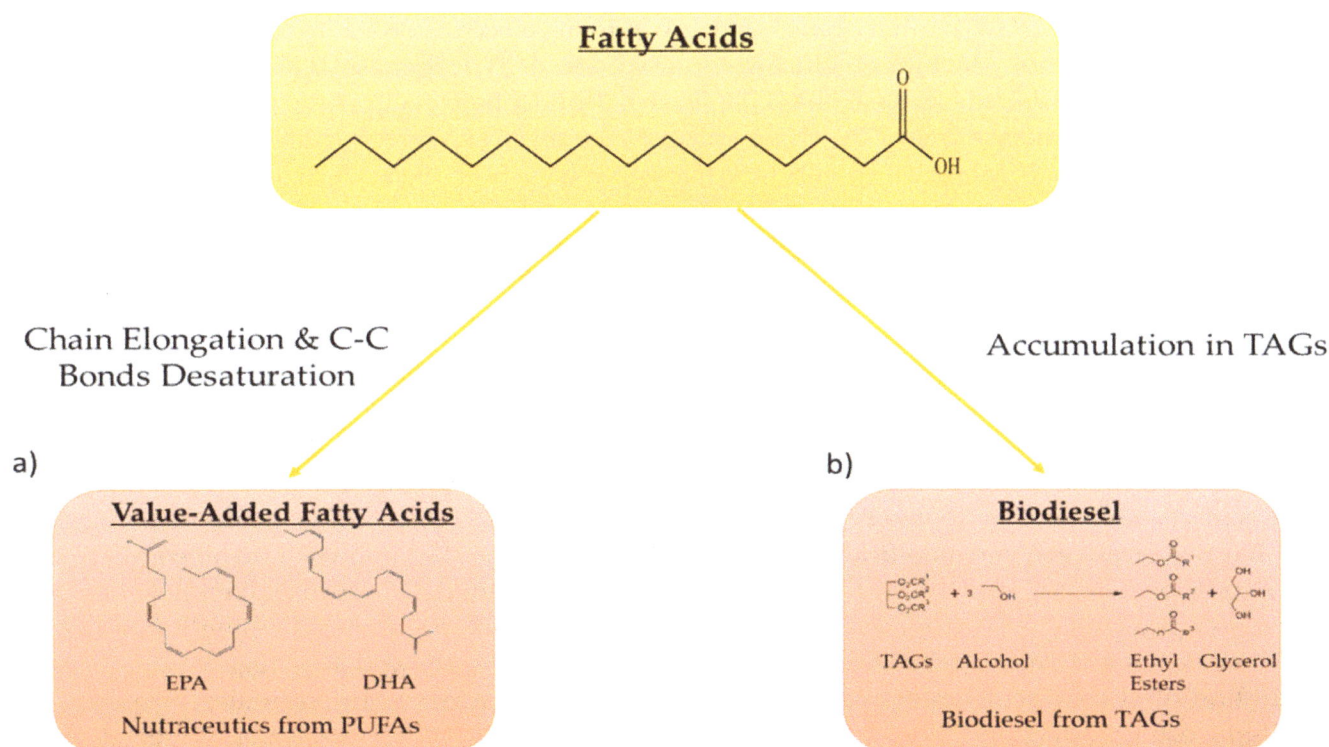

Figure 2. Examples of fatty acids of biotechnological interest. (**a**) Through various reactions of elongation and formation of double C-C bonds, poly-unsaturated fatty acids (PUFA) can be synthetized, such as eicosapentaenoic acid (EPA) and docosahexaenoic acid (DHA) with nutraceutical or food applications; (**b**) Accumulation in triglycerides (TAGs) and biodiesel formation via chemical transesterification.

The most frequently studied enzyme involved in lipid synthesis is acyl-CoA diacylglycerol acyltransferase (DGAT), involved in the final reaction of the TAG biosynthetic pathway [23,24]. Three independent groups of enzymes, referred to as acyl-CoA diacylglycerol acyltransferases type 1, 2, and 3 (DGATs 1-2-3), take part in the acyl-CoA-dependent formation of TAGs from its precursor sn-1,2-diacylglycerol (DAG) [25]. The individual contribution of each DGAT isoenzyme to the fatty acid profile of TAG differs between species [24,26].

A gene encoding DGAT1 was initially discovered in the green alga *Chlorella ellipsoidea* by Guo et al. [27], and an experiment involving overexpression of DGAT1 was subsequently performed in the oleaginous microalgae *Nannochloropsis oceanica* [28]. The first DGAT2 sequence was obtained from the green alga *Ostreococcus tauri* [29], and different studies involving overexpression of DGAT2 were performed. In particular, DGAT2 overexpression led to an increase in TAG production in the diatoms *Phaeodactylum tricornutum* [30] and *Thalassiosira pseudonana* [31], and in the oleaginous microalgae *Neochloris oleoabundans* [32] and *N. oceanica* [33]. Different isoforms of DGAT2 (NoDGAT2A, 2C, 2D) have successively been identified in *N. oceanica* and different combinations of either overexpression or under-expression have been analyzed. These combinations gave different fatty acid-production profiles, with some optimized for nutritional applications and others for biofuel purposes [34]. Even if the green alga *Chlamydomonas reinhardtii* is considered a common biofuel feedstock, it showed no clear trends following overexpression of different DGAT2 isoforms, with increased levels of TAG in some reports [35], while levels were not increased in others [36]. Recently, Cui and coworkers [37] characterized a dual-function wax ester synthase (WS)/DGAT enzyme in *P. tricornutum*, whose overexpression led to an accumulation of both TAGs and wax esters. This was the first report of this particular enzyme in a microalga, and a patent involving the enzyme was subsequently filed (Patent Code: CN107299090A, 2017).

In addition to DGAT, other genes have been targeted in order to increase high-value added lipid production, including glucose-6-phosphate dehydrogenase (G6PD), Δ6-desaturase, 6-phosphogluconate dehydrogenase (6PGD), glycerol-3-phosphate acyltransferase (GPAT1-GPAT2), and acetyl-CoA synthetase 2 (ACS2). Overexpression of these enzymes resulted in increased lipid contents [38–42]. In particular, two patents for desaturases have been filed. One covers a Δ6-desaturase from *Nannochloropsis* spp., which converts linoleic acid to γ-linolenic acid (GLA) and α-linolenic acid (ALA) to stearidnoic acid (Patent Code: CN101289659A, 2010). The other covers a Ω6-desaturase from *Arctic chlamydomonas* sp. *ArF0006*, which converts oleic acid to linoleic acid (Patent Code: KR101829048B1, 2018).

Other approaches to increase lipid production and/or alter lipid profiles via gene disruption have been employed. Examples include the knock-out of a phospholipase A2 (PLA2) gene via CRISPR/Cas9 ribonucleoproteins in *C. reinhardtii* [43], microRNA silencing of the stearoyl-ACP desaturase (that forms oleic acid via addition of a double-bond in a lipid chain [44]) in *C. reinhardtii* [45], and meganuclease and TALE nuclease genome modification in *P. tricornutum* [46]. This last approach involved modifying the expression of seven genes, potentially affecting the lipid content (UDP-glucose pyrophosphorylase, glycerol-3-phosphate dehydrogenase, and enoyl-ACP reductase), the acyl chain length (long chain acyl-CoA elongase and a putative palmitoyl-protein thioesterase), and the degree of fatty acid saturation (Ω-3 fatty acid desaturase and Δ-12-fatty acid desaturase). In particular, a mutant for UDP-glucose pyrophosphorylase showed a 45-fold increase in TAG accumulation under nitrogen starvation conditions. Figure 3 provides an overview of the subcellular localization of metabolic pathways and engineered enzymes in the aforementioned examples.

Finally, Sorigué and coworkers [47] reported, for the first time, the presence of a photoenzyme named fatty acid photodecarboxylase (FAP) in *Chlorella variabilis* str microalgae. NC64A. FAP converts fatty acids to hydrocarbons and may be useful in light-driven production of hydrocarbons. It is worth mentioning that Misra et al. [48] have developed a database to catalogue the enzymes which have been identified as being responsible for lipid synthesis from available microalgal genomes (e.g., *C. reinhardtii*, *P. tricornutum*, *Volvox carteri*), called dEMBF (website: http://bbprof.immt.res.in/embf/). To date, the database has collected 316 entries from 16 organisms, while providing different browsing options (Search by: "Enzyme Classification", "Organism", and "Enzyme Class") and different web-based tools (NCBI's Blast software integrated, sequence comparison, Motif prediction via the MEME software). The enzymes discussed in this section are reported in Table 1.

Table 1. Enzymes from Microalgae for Lipid and Biodiesel Production. Marine and freshwater ecological strain sources are abbreviated as M or F, respectively. Algal classes of *Bacillariophyceae, Chlorophyceae, Trebouxiophyceae, Eustigmatophyceae, Mamiellophyceae, Coscinodiscophyceae,* and *Cyanidiophyceae* are abbreviated as BA, CH, TR, EU, MA, CO, and CY, respectively.

Ref.	Enzymes	Microalgae	Strain Source	Microalgal Class	Main Results
[39]	Δ6-Desaturase	*Phaeodactylum tricornutum*	M	BA	Neutral lipid production enhanced and increase of EPA content
[41]	acetyl-CoA synthetase	*Chlamydomonas reinhardtii*	F	CH	Increase in neutral lipid production
[27]	acyl-CoA diacylglycerol acyltransferase 1	*Chlorella ellipsoidea*	F	TR	Sequence identification and function of TAG accumultation characterized
[28]	acyl-CoA diacylglycerol acyltransferase 1A	*Nannochloropsis oceanica*	M	EU	Increase in TAGs production both in nitrogen-replete and -deplete conditions
[36]	acyl-CoA diacylglycerol acyltransferase 2	*Chlamydomonas reinhardtii*	F	CH	No TAGs overproduction
[35]	acyl-CoA diacylglycerol acyltransferase 2	*Chlamydomonas reinhardtii*	F	CH	Five DGAT2 homologous genes identification and the overexpression of CrDGAT2-1 and CrDGAT2-5 resulting in a significant increase in lipid production
[33]	acyl-CoA diacylglycerol acyltransferase 2	*Nannochloropsis oceanica*	M	EU	Increase in neutral lipid production
[32]	acyl-CoA diacylglycerol acyltransferase 2	*Neochloris oleoabundans*	F	CH	Change of lipid profile
[29]	acyl-CoA diacylglycerol acyltransferase 2	*Ostreococcus tauri*	M	MA	Gene identification and enzyme characterization in heterologous systems
[30]	acyl-CoA diacylglycerol acyltransferase 2	*Phaeodactylum tricornutum*	M	BA	Increase in neutral lipid production with enrichment EPA-PUFAs content
[31]	acyl-CoA diacylglycerol acyltransferase 2	*Thalassiosira pseudonana*	M	CO	Increase in TAGs production with focus on the intracellular enzyme localization
[34]	acyl-CoA diacylglycerol acyltransferase 2A, 2C, 2D	*Nannochloropsis oceanica*	M	EU	Differential DGAT2 isoforms expression in different engineered strains with individual specialized lipid profiles

Table 1. *Cont.*

Ref.	Enzymes	Microalgae	Strain Source	Microalgal Class	Main Results
[47]	fatty acid photodecarboxylase	*Chlorella variabilis*	F	TR	Enzyme identification and alkane synthase activity tested
[38]	glucose-6-phosphate dehydrogenase	*Phaeodactylum tricornutum*	M	BA	Modest increase in neutral lipid production with a lipid composition switch from polyunsaturated to monounsaturated
[42]	glucose-6-phosphate dehydrogenase; phosphogluconate dehydrogenase	*Fistulifera solaris*	M	BA	Slight increase in TAGs production
[40]	glycerol-3-phosphate acyltransferase 1, 2	*Cyanidioschyzon merolae*	F	CY	Significant increase in TAGs production
[43]	phospholipase A2	*Chlamydomonas reinhardtii*	F	CH	Increase in TAGs production
[45]	stearoyl-ACP desaturase	*Chlamydomonas reinhardtii*	F	CH	Production of TAGs enriched in stearic acid
[46]	UDP-glucose pyrophosphorylase, glycerol-3-phosphate dehydrogenase, enoyl-ACP reductase, long chain acyl-CoA elongase, putative palmitoyl-protein thioesterase, Ω-3 fatty acid desaturase and Δ-12-fatty acid desaturase	*Phaeodactylum tricornutum*	M	BA	Significant increase in TAGs production (45-fold increase for UDP-glucose pyrophosphorylase mutant)
[37]	wax esther synthase/acyl-CoA diacylglycerol acyltransferase	*Phaeodactylum tricornutum*	M	BA	Increase in neutral lipids and wax esters production
Patent Code (Year)	**Enzymes**	**Microalgae**	**Strain Source**	**Microalgal Class**	**Notes**
CN107299090A (2017)	wax esther synthase/acyl-CoA diacylglycerol acyltransferase	*Phaeodactylum tricornutum*	M	BA	Neutral lipids and wax esters production enhanced
CN101289659A (2010)	Δ6-Desaturase	*Nannochloropsis* spp.	M	EU	The enzyme sequence was identified and the enzyme characterized in bacterial systems
KR101829048B1 (2018)	Ω6-Desaturase	*Arctic Chlamydomonas* sp. *ArF0006*	F	CH	The enzyme sequence was identified and the enzyme characterized in bacterial systems

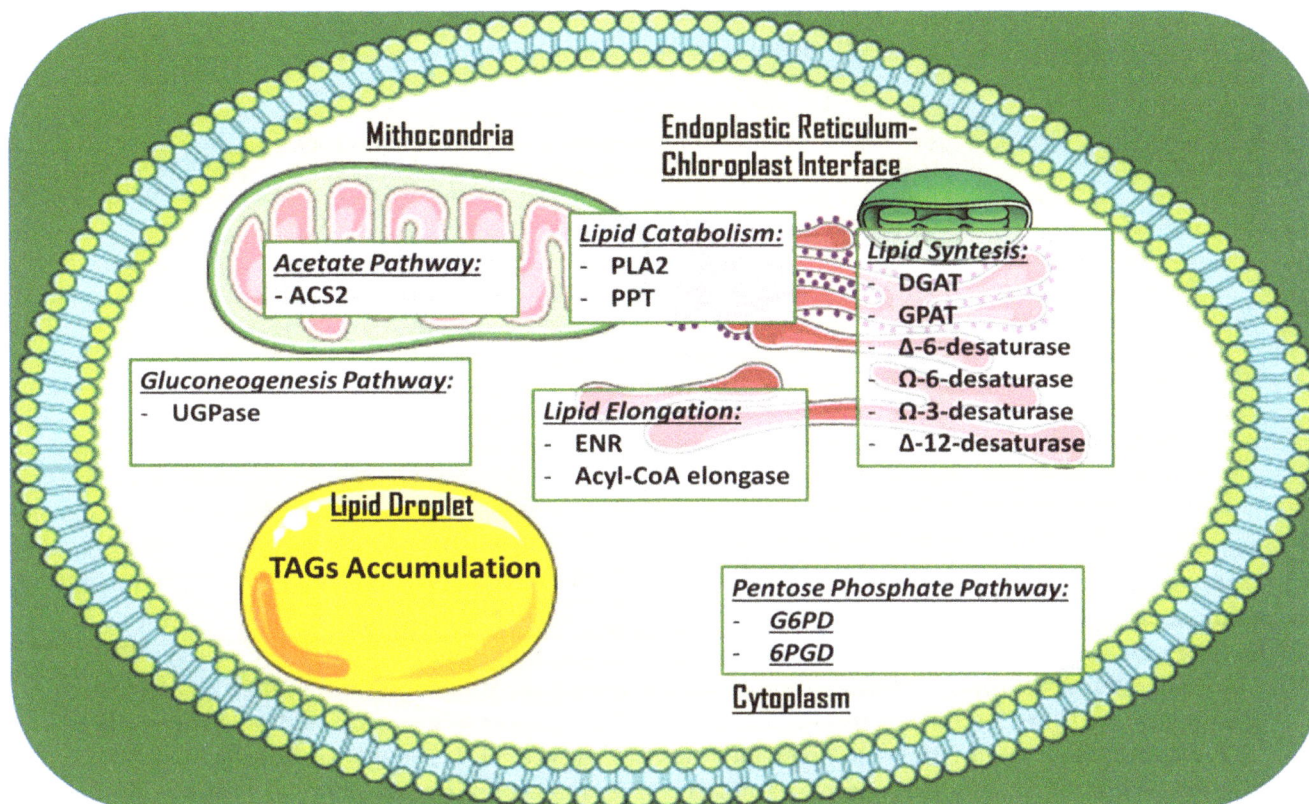

Figure 3. Main studied and engineered enzymes for TAGs and PUFAs in microalgae for the production of high value-added lipids. Enzymes are roughly divided in subcellular compartments. A single lipid droplet where TAGs are accumulated is added. Abbreviations: DGAT: Acyl-CoA diacylglycerol acyltransferase; G6PD: Glucose-6-phosphate dehydrogenase; 6PGD: 6-phosphogluconate dehydrogenase; GPAT: Glycerol-3-phosphate acyltransferase; ACS2: acetyl-CoA synthetase 2; PLA2: Phospholipase A2; Δ-6/Δ-12-Desaturase: delta-6/delta-12 fatty acid desaturase; Ω-3/Ω-6-desaturase: omega-2/omega-6 fatty acid desaturase; ENR: Enoyl-acyl carrier protein reductase; UGPase: UDP-glucose pyrophosphorylase; TAG: Triglyceride.

2.2. Enzymes for Healthcare Application

Enzymes for healthcare applications can include: (1) Enzymes used directly as "drugs", or (2) enzymes involved in the biosynthetic pathway of bioactive compounds (Figure 4). Regarding the first group, the most studied enzyme is ʟ-asparaginase. ʟ-asparaginase is an ʟ-asparagine amidohydrolase enzyme used for the treatment of acute lymphoblastic leukemia, acute myeloid leukemia, and non-Hodgkin's lymphoma [49].

Its hydrolytic effect reduces asparagine availability for cancer cells that are unable to synthesize ʟ-asparaginase autonomously [50] ʟ-asparaginase was historically first discovered and then produced in bacteria (e.g., *Escherichia coli, Erwinia aroideae, Bacillus cereus*) [51–53].

However, in order to overcome some of the economical and safety limits associated with marketing the enzyme [54,55], increased efforts began to focus on the identification and characterization of the enzyme in microalgae strains.

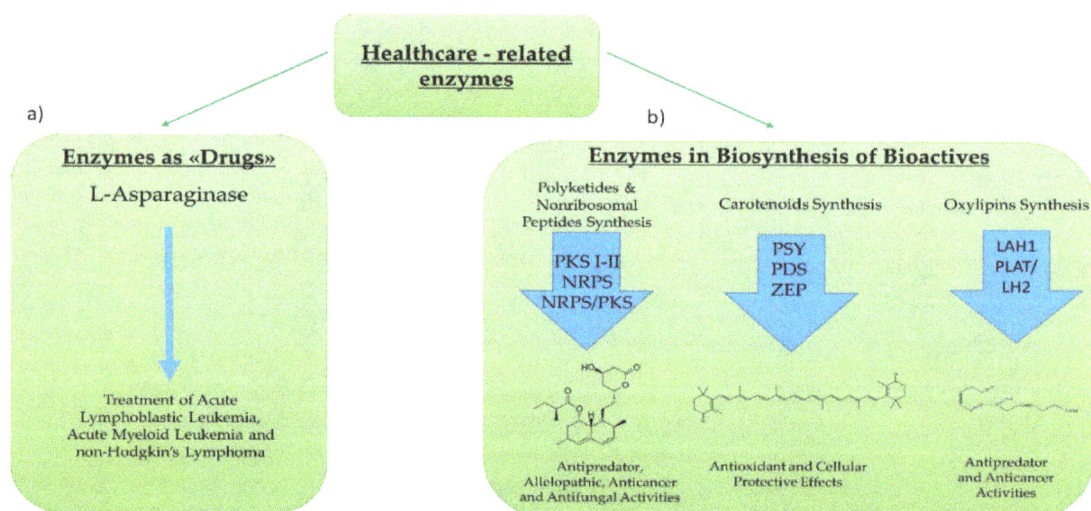

Figure 4. Enzymes for Healthcare Applications. Enzymes for healthcare applications can include: (**a**) Enzymes used directly as "drugs", such as the L-asparaginase (**b**) enzymes involved in the biosynthetic pathway of active compounds, such as polyketides, carotenoids, or oxylipins. In the synthesis of polyketides, the enzymes studied are polyketide synthases and nonribosomal peptide synthases. For the synthesis of carotenoids, the most studied enzymes are phytoene synthase (PSY), phytoene decarboxylase (PDS) and zeaxanthin epoxidase (ZEP). For the synthesis of oxylipins the studied enzymes are lipoic acid hydrolases (LAH) and PLAT (Polycystin-1, Lipoxygenase, Alpha-Toxin)/LH2 (Lipoxygenase homology). An example of molecules and their roles for each pathway is also outlined.

Paul [56] first purified an L-asparaginase in *Chlamidomonas* spp. with limited anticancer activity, and tested it in an in vivo anti-lymphoma assay. Ebrahiminezhad and coworkers screened 40 microalgal isolates via activity assays and reported on *Chlorella vulgaris* as a novel potential feedstock for L-asparaginase production [57].

Regarding enzymatic pathways involved in the synthesis of bioactive compounds, many studies have focused on polyketide synthases (PKS) and nonribosomal peptide synthetases (NRPS). PKS produce polyketides, while NRPS produce nonribosomal peptides. Both classes of secondary metabolites are formed by sequential reactions operated by these "megasynthase" enzymes [58,59]. Polyketides and nonribosomal peptides have been reported to have antipredator, allelopathic, anticancer, and antifungal activities [58,60–62]. PKS can be multi-domain enzymes (Type I PKS), large enzyme complexes (Type II), or homodimeric complexes (Type III) [63]. Genes potentially encoding these first two types' of PKSs have been identified in several microalgae (e.g., *Amphidinium carterae*, *Azadinium spinosum*, *Gambierdiscus* spp., *Karenia brevis* [64–67]). Similarly, NRPSs have a modular organization similar to type I PKSs, and genes potentially encoding NRPSs have been found in different microalgae [68]. Moreover, metabolites that are likely to derive from hybrid NRPS/PKS gene clusters have been reported from *Karenia brevis* [69]. However, to our knowledge, there are no studies reporting the direct correlation of a PKS or NRPS gene from a microalga with the production of a bioactive compound.

Other microalgal enzymes which have been widely studied are those involved in the synthesis of compounds with nutraceutical and cosmeceutical applications, such as those involved in carotenoid synthesis (e.g., astaxanthin, β-carotene, lutein, and canthaxanthin). Carotenoids are isoprenoid pigments, which have many cellular protective effects, such as antioxidant effects occurring via the chemical quenching of O_2 and other reactive oxygen species [70–72]. Their antioxidant properties can potentially protect humans from a compromised immune response, premature aging, arthritis, cardiovascular diseases, and/or certain cancers [72]. Among microalgae, the most studied for the industrial production of carotenoids are the halophile microalga *Dunaliella salina* and the green alga *Haematococcus pluvialis*, which naturally produce high amounts of carotenoids [73]. Moreover, *D.*

salina is a particularly versatile feedstock, and many researchers have focused on obtaining maximum carotenoid yields without impeding its growth [74–76]. In addition, *D. salina* has been successfully transformed via different approaches, such as microparticle bombardment [77] or via *Agrobacterium tumefaciens* [78], increasing the feasibility of its use for biotechnological applications.

The most studied enzymes involved in carotenoid synthesis are: β-carotene oxygenase, lycopene-β-cyclase, phytoene synthase, phytoene desaturase, β-carotene hydroxylase, and zeaxanthin epoxidase [79]. In order to improve the production of carotenoids, different metabolic engineering approaches have been employed. The initial method used was to induce random or site directed mutations in an attempt to improve the activity of enzymes involved in the carotenoid metabolic pathway. Increased production of carotenoids can also be achieved by changing culturing conditions or by employing genetic modifications [79]. For example, mRNA levels of β-carotene oxygenase, involved in the biosynthesis of ketocarotenoids [80], increased in *Chlorella zofengiensis* under combined nitrogen starvation and high-light irradiation, and an increase canthaxanthin, zeaxanthin, and astaxanthin was observed [81]. Couso et al. [82] reported an upregulation in lycopene-β-cyclase, which converts lycopene to β-carotene [83] in *C. reinhardtii* under conditions of high light.

Regarding genetic modifications, Cordero [84] transformed the green microalga *C. reinhardtii* by overexpressing a phytoene synthase (which converts geranylgeranyl pyrophosphate to phytoene) isolated from *Chlorella zofingiensis*, resulting in a 2.0- and 2.2-fold increase in violaxanthin and lutein production, respectively. A phytoene desaturase, which transforms the colorless phytoene into the red-colored lycopene [85], was mutated in *H. pluvialis* by Steinbrenner and Sandmann [86], resulting in the upregulation of the enzyme and an increase in astaxanthin production. Galarza and colleagues expressed a nuclear phytoene desaturase in the plastidial genome of *H. pluvialis*, resulting in a 67% higher astaxanthin accumulation when the strain was grown under stressful conditions [87]. The insertion of a β-carotene hydroxylase from *C. reinhardtii* in *Dunaliella salina* resulted in a 3-fold increase of violaxanthin and a 2-fold increase of zeaxanthin [78]. The inhibition of *D. salina* phytoene desaturase using RNAi technology [88] resulted in an increase in phytoene content, but also a decrease in photosynthetic efficiency and growth rate.

More modern methods which have been used include the use of CRISPR/Cas9 (clustered regularly interspaced short palindromic repeats/CRISPR-associated protein 9) for precise and highly efficient "knock-out" of key genes [89]. For example, Baek et al. have used CRISPR/Cas9 to knock-out the zeaxanthin epoxidase (ZEP) gene in *C. reinhardtii* [90]. This enzyme is involved in the conversion of zeaxantin to violaxantin [91], and with its knock-out they obtained a 47-fold increase in zeaxanthin productivity. The current state-of-art involved in metabolic engineering for carotenoid production in microalgae is further discussed in other reviews [72,92].

Other studies have focused on enzymes involved in the synthesis of oxylipins, which are secondary metabolites that have previously been shown to have antipredator and anticancer activities [93–95]. Although oxylipin chemistry and putative biosynthetic pathways have been extensively studied in both plants and microalgae [96–98], the related enzymes and genes have only recently been identified and characterized in microalgae. Adelfi and coworkers have studied genes involved in the biosynthesis of oxylipins in *Pseudo-nitzchia multistriata* and performed transcriptome analysis on these genes in *Pseudo-nitzchia arenysensis* [99]. In diatoms, they characterized, for the first time, two patatin-like lypolitic acid hydrolases (LAH1) involved in the release of the fatty acid precursors of oxylipins and tested their galactolipase activity in vitro. Transcriptomic analysis also revealed three of seven putative patatin genes (g9879, g2582, and g3354) in *N. oceanica* and demonstrated that they were u-regulated under nitrogen-starvation conditions [100]. Similarly, Lauritano and coworkers analyzed the transcriptome of the green alga *Tetraselmis suecica* and reported three PLAT (Polycystin-1, Lipoxygenase, Alpha-Toxin)/LH2 (Lipoxygenase homology) domain transcripts [68]. The group also performed in silico domain assessment and structure predictions. The enzymes discussed in this section are described in Table 2.

Table 2. Enzymes from Microalgae for Healthcare Applications. Marine, freshwater, and soil strain sources are abbreviated as M, F, or S, respectively. Algal classes of *Chlorophyceae*, *Trebouxiophyceae*, *Bacillariophyceae*, *Dinophyceae*, and *Chlorodendrophyceae*, are abbreviated as CH, TR, BA, DY, and CR respectively.

Reference	Enzymes	Microalgae	Strain Source	Microalgal Class	Main Results
[78]	β-carotene hydroxylase	*Dunaliella salina*	M	CH	Increase in violaxanthin and zeaxanthin production
[81]	β-carotene oxygenase	*Chlorella zofingiensis*	S	TR	Increase in canthaxanthin, zeaxanthin and astaxanthin production under combined nitrogen starvation and high light stress
[56]	L-asparaginase	*Chlamidomonas* spp.	F	CH	Enzyme purified and tested
[57]	L-asparaginase	*Chlorella vulgaris*	F, S	TR	Screening of 40 microalgal isolates searching for new L-asparaginase sources
[82]	lycopene-β-cyclase	*Chlamidomonas reinhardtii*	F	CH	Increased gene expression under high light stress
[99]	lypolitic acid hydrolase 1	*Pseudo-nitzschia multistrata*, *Pseudo-nitzschia arenysensis*	M	BA	Enzyme finding, characterization and retrieval of homologous sequences in other diatoms
[69]	non-ribosomal peptide synthase	*Karenia brevis*	M	DY	Gene cluster identification and chloroplastic localization identification
[68]	polycystin-1, Lipoxygenase, Alpha-Toxin/ lipoxygenase homology 2	*Tetraselmis suecica*	M	CR	Three putative enzyme sequences identification and in silico domain assessment and structure prediction
[88]	phytoene desaturase	*Dunaliella salina*	M	CH	Increase in phytoene production
[84]	phytoene synthase	*Chlamidomonas reinhardtii*	F	CH	Increase in violaxanthin (2.0 fold) and lutein (2.2-fold) production
[86]	phytoene desaturase	*Haematococcus pluvialis*	F	CH	Increase in astaxanthin production
[87]	phytoene desaturase	*Haematococcus pluvialis*	F	CH	Increase in astaxanthin production
[64]	polyketide synthase	*Amphidinium carterae*	M	DY	Identification of a transcript coding for type I PKS β-ketosynthase domain

Table 2. *Cont.*

Reference	Enzymes	Microalgae	Strain Source	Microalgal Class	Main Results
[65]	polyketide synthase	*Azadinium spinosum*	M	DY	Identification of type I PKS domains using a combination of genomic and transcriptomic anayses
[66]	polyketide synthase	*Gamberdiscus polynesiensis, Gamberdiscus excentricus*	M	CH	Identification of transcripts coding for type I and type II PKS domains
[67]	polyketide synthase	*Karenia brevis*	M	DY	Identification of eight transcripts, six of which coding for type I PKS catalytic domains
[90]	zeaxanthin epoxidase	*Chlamydomonas reinhardtii*	F	CH	Increase in zeaxanthin production of 47-fold

2.3. Enzymes for Bioremediation

Bioremediation is the use of microorganisms and their enzymes for the degradation and/or transformation of toxic pollutants into less dangerous metabolites/moieties. The potential, which microalgae possess to proliferate in environments that are rich in nutrients (e.g., eutrophic environments) and to biosequestrate heavy metal ions, makes them ideal candidate organisms for bioremediation strategies [101,102]. The optimal goal in this area is to combine bioremediation activities with the possibility of extracting lipids and other high-value added compounds from the biomass that is produced [103–106] in order to reduce overall costs and to recycle materials. In this section, the focus will be on enzymatic bioremediation, which is a novel approach involving the direct use of purified or partially purified enzymes from microorganisms, and in this case, from microalgae, in order to detoxify a specific toxicant/pollutant [107]. This method has recently started to demonstrate promising results through the use of bacterial enzymes [108,109]. Examples are the use of enzymes for the bioremediation of industrial waste and, in particular, the recent use of chromate reductases found in chromium resistant bacteria, known to detoxify the highly toxic chromium Cr(VI) to the less-toxic Cr(III) [110].

In microalgae, a recent study focused on Cr(VI) reduction involving *C. vulgaris* [111]. This activity was suggested to involve both a biological route, through the putative enzyme chromium reductase, and a nonbiological route: Using the scavenger molecule glutathione (GSH). With respect to chromium removal, several strains of microalgae have been reported to be capable of achieving Cr(IV) removal from water bodies, including *Scenedesmus* and *Chlorella* species [112–114]. In the aforementioned transcriptome study on the green algae *Tetraselmis suecica*, a transcript for a putative nitrilase was reported [68]. Given that nitrilases are enzymes that catalyze the hydrolysis of nitriles to carboxylic acids and ammonia [115] and that this enzyme has recently been used for cyanide bioremediation in wastewaters [116], this nitrilase in *T. suecica* may prove to be useful in the treatment of cyanide contaminated water bodies.

Other enzymes have been reported to be overexpressed in microalgae when they are exposed to contaminants, but it is not clear whether or not they are directly involved in their degradation or whether they are produced as a stress defensive response in the cell in order to help balance cellular homeostasis (e.g., to detoxify reactive oxygen/nitrogen species produced after exposure to contaminants). Examples of these enzymes include peroxidases (Px), superoxide dismutase (SOD), catalase (CAT), and glutathione reductase (GR). SOD, Px, and CAT typically function in helping detoxify the cell from oxygen reactive species [117,118], while GR replenishes bioavailable glutathione, catalyzing the reduction of glutathione disulfide (GSSG) to the sulfhydryl form (GSH) [119]. Regarding the detoxification of reactive nitrogen species, the most studied enzymes in microalgae are the nitrate and nitrite reductases. The first enzyme reduces nitrate (NO_3^-) to nitrite (NO_2^-), while the second subsequently reduces nitrite to ammonia (NH_4^+). NH_4^+ is then assimilated into amino acids via the glutamine synthetase/glutamine-2-oxoglutarate amino-transferase cycle [120] (Figure 5).

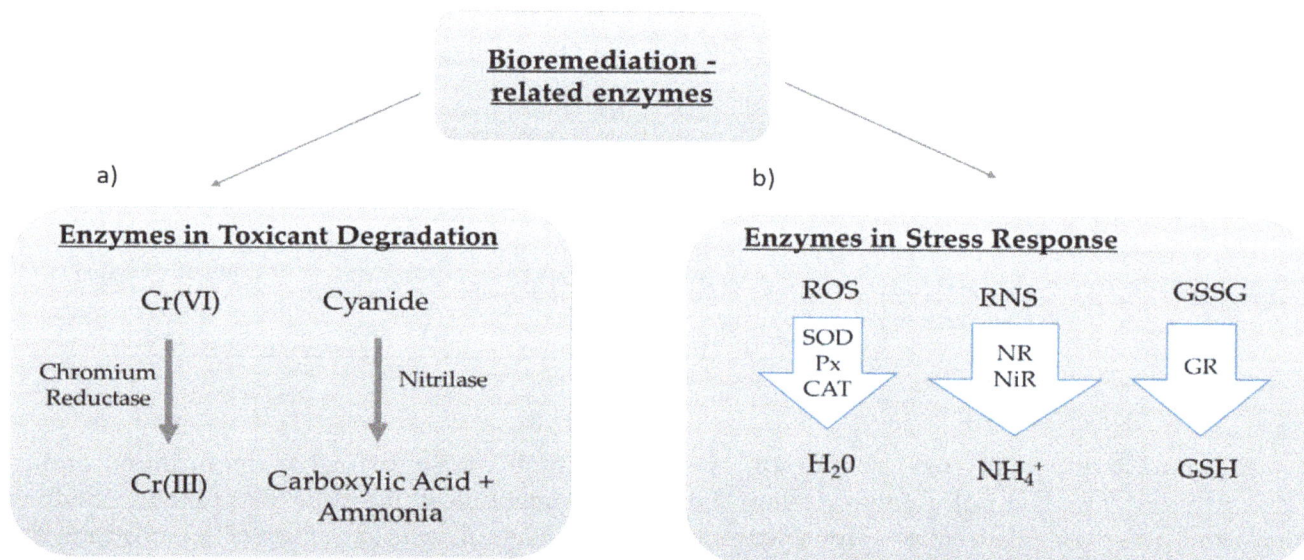

Figure 5. Enzymes for Bioremediation. Enzymes for Bioremediation can be: (**a**) Enzymes directly used for the degradation of toxicant compounds to less or non toxic versions (e.g., the hexavalent Chromium is converted to the less toxic trivalent Chromium due to the activity of Chromium Reductase); (**b**) Enzymes involved in cellular stress response mechanisms, such as peroxidases (Px), superoxide dismutase (SOD), and catalase (CAT) that detoxify reactive oxygen species (ROS), nitrate reductase (NR), and nitrite reductase (NiR) that detoxify reactive nitrogen species (RNS) in ammonium, and GR, that catalyzes the reduction of glutathione disulfide (GSSG) to glutathione (GSH).

For example, peroxidase activity has been reported in extracts from the green alga *Selenastrum capricornutum* (now named *Raphidocelis subcapitata* [121]), which was highly sensitive to very small concentrations of copper (Cu) (0.1 mM), and the authors proposed that the enzyme could be employed as a sensitive bioindicator of copper contamination in fresh waters [122]. Levels of Px, SOD, CAT, and GR have been reported to be upregulated following Cu contamination in *P. tricornutum* and following lead (Pb) contamination in two lichenic microalgal strains from the *Trebouxia* genus (prov. names, TR1 and TR9) [123,124]. In Morelli's work, an increase of 200% in CAT activity indicated its important role in Cu detoxification. In contrast, Alvarez and coworkers reported that Px, SOD, CAT, and GR activity was higher in TR1 than in TR9 under control conditions (with the exception of CAT), while prolonged exposure to Pb resulted in the enzymatic activities of the two microalgae changing to similar levels, reflecting the different physiological and anatomical adaptations of the two organisms. TR1 possesses a thinner cell wall, thereby requiring it to have a more efficient basal enzymatic defence system, while TR9 has a thicker cell wall and induces the expression of intracellular defense mechanisms when the contaminant concentrations are high and physical barriers are no longer effective. Further studies will be required to assess whether these TR1 enzymes are more efficient than enzymes from other microalgal sources and the potential applications that these enzymes may have. All of the enzymes discussed in this section are reported in Table 3.

Table 3. Enzymes from Microalgae with utility in Bioremediation applications Marine, freshwater, and lichenic strain sources are abbreviated as M, F, and L respectively. Algal classes of *Trebouxiophyceae*, *Chlorodendrophyceae*, *Chlorophyceae*, and *Bacillariophyceae* are abbreviated as TR, CR, CH, and BA, respectively.

Reference	Enzymes	Microalgae	Strain Source	Microalgal Class	Main Results
[111]	Putative Cr Reductase	*Chlorella vulgaris*	F	TR	Enzymatic Cr conversion (from Cr(VI) to Cr(III)) detected
[68]	Nitrilase	*Tetraselmis suecica*	M	CR	Putative enzyme sequence identification
[122]	Putative ascorbate peroxidase	*Selenastrum capricornutum*	F	CH	High sensitivity to Cu concentration activity
[123]	superoxide-dismutase, catalase, glutathione reductase	*Phaeodactylum tricornutum*	M	BA	Higher detected enzymatic activity after Cu accumulation
[124]	superoxide-dismutase, catalase, glutathione reductase, ascorbate peroxidase	*Trebouxia 1 (TR1), Trebouxia 9 (TR9)*	L	TR	Constitutive higher enzymatic activity detected in TR1, while exposed to Pb brings TR1 and TR9 enzymatic activities to comparable levels

3. Conclusions and Future Perspectives

Among aquatic organisms that have recently received attention as potential sources of industrially relevant enzymes [125,126], microalgae, in particular, stand out as a new sustainable and ecofriendly source of biological products (e.g., lipids, carotenoids, oxylipins, and polyketides). This review summarized the available information on enzymes from microalgae with possible biotechnological applications, with a particular focus on value-added lipid production, together with healthcare and bioremediation applications.

The promise of microalgae as potential sources of novel enzymes of interest is reflected in the abundance of recent reports in the literature in this area. However, the biotechnological exploitation of their enzymes in comparison to other potential sources has only become more feasible quite recently, primarily due to the implementation of novel isolation and culturing procedures, together with an increase in the availability of -omics data. This data has facilitated the use of a broader array of approaches, such as site-specific mutagenesis, bioinformatics-based searches for genes of interest, and/or the use of genome editing tools (e.g., CRISPR/Cas9 and TILLING), resulting in promising results particularly with respect to high-performance lipid [46] and carotenoid [89] production in different microalgae.

The majority of studies to date have focused on enzymes involved in pathways for lipid synthesis in order to increase their total production or to direct cellular production to lipid classes with applications as nutraceuticals, cosmeceuticals, or as a feedstock for biodiesel production. For this reason, several recent studies have focused on the improvement of lipid production in oleaginous microalgae. In addition, algal biomass is often used for the extraction of both lipids and other value-added products, such as pigments and proteins, in order to maximize the production of useful products such as these at the lowest possible cost [127–129].

Future approaches to maximize the enzymatic potential of microalgae are likely to focus on three different approaches: (1) The use of ever-increasing amounts of available -omics data to optimize microalgal strains for the production of valuable products, through the overexpression of one or more enzymes through the use of genome editing tools; (2) identification and subsequent characterization of metabolic pathways involved in the production of specific bioactives (e.g., polyketides), many of which are still poorly characterized; (3) the search for genes with direct biotechnological applications (e.g., L-asparaginase, chromate reductase, nitrilase) in microalgal genomes and transcriptomes datasets. A common element in all three approaches is the potential use of next generation sequencing based approaches (NGS) [130], the price of which is declining rapidly [131].

The feasibility of employing any of the aforementioned three approaches will be directly influenced by progress in methods to decrease the costs of growth and genetic manipulation of microalgae. The ultimate aim would be to mimic what has happened in the area of bacterial enzymology, where robust pipelines for enzyme discovery have been established. If this could be achieved, then it is clear that microalgae are likely to meet our expectations as a promising source of novel enzymes with utility in a variety of different biotechnological applications.

Author Contributions: G.M.V., P.D.L., A.I., A.D.W.D. and C.L. co-wrote the review.

Acknowledgments: Authors thank Servier Medical Art (SMART) website (https://smart.servier.com/) by Servier for the elements of Figure 3. SMART is licensed under a Creative Commons Attribution 3.0 Unported License.

References

1. Schneider, S.H.; Root, T.L.; Mastrandrea, M.D. *Encyclopedia of climate and weather*; Oxford University Press: Oxford, UK, 2011.
2. Bagchi, B. *Water in Biological and Chemical Processes*; Cambridge University Press: Cambridge, UK, 2013.
3. Faber, K. *Biotransformations in Organic Chemistry*; Springer: Heidelberg, Germany, 2011.
4. Cao, Y.; Tan, H. Effects of cellulase on the modification of cellulose. *Carbohydr. Res.* **2002**, *337*, 1291–1296. [CrossRef]
5. Nigam, P.S. Microbial enzymes with special characteristics for biotechnological applications. *Biomolecules* **2013**, *3*, 597–611. [CrossRef] [PubMed]
6. Fernandes, P. Enzymes in food processing: a condensed overview on strategies for better biocatalysts. *Enzyme Res.* **2010**, *2010*, 1–19. [CrossRef] [PubMed]
7. Vellard, M. The enzyme as drug: application of enzymes as pharmaceuticals. *Curr. Opin. Biotechnol.* **2003**, *14*, 444–450. [CrossRef]
8. Andersen, R.A. *Algal culturing techniques*; Academic Press: Cambridge, MA, USA, 2005.
9. Khan, M.I.; Shin, J.H.; Kim, J.D. The promising future of microalgae: current status, challenges, and optimization of a sustainable and renewable industry for biofuels, feed, and other products. *Microb. Cell Fact.* **2018**, *17*, 36. [CrossRef] [PubMed]
10. Martínez Andrade, K.; Lauritano, C.; Romano, G.; Ianora, A. Marine Microalgae with Anti-Cancer Properties. *Mar. Drugs* **2018**, *16*, 165. [CrossRef]
11. Bhalamurugan, G.L.; Valerie, O.; Mark, L. Valuable bioproducts obtained from microalgal biomass and their commercial applications: A review. *Environ. Eng. Res.* **2018**, *23*, 229–241. [CrossRef]
12. Doron, L.; Segal, N.; Shapira, M. Transgene Expression in Microalgae-From Tools to Applications. *Front. Plant Sci.* **2016**, *7*, 505. [CrossRef]
13. Brasil, B.d.S.A.F.; de Siqueira, F.G.; Salum, T.F.C.; Zanette, C.M.; Spier, M.R. Microalgae and cyanobacteria as enzyme biofactories. *Algal Res.* **2017**, *25*, 76–89. [CrossRef]
14. Lauritano, C.; Ianora, A. Grand Challenges in Marine Biotechnology: Overview of Recent EU-Funded Projects. In *Grand Challenges in Marine Biotechnology*; Rampellotto, P.H., Trincone, A., Eds.; Springer: Heidelberg, Germany, 2018; pp. 425–449.
15. Mogharabi, M.; Faramarzi, M.A. Are Algae the Future Source of Enzymes? *Trends Pept. Protein Sci.* **2016**, *1*, 1–6.
16. Chisti, Y. Biodiesel from microalgae. *Biotechnol. Adv.* **2007**, *25*, 294–306. [CrossRef]
17. Sanghvi, A.M.; Martin Lo, Y. Present and Potential Industrial Applications of Macro- and Microalgae. *Recent Patents Food, Nutr. Agric.* **2010**, *2*, 187–194.
18. Bellou, S.; Baeshen, M.N.; Elazzazy, A.M.; Aggeli, D.; Sayegh, F.; Aggelis, G. Microalgal lipids biochemistry and biotechnological perspectives. *Biotechnol. Adv.* **2014**, *32*, 1476–1493. [CrossRef]

19. Moriyama, T.; Toyoshima, M.; Saito, M.; Wada, H.; Sato, N. Revisiting the Algal "Chloroplast Lipid Droplet": The Absence of an Entity That Is Unlikely to Exist. *Plant Physiol.* **2018**, *176*, 1519–1530. [CrossRef]

20. Fukuda, H.; Kondo, A.; Noda, H. Biodiesel fuel production by transesterification of oils. *J. Biosci. Bioeng.* **2001**, *92*, 405–416. [CrossRef]

21. Wells, M.L.; Potin, P.; Craigie, J.S.; Raven, J.A.; Merchant, S.S.; Helliwell, K.E.; Smith, A.G.; Camire, M.E.; Brawley, S.H. Algae as nutritional and functional food sources: revisiting our understanding. *J. Appl. Phycol.* **2017**, *29*, 949–982. [CrossRef]

22. Caporgno, M.P.; Mathys, A. Trends in microalgae incorporation into innovative food products with potential health benefits. *Front. Nutr.* **2018**, *5*, 1–10. [CrossRef]

23. Merchant, S.S.; Kropat, J.; Liu, B.; Shaw, J.; Warakanont, J. TAG, You're it! *Chlamydomonas* as a reference organism for understanding algal triacylglycerol accumulation. *Curr. Opin. Biotechnol.* **2012**, *23*, 352–363. [CrossRef]

24. Xu, Y.; Caldo, K.M.P.; Pal-Nath, D.; Ozga, J.; Lemieux, M.J.; Weselake, R.J.; Chen, G. Properties and Biotechnological Applications of Acyl-CoA:diacylglycerol Acyltransferase and Phospholipid:diacylglycerol Acyltransferase from Terrestrial Plants and Microalgae. *Lipids* **2018**, *53*, 663–688. [CrossRef]

25. Lung, S.-C.; Weselake, R.J. Diacylglycerol acyltransferase: a key mediator of plant triacylglycerol synthesis. *Lipids* **2006**, *41*, 1073–1088. [CrossRef]

26. Shockey, J.M.; Gidda, S.K.; Chapital, D.C.; Kuan, J.-C.; Dhanoa, P.K.; Bland, J.M.; Rothstein, S.J.; Mullen, R.T.; Dyer, J.M. Tung tree DGAT1 and DGAT2 have nonredundant functions in triacylglycerol biosynthesis and are localized to different subdomains of the endoplasmic reticulum. *Plant Cell* **2006**, *18*, 2294–2313. [CrossRef]

27. Guo, X.; Fan, C.; Chen, Y.; Wang, J.; Yin, W.; Wang, R.R.C.; Hu, Z. Identification and characterization of an efficient acyl-CoA: Diacylglycerol acyltransferase 1 (DGAT1) gene from the microalga *Chlorella ellipsoidea*. *BMC Plant Biol.* **2017**, *17*, 1–16. [CrossRef]

28. Wei, H.; Shi, Y.; Ma, X.; Pan, Y.; Hu, H.; Li, Y.; Luo, M.; Gerken, H.; Liu, J. A type-I diacylglycerol acyltransferase modulates triacylglycerol biosynthesis and fatty acid composition in the oleaginous microalga, *Nannochloropsis oceanica*. *Biotechnol. Biofuels* **2017**, *10*, 1–18. [CrossRef]

29. Wagner, M.; Hoppe, K.; Czabany, T.; Heilmann, M.; Daum, G.; Feussner, I.; Fulda, M. Identification and characterization of an acyl-CoA:diacylglycerol acyltransferase 2 (DGAT2) gene from the microalga *Ostreococcus tauri*. *Plant Physiol. Biochem.* **2010**, *48*, 407–416. [CrossRef]

30. Niu, Y.-F.; Zhang, M.-H.; Li, D.-W.; Yang, W.-D.; Liu, J.-S.; Bai, W.-B.; Li, H.-Y. Improvement of Neutral Lipid and Polyunsaturated Fatty Acid Biosynthesis by Overexpressing a Type 2 Diacylglycerol Acyltransferase in Marine Diatom *Phaeodactylum tricornutum*. *Mar. Drugs* **2013**, *11*, 4558–4569. [CrossRef]

31. Manandhar-Shrestha, K.; Hildebrand, M. Characterization and manipulation of a DGAT2 from the diatom *Thalassiosira pseudonana*: Improved TAG accumulation without detriment to growth, and implications for chloroplast TAG accumulation. *Algal Res.* **2015**, *12*, 239–248. [CrossRef]

32. Klaitong, P.; Fa-aroonsawat, S.; Chungjatupornchai, W. Accelerated triacylglycerol production and altered fatty acid composition in oleaginous microalga *Neochloris oleoabundans* by overexpression of diacylglycerol acyltransferase 2. *Microb. Cell Fact.* **2017**, *16*, 1–10. [CrossRef]

33. Li, D.-W.; Cen, S.-Y.; Liu, Y.-H.; Balamurugan, S.; Zheng, X.-Y.; Alimujiang, A.; Yang, W.-D.; Liu, J.-S.; Li, H.-Y. A type 2 diacylglycerol acyltransferase accelerates the triacylglycerol biosynthesis in heterokont oleaginous microalga *Nannochloropsis oceanica*. *J. Biotechnol.* **2016**, *229*, 65–71. [CrossRef]

34. Xin, Y.; Lu, Y.; Lee, Y.-Y.; Wei, L.; Jia, J.; Wang, Q.; Wang, D.; Bai, F.; Hu, H.; Hu, Q.; et al. Producing Designer Oils in Industrial Microalgae by Rational Modulation of Co-evolving Type-2 Diacylglycerol Acyltransferases. *Mol. Plant* **2017**, *10*, 1523–1539. [CrossRef]

35. Deng, X.-D.; Gu, B.; Li, Y.-J.; Hu, X.-W.; Guo, J.-C.; Fei, X.-W. The roles of acyl-CoA: diacylglycerol acyltransferase 2 genes in the biosynthesis of triacylglycerols by the green algae *Chlamydomonas reinhardtii*. *Mol. Plant* **2012**, *5*, 945–947. [CrossRef]

36. La Russa, M.; Bogen, C.; Uhmeyer, A.; Doebbe, A.; Filippone, E.; Kruse, O.; Mussgnug, J.H. Functional analysis of three type-2 DGAT homologue genes for triacylglycerol production in the green microalga *Chlamydomonas reinhardtii*. *J. Biotechnol.* **2012**, *162*, 13–20. [CrossRef]

37. Cui, Y.; Zhao, J.; Wang, Y.; Qin, S.; Lu, Y. Characterization and engineering of a dual-function diacylglycerol acyltransferase in the oleaginous marine diatom *Phaeodactylum tricornutum*. *Biotechnol. Biofuels* **2018**, *11*, 32. [CrossRef]

38. Xue, J.; Balamurugan, S.; Li, D.-W.; Liu, Y.-H.; Zeng, H.; Wang, L.; Yang, W.-D.; Liu, J.-S.; Li, H.-Y. Glucose-6-phosphate dehydrogenase as a target for highly efficient fatty acid biosynthesis in microalgae by enhancing NADPH supply. *Metab. Eng.* **2017**, *41*, 212–221. [CrossRef]

39. Zhu, B.-H.; Tu, C.-C.; Shi, H.-P.; Yang, G.-P.; Pan, K.-H. Overexpression of endogenous delta-6 fatty acid desaturase gene enhances eicosapentaenoic acid accumulation in *Phaeodactylum tricornutum*. *Process Biochem.* **2017**, *57*, 43–49. [CrossRef]

40. Fukuda, S.; Hirasawa, E.; Takemura, T.; Takahashi, S.; Chokshi, K.; Pancha, I.; Tanaka, K.; Imamura, S. Accelerated triacylglycerol production without growth inhibition by overexpression of a glycerol-3-phosphate acyltransferase in the unicellular red alga *Cyanidioschyzon merolae*. *Sci. Rep.* **2018**, *8*, 1–12. [CrossRef]

41. Rengel, R.; Smith, R.T.; Haslam, R.P.; Sayanova, O.; Vila, M.; León, R. Overexpression of acetyl-CoA synthetase (ACS) enhances the biosynthesis of neutral lipids and starch in the green microalga *Chlamydomonas reinhardtii*. *Algal Res.* **2018**, *31*, 183–193. [CrossRef]

42. Osada, K.; Maeda, Y.; Yoshino, T.; Nojima, D.; Bowler, C.; Tanaka, T. Enhanced NADPH production in the pentose phosphate pathway accelerates lipid accumulation in the oleaginous diatom *Fistulifera solaris*. *Algal Res.* **2017**, *23*, 126–134. [CrossRef]

43. Shin, Y.S.; Jeong, J.; Nguyen, T.H.T.; Kim, J.Y.H.; Jin, E.; Sim, S.J. Targeted knockout of phospholipase A2 to increase lipid productivity in *Chlamydomonas reinhardtii* for biodiesel production. *Bioresour. Technol.* **2019**, *271*, 368–374. [CrossRef]

44. Los, D.A.; Murata, N. Structure and expression of fatty acid desaturases. *Biochim. Biophys. Acta Lipids Lipid Metab.* **1998**, *1394*, 3–15. [CrossRef]

45. De Jaeger, L.; Springer, J.; Wolbert, E.J.H.; Martens, D.E.; Eggink, G.; Wijffels, R.H. Gene silencing of stearoyl-ACP desaturase enhances the stearic acid content in *Chlamydomonas reinhardtii*. *Bioresour. Technol.* **2017**, *245*, 1616–1626. [CrossRef]

46. Daboussi, F.; Leduc, S.; Maréchal, A.; Dubois, G.; Guyot, V.; Perez-Michaut, C.; Amato, A.; Falciatore, A.; Juillerat, A.; Beurdeley, M.; et al. Genome engineering empowers the diatom *Phaeodactylum tricornutum* for biotechnology. *Nat. Commun.* **2014**, *5*, 3831. [CrossRef]

47. Sorigué, D.; Légeret, B.; Cuiné, S.; Blangy, S.; Moulin, S.; Billon, E.; Richaud, P.; Brugière, S.; Couté, Y.; Nurizzo, D.; et al. An algal photoenzyme converts fatty acids to hydrocarbons. *Science* **2017**, *357*, 903–907. [CrossRef]

48. Misra, N.; Panda, P.K.; Parida, B.K.; Mishra, B.K. dEMBF: A comprehensive database of enzymes of microalgal biofuel feedstock. *PLoS One* **2016**, *11*, 146–158. [CrossRef]

49. Batool, T.; Makky, E.A.; Jalal, M.; Yusoff, M.M. A comprehensive review on L-asparaginase and its applications. *Appl. Biochem. Biotechnol.* **2016**, *178*, 900–923. [CrossRef]

50. Ali, U.; Naveed, M.; Ullah, A.; Ali, K.; Shah, S.A.; Fahad, S.; Mumtaz, A.S. L-asparaginase as a critical component to combat Acute Lymphoblastic Leukemia (ALL): A novel approach to target ALL. *Eur. J. Pharmacol.* **2016**, *771*, 199–210. [CrossRef]

51. Roberts, J.; Prager, M.D.; Bachynsky, N. The antitumor activity of *Escherichia coli* L-asparaginase. *Cancer Res.* **1966**, *26*, 2213–2217.

52. Peterson, R.E.; Ciegler, A. L-asparaginase production by various bacteria. *Appl. Microbiol.* **1969**, *17*, 929–930.

53. Thenmozhi, C.; Sankar, R.; Karuppiah, V.; Sampathkumar, P. L-asparaginase production by mangrove derived *Bacillus cereus* MAB5: Optimization by response surface methodology. *Asian Pac. J. Trop. Med.* **2011**, *4*, 486–491. [CrossRef]

54. Ahmad, N.; Pandit, N.; Maheshwari, S. L-asparaginase gene-a therapeutic approach towards drugs for cancer cell. *Int. J. Biosci.* **2012**, *2*, 1–11.

55. Vidya, J.; Sajitha, S.; Ushasree, V.; Sindhu, R.; Binod, P.; Madhavan, A.; Pandey, A. Genetic and metabolic engineering approaches for the production and delivery of L-asparaginases: An overview. *Bioresour. Technol.* **2017**, *245*, 1775–1781. [CrossRef]

56. Paul, J.H. Isolation and characterization of a *Chlamydomonas* L-asparaginase. *Biochem. J.* **1982**, *203*, 109–115. [CrossRef]

57. Ebrahiminezhad, A.; Rasoul-Amini, S.; Ghoshoon, M.B.; Ghasemi, Y. *Chlorella vulgaris*, a novel microalgal source for L-asparaginase production. *Biocatal. Agric. Biotechnol.* **2014**, *3*, 214–217. [CrossRef]

58. Sasso, S.; Pohnert, G.; Lohr, M.; Mittag, M.; Hertweck, C. Microalgae in the postgenomic era: a blooming reservoir for new natural products. *FEMS Microbiol. Rev.* **2012**, *36*, 761–785. [CrossRef]

59. Berry, J. Marine and freshwater microalgae as a potential source of novel herbicides. In *Herbicides and Environment, Kortekamp A.*; InTechOpen: London, UK, 2011.

60. Kobayashi, J. Amphidinolides and its related macrolides from marine dinoflagellates. *J. Antibiot. (Tokyo)* **2008**, *61*, 271–284. [CrossRef]

61. Kellmann, R.; Stüken, A.; Orr, R.J.S.; Svendsen, H.M.; Jakobsen, K.S. Biosynthesis and molecular genetics of polyketides in marine dinoflagellates. *Mar. Drugs* **2010**, *8*, 1011–1048. [CrossRef]

62. Kohli, G.S.; John, U.; Van Dolah, F.M.; Murray, S.A. Evolutionary distinctiveness of fatty acid and polyketide synthesis in eukaryotes. *ISME J.* **2016**, *10*, 1877–1890. [CrossRef]

63. Jenke-Kodama, H.; Sandmann, A.; Müller, R.; Dittmann, E. Evolutionary implications of bacterial polyketide synthases. *Mol. Biol. Evol.* **2005**, *22*, 2027–2039. [CrossRef]

64. Lauritano, C.; De Luca, D.; Ferrarini, A.; Avanzato, C.; Minio, A.; Esposito, F.; Ianora, A. De novo transcriptome of the cosmopolitan dinoflagellate *Amphidinium carterae* to identify enzymes with biotechnological potential. *Sci. Rep.* **2017**, *7*, 11701. [CrossRef]

65. Meyer, J.M.; Rödelsperger, C.; Eichholz, K.; Tillmann, U.; Cembella, A.; McGaughran, A.; John, U. Transcriptomic characterisation and genomic glimps into the toxigenic dinoflagellate *Azadinium spinosum*, with emphasis on polykeitde synthase genes. *BMC Genomics* **2015**, *16*, 27. [CrossRef]

66. Kohli, G.S.; Campbell, K.; John, U.; Smith, K.F.; Fraga, S.; Rhodes, L.L.; Murray, S.A. Role of modular polyketide synthases in the production of polyether ladder compounds in ciguatoxin-producing *Gambierdiscus polynesiensis* and *G. excentricus* (Dinophyceae). *J. Eukaryot. Microbiol.* **2017**, *64*, 691–706. [CrossRef]

67. Monroe, E.A.; Van Dolah, F.M. The toxic dinoflagellate *Karenia brevis* encodes novel Type I-like polyketide synthases containing discrete catalytic domains. *Protist* **2008**, *159*, 471–482. [CrossRef]

68. Lauritano, C.; De Luca, D.; Amoroso, M.; Benfatto, S.; Maestri, S.; Racioppi, C.; Esposito, F.; Lanora, A. New molecular insights on the response of the green alga *Tetraselmis suecica* to nitrogen starvation. *Sci. Rep.* **2019**, *9*, 3336. [CrossRef]

69. López-Legentil, S.; Song, B.; DeTure, M.; Baden, D.G. Characterization and localization of a hybrid non-ribosomal peptide synthetase and polyketide synthase gene from the toxic dinoflagellate *Karenia brevis*. *Mar. Biotechnol.* **2010**, *12*, 32–41. [CrossRef]

70. Fiedor, J.; Burda, K. Potential role of carotenoids as antioxidants in human health and disease. *Nutrients* **2014**, *6*, 466–488. [CrossRef]

71. Musser, A.J.; Maiuri, M.; Brida, D.; Cerullo, G.; Friend, R.H.; Clark, J. The nature of singlet exciton fission in carotenoid aggregates. *J. Am. Chem. Soc.* **2015**, *137*, 5130–5139. [CrossRef]

72. Gong, M.; Bassi, A. Carotenoids from microalgae: A review of recent developments. *Biotechnol. Adv.* **2016**, *34*, 1396–1412. [CrossRef]

73. Rammuni, M.N.; Ariyadasa, T.U.; Nimarshana, P.H.V.; Attalage, R.A. Comparative assessment on the extraction of carotenoids from microalgal sources: Astaxanthin from *H. pluvialis* and β-carotene from *D. salina*. *Food Chem.* **2019**, *277*, 128–134. [CrossRef]

74. Lamers, P.P.; Janssen, M.; De Vos, R.C.H.; Bino, R.J.; Wijffels, R.H. Exploring and exploiting carotenoid accumulation in *Dunaliella salina* for cell-factory applications. *Trends Biotechnol.* **2008**, *26*, 631–638. [CrossRef]

75. Prieto, A.; Pedro Cañavate, J.; García-González, M. Assessment of carotenoid production by *Dunaliella salina* in different culture systems and operation regimes. *J. Biotechnol.* **2011**, *151*, 180–185. [CrossRef]

76. Besson, A.; Formosa-Dague, C.; Guiraud, P. Flocculation-flotation harvesting mechanism of *Dunaliella salina*: From nanoscale interpretation to industrial optimization. *Water Res.* **2019**, *155*, 352–361. [CrossRef]

77. Tan, C.; Qin, S.; Zhang, Q.; Jiang, P.; Zhao, F. Establishment of a micro-particle bombardment transformation system for *Dunaliella salina*. *J. Microbiol.* **2005**, *43*, 361–365.

78. Simon, D.P.; Anila, N.; Gayathri, K.; Sarada, R. Heterologous expression of β-carotene hydroxylase in *Dunaliella salina* by *Agrobacterium*-mediated genetic transformation. *Algal Res.* **2016**, *18*, 257–265. [CrossRef]

79. Saini, D.K.; Chakdar, H.; Pabbi, S.; Shukla, P. Enhancing production of microalgal biopigments through metabolic and genetic engineering. *Crit. Rev. Food Sci. Nutr.* **2019**, 1–15. [CrossRef]

80. Huang, J.C.; Wang, Y.; Sandmann, G.; Chen, F. Isolation and characterization of a carotenoid oxygenase gene from *Chlorella zofingiensis* (Chlorophyta). *Appl. Microbiol. Biotechnol.* **2006**, *71*, 473–479. [CrossRef]

81. Cordero, B.F.; Couso, I.; Leon, R.; Rodriguez, H.; Vargas, M.A. Isolation and characterization of a lycopene ε-cyclase gene of *Chlorella* (*Chromochloris*) *zofingiensis*. Regulation of the carotenogenic pathway by nitrogen and light. *Mar. Drugs* **2012**, *10*, 2069–2088. [CrossRef]

82. Couso, I.; Vila, M.; Vigara, J.; Cordero, B.F.; Vargas, M.Á.; Rodríguez, H.; León, R. Synthesis of carotenoids and regulation of the carotenoid biosynthesis pathway in response to high light stress in the unicellular microalga *Chlamydomonas reinhardtii*. *Eur. J. Phycol.* **2012**, *47*, 223–232. [CrossRef]

83. Cunningham, F.X.; Pogson, B.; Sun, Z.; McDonald, K.A.; DellaPenna, D.; Gantt, E.; Gantt, E. Functional analysis of the beta and epsilon lycopene cyclase enzymes of *Arabidopsis* reveals a mechanism for control of cyclic carotenoid formation. *Plant Cell* **1996**, *8*, 1613–1626. [CrossRef]

84. Cordero, B.F.; Couso, I.; León, R.; Rodríguez, H.; Vargas, M.Á. Enhancement of carotenoids biosynthesis in *Chlamydomonas reinhardtii* by nuclear transformation using a phytoene synthase gene isolated from *Chlorella zofingiensis*. *Appl. Microbiol. Biotechnol.* **2011**, *91*, 341–351. [CrossRef]

85. Fraser, P.D.; Misawa, N.; Linden, H.; Yamano, S.; Kobayashi, K.; Sandmann, G. Expression in *Escherichia coli*, purification, and reactivation of the recombinant *Erwinia uredovora* phytoene desaturase. *J. Biol. Chem.* **1992**, *267*, 19891–19895.

86. Steinbrenner, J.; Sandmann, G. Transformation of the green alga *Haematococcus pluvialis* with a phytoene desaturase for accelerated astaxanthin biosynthesis. *Appl. Environ. Microbiol.* **2006**, *72*, 7477–7484. [CrossRef]

87. Galarza, J.I.; Gimpel, J.A.; Rojas, V.; Arredondo-Vega, B.O.; Henríquez, V. Over-accumulation of astaxanthin in *Haematococcus pluvialis* through chloroplast genetic engineering. *Algal Res.* **2018**, *31*, 291–297. [CrossRef]

88. Srinivasan, R.; Babu, S.; Gothandam, K.M. Accumulation of phytoene, a colorless carotenoid by inhibition of phytoene desaturase (PDS) gene in *Dunaliella salina* V-101. *Bioresour. Technol.* **2017**, *242*, 311–318. [CrossRef]

89. Cong, L.; Ran, F.A.; Cox, D.; Lin, S.; Barretto, R.; Habib, N.; Hsu, P.D.; Wu, X.; Jiang, W.; Marraffini, L.A.; et al. Multiplex genome engineering using CRISPR/Cas systems. *Science.* **2013**, *339*, 819–823. [CrossRef]

90. Baek, K.; Yu, J.; Jeong, J.; Sim, S.J.; Bae, S.; Jin, E. Photoautotrophic production of macular pigment in a *Chlamydomonas reinhardtii* strain generated by using DNA-free CRISPR-Cas9 RNP-mediated mutagenesis. *Biotechnol. Bioeng.* **2018**, *115*, 719–728. [CrossRef]

91. Frommolt, R.; Goss, R.; Wilhelm, C. The de-epoxidase and epoxidase reactions of *Mantoniella squamata* (Prasinophyceae) exhibit different substrate-specific reaction kinetics compared to spinach. *Planta* **2001**, *213*, 446–456. [CrossRef]

92. Gimpel, J.A.; Henríquez, V.; Mayfield, S.P. In Metabolic Engineering of Eukaryotic Microalgae: Potential and Challenges Come with Great Diversity. *Front. Microbiol.* **2015**, *6*, 1376. [CrossRef]

93. De los Reyes, C.; Ávila-Román, J.; Ortega, M.J.; de la Jara, A.; García-Mauriño, S.; Motilva, V.; Zubía, E. Oxylipins from the microalgae *Chlamydomonas debaryana* and *Nannochloropsis gaditana* and their activity as TNF-α inhibitors. *Phytochemistry* **2014**, *102*, 152–161. [CrossRef]

94. Lauritano, C.; Romano, G.; Roncalli, V.; Amoresano, A.; Fontanarosa, C.; Bastianini, M.; Braga, F.; Carotenuto, Y.; Ianora, A. New oxylipins produced at the end of a diatom bloom and their effects on copepod reproductive success and gene expression levels. *Harmful Algae* **2016**, *55*, 221–229. [CrossRef]

95. Ávila-Román, J.; Talero, E.; de los Reyes, C.; García-Mauriño, S.; Motilva, V. Microalgae-derived oxylipins decrease inflammatory mediators by regulating the subcellular location of NFκB and PPAR-γ. *Pharmacol. Res.* **2018**, *128*, 220–230. [CrossRef]

96. Pohnert, G. Phospholipase A2 activity triggers the wound-activated chemical defense in the diatom *Thalassiosira rotula*. *Plant Physiol.* **2002**, *129*, 103–111. [CrossRef]

97. Matos, A.R.; Pham-Thi, A.T. Lipid deacylating enzymes in plants: Old activities, new genes. *Plant Physiol. Biochem.* **2009**, *47*, 491–503. [CrossRef]

98. Cutignano, A.; Lamari, N.; D'ippolito, G.; Manzo, E.; Cimino, G.; Fontana, A. Lipoxygenase products in marine diatoms: A concise analytical method to explore the functional potential of oxylipins. *J. Phycol.* **2011**, *47*, 233–243. [CrossRef]

99. Adelfi, M.G.; Vitale, R.M.; d'Ippolito, G.; Nuzzo, G.; Gallo, C.; Amodeo, P.; Manzo, E.; Pagano, D.; Landi, S.; Picariello, G.; et al. Patatin-like lipolytic acyl hydrolases and galactolipid metabolism in marine diatoms of the genus *Pseudo-nitzschia*. *Biochim. Biophys. Acta Mol. Cell Biol. Lipids* **2019**, *1864*, 181–190. [CrossRef]

100. Li, J.; Han, D.; Wang, D.; Ning, K.; Jia, J.; Wei, L.; Jing, X.; Huang, S.; Chen, J.; Li, Y.; et al. Choreography of Transcriptomes and Lipidomes of *Nannochloropsis* Reveals the Mechanisms of Oil Synthesis in Microalgae. *Plant Cell* **2014**, *26*, 1645–1665. [CrossRef]

101. De la Noüe, J.; Laliberté, G.; Proulx, D. Algae and waste water. *J. Appl. Phycol.* **1992**, *4*, 247–254. [CrossRef]

102. Mathimani, T.; Pugazhendhi, A. Utilization of algae for biofuel, bio-products and bio-remediation. *Biocatal. Agric. Biotechnol.* **2019**, *17*, 326–330. [CrossRef]

103. Kuo, C.-M.; Chen, T.-Y.; Lin, T.-H.; Kao, C.-Y.; Lai, J.-T.; Chang, J.-S.; Lin, C.-S. Cultivation of *Chlorella* sp. GD using piggery wastewater for biomass and lipid production. *Bioresour. Technol.* **2015**, *194*, 326–333. [CrossRef]

104. Kim, H.-C.; Choi, W.J.; Chae, A.N.; Park, J.; Kim, H.J.; Song, K.G. Evaluating integrated strategies for robust treatment of high saline piggery wastewater. *Water Res.* **2016**, *89*, 222–231. [CrossRef]

105. Hemalatha, M.; Sravan, J.S.; Yeruva, D.K.; Venkata Mohan, S. Integrated ecotechnology approach towards treatment of complex wastewater with simultaneous bioenergy production. *Bioresour. Technol.* **2017**, *242*, 60–67. [CrossRef]

106. Rugnini, L.; Costa, G.; Congestri, R.; Antonaroli, S.; Sanità di Toppi, L.; Bruno, L. Phosphorus and metal removal combined with lipid production by the green microalga *Desmodesmus sp.*: An integrated approach. *Plant Physiol. Biochem.* **2018**, *125*, 45–51. [CrossRef]

107. Sharma, B.; Dangi, A.K.; Shukla, P. Contemporary enzyme based technologies for bioremediation: a review. *J. Environ. Manage.* **2018**, *210*, 10–22. [CrossRef]

108. Thatoi, H.; Das, S.; Mishra, J.; Rath, B.P.; Das, N. Bacterial chromate reductase, a potential enzyme for bioremediation of hexavalent chromium: a review. *J. Environ. Manage.* **2014**, *146*, 383–399. [CrossRef]

109. Sivaperumal, P.; Kamala, K.; Rajaram, R. Bioremediation of industrial waste through enzyme producing marine microorganisms. In *Advances in Food and Nutrition Research*; Academic Press: Cambridge, MA, USA, 2017; Volume 80, pp. 165–179, ISBN 9780128095874.

110. Joutey, N.T.; Sayel, H.; Bahafid, W.; El Ghachtouli, N. Mechanisms of hexavalent chromium resistance and removal by microorganisms. *Rev. Environ. Contam. Toxicol.* **2015**, *233*, 45–69.

111. Yen, H.-W.; Chen, P.-W.; Hsu, C.-Y.; Lee, L. The use of autotrophic *Chlorella vulgaris* in chromium (VI) reduction under different reduction conditions. *J. Taiwan Inst. Chem. Eng.* **2017**, *74*, 1–6. [CrossRef]

112. Han, X.; Wong, Y.S.; Wong, M.H.; Tam, N.F.Y. Biosorption and bioreduction of Cr(VI) by a microalgal isolate, *Chlorella miniata*. *J. Hazard. Mater.* **2007**, *146*, 65–72. [CrossRef]

113. Jácome-Pilco, C.R.; Cristiani-Urbina, E.; Flores-Cotera, L.B.; Velasco-García, R.; Ponce-Noyola, T.; Cañizares-Villanueva, R.O. Continuous Cr(VI) removal by *Scenedesmus incrassatulus* in an airlift photobioreactor. *Bioresour. Technol.* **2009**, *100*, 2388–2391. [CrossRef]

114. Pradhan, D.; Sukla, L.B.; Mishra, B.B.; Devi, N. Biosorption for removal of hexavalent chromium using microalgae *Scenedesmus sp.* *J. Clean. Prod.* **2019**, *209*, 617–629. [CrossRef]

115. Raczynska, J.E.; Vorgias, C.E.; Antranikian, G.; Rypniewski, W. Crystallographic analysis of a thermoactive nitrilase. *J. Struct. Biol.* **2011**, *173*, 294–302. [CrossRef]

116. Park, J.M.; Trevor Sewell, B.; Benedik, M.J. Cyanide bioremediation: the potential of engineered nitrilases. *Appl. Microbiol. Biotechnol.* **2017**, *101*, 3029–3042. [CrossRef]

117. Cirulis, J.T.; Scott, J.A.; Ross, G.M. Management of oxidative stress by microalgae. *Can. J. Physiol. Pharmacol.* **2013**, *91*, 15–21. [CrossRef]

118. Lauritano, C.; Orefice, I.; Procaccini, G.; Romano, G.; Ianora, A. Key genes as stress indicators in the ubiquitous diatom *Skeletonema marinoi*. *BMC Genomics* **2015**, *16*, 411. [CrossRef]

119. Couto, N.; Wood, J.; Barber, J. The role of glutathione reductase and related enzymes on cellular redox homoeostasis network. *Free Radic. Biol. Med.* **2016**, *95*, 27–42. [CrossRef]

120. Rogato, A.; Amato, A.; Iudicone, D.; Chiurazzi, M.; Ferrante, M.I.; d'Alcalà, M.R. The diatom molecular toolkit to handle nitrogen uptake. *Mar. Genomics* **2015**, *24*, 95–108. [CrossRef]

121. Suzuki, S.; Yamaguchi, H.; Nakajima, N.; Kawachi, M. *Raphidocelis subcapitata (=Pseudokirchneriella subcapitata)* provides an insight into genome evolution and environmental adaptations in the *Sphaeropleales*. *Sci. Rep.* **2018**, *8*, 8058. [CrossRef]

122. Sauser, K.R.; Liu, J.K.; Wong, T.-Y. Identification of a copper-sensitive ascorbate peroxidase in the unicellular green alga *Selenastrum capricornutum*. *Biometals* **1997**, *10*, 163–168. [CrossRef]

123. Morelli, E.; Scarano, G. Copper-induced changes of non-protein thiols and antioxidant enzymes in the marine microalga *Phaeodactylum tricornutum*. *Plant Sci.* **2004**, *167*, 289–296. [CrossRef]

124. Álvarez, R.; del Hoyo, A.; García-Breijo, F.; Reig-Armiñana, J.; del Campo, E.M.; Guéra, A.; Barreno, E.; Casano, L.M. Different strategies to achieve Pb-tolerance by the two *Trebouxia* algae coexisting in the lichen *Ramalina farinacea*. *J. Plant Physiol.* **2012**, *169*, 1797–1806. [CrossRef]

125. Kennedy, J.; Margassery, L.M.; Morrissey, J.P.; O'Gara, F.; Dobson, A.D.W. Metagenomic strategies for the discovery of novel enzymes with biotechnological application from marine ecosystems. *Mar. Enzym. Biocatal.* **2013**, 109–130.

126. Trincone, A. Enzymatic processes in marine biotechnology. *Mar. Drugs* **2017**, *15*, 93. [CrossRef]

127. Raut, N.; Al-Balushi, T.; Panwar, S.; Vaidya, R.S.; Shinde, G.B. Microalgal Biofuel. In *Biofuels-Status and Perspective*; InTech: London, UK, 2015.

128. Ruiz, J.; Olivieri, G.; de Vree, J.; Bosma, R.; Willems, P.; Reith, J.H.; Eppink, M.H.M.; Kleinegris, D.M.M.; Wijffels, R.H.; Barbosa, M.J. Towards industrial products from microalgae. *Energy Environ. Sci.* **2016**, *9*, 3036–3043. [CrossRef]

129. Gifuni, I.; Pollio, A.; Safi, C.; Marzocchella, A.; Olivieri, G. Current Bottlenecks and Challenges of the Microalgal Biorefinery. *Trends Biotechnol.* **2018**, *37*, 242–252. [CrossRef]

130. Lauritano, C.; Ferrante, M.I.; Rogato, A. Marine Natural Products from Microalgae: An -Omics Overview. *Mar. Drugs* **2019**, *17*, 269. [CrossRef]

131. Mardis, E.R. A decade's perspective on DNA sequencing technology. *Nature* **2011**, *470*, 198–203. [CrossRef]

9

Cloning, Expression and Characterization of a Novel Cold-Adapted β-galactosidase from the Deep-Sea Bacterium *Alteromonas* sp. ML52

Jingjing Sun [1,2,*], Congyu Yao [1,3], Wei Wang [1,2], Zhiwei Zhuang [4], Junzhong Liu [1,2], Fangqun Dai [1,2] and Jianhua Hao [1,2,5,*]

[1] Key Laboratory of Sustainable Development of Polar Fishery, Ministry of Agriculture and Rural Affairs, Yellow Sea Fisheries Research Institute, Chinese Academy of Fishery Sciences, Qingdao 266071, China; yaocongyv@foxmail.com (C.Y.); weiwang@ysfri.ac.cn (W.W.); qdjz99@163.com (J.L.); dai@ysfri.ac.cn (F.D.)

[2] Laboratory for Marine Drugs and Bioproducts, Laboratory for Marine Fisheries Science and Food Production Processes, Qingdao National Laboratory for Marine Science and Technology, Qingdao 266071, China

[3] College of Food Science and Technology, Shanghai Ocean University, Shanghai 201306, China

[4] New Hope Liuhe Co. Ltd., Qingdao 266071, China; zzw19680@163.com

[5] Jiangsu Collaborative Innovation Center for Exploitation and Utilization of Marine Biological Resource, Lianyungang 222005, China

* Correspondence: sunjj@ysfri.ac.cn (J.S.); haojh@ysfri.ac.cn (J.H.).

Abstract: The bacterium *Alteromonas* sp. ML52, isolated from deep-sea water, was found to synthesize an intracellular cold-adapted β-galactosidase. A novel β-galactosidase gene from strain ML52, encoding 1058 amino acids residues, was cloned and expressed in *Escherichia coli*. The enzyme belongs to glycoside hydrolase family 2 and is active as a homotetrameric protein. The recombinant enzyme had maximum activity at 35 °C and pH 8 with a low thermal stability over 30 °C. The enzyme also exhibited a K_m of 0.14 mM, a V_{max} of 464.7 U/mg and a k_{cat} of 3688.1 S^{-1} at 35 °C with 2-nitrophenyl-β-D-galactopyranoside as a substrate. Hydrolysis of lactose assay, performed using milk, indicated that over 90% lactose in milk was hydrolyzed after incubation for 5 h at 25 °C or 24 h at 4 °C and 10 °C, respectively. These properties suggest that recombinant *Alteromonas* sp. ML52 β-galactosidase is a potential biocatalyst for the lactose-reduced dairy industry.

Keywords: *Alteromonas*; deep sea; cold-adapted enzyme; β-galactosidase; lactose-free milk

1. Introduction

Beta-galactosidase (EC 3.2.1.23), a glycoside hydrolase enzyme, catalyzes the hydrolysis of terminal non-reducing β-D-galactose residues into β-D-galactosides and also catalyzes transgalactosylation reactions [1–3]. Beta-galactosidases exist naturally in many organisms, including microorganisms, plants and animals [4,5]. Most industrial β-galactosidases are obtained from microorganisms. For example, the enzymes isolated from bacteria [6] and yeast [7], with neutral optimum pH, were used in milk products, and fungal [8] enzymes with an acid optimum pH were used in acid whey products. The main application of β-galactosidase is to hydrolyze lactose in milk in the dairy industry to provide lactose-free milk for lactose-intolerant consumers [9]. Another application of β-galactosidase is to transfer lactose and monosaccharide to a series of galacto-oligosaccharides (GOS) which are functional galactosylated products [10–12]. However, β-galactosidase catalyzed at moderate temperatures may cause some issues, e.g., increased production costs, wasted energy and producing undesirable microbial contamination [13]. Cold-adapted β-galactosidases, with low optimum temperatures, could catalyze hydrolysis or transgalactosylation reactions at refrigerating temperatures (4–10 °C), thus potentially overcoming these shortcomings. It may be especially beneficial to the dairy industry which could improve the hydrolysis of lactose at low temperatures.

While a minority of β-galactosidases from fungus are secreted to the extracellular medium, e.g., an acid β-galactosidase from *Aspergillus* spp. [14], β-galactosidases are generally intracellular enzymes in yeast and bacteria. Most reported β-galactosidases are recombinant enzymes derived from heterologous expression than from a natural source.

In recent years, the number of cold-adapted β-galactosidases were isolated from psychrophilic and psychrotrophic microorganisms obtained from isothermal cold environments such as polar [15–17], deep-sea [18] and high mountainous regions [19]. The main source of enzymes has been obtained from bacterial strains such as *Arthrobacter psychrolactophilus* strain F2 [20], *Arthrobacter* sp. 32c [21], *Halomonas* sp. S62 [22], *Paracoccus* sp. 32d [23], *Pseudoalteromonas haloplanktis* [24] and *Rahnella* sp. R3 [19].

Only a few cold-adapted β-galactosidases have been discovered from other sources, including psychrophilic-basidiomycetous yeast *Guehomyces pullulan* [25] and Antarctic haloarchaeon *Halorubrum lacusprofundi* [26]. The β-galactosidase from *Arthrobacter psychrolactophilus* strain F2 showed the lowest optimum temperature at 10 °C with an optimum pH of 8.

Based on the specific features of sequence, structure, substrate specificity and reaction mechanism, β-galactosidases have been classified into GH1, GH2, GH35 and GH42 families [27]. Most reported microorganism β-galactosidases belong to the GH2 [15,20,23,24,28] and GH42 [19,21,29] families. A typical GH2 β-galactosidase from *E. coli* is made up of five sequential domains and forms a functional tetramer [30].

Most of the characterized cold-adapted β-galactosidases from the GH2 family are tetrameric enzymes, except for a dimeric enzyme from *Paracoccus* sp. 32d and a hexameric enzyme from *Arthrobacter* sp. C2-2 [31]. Hitherto, three crystal structures of cold-adapted β-galactosidases have been obtained: GH42 β-galactosidase from *Planococcus* sp. L4 [32] and two GH2 β-galactosidases from Antarctic bacteria *Arthrobacter* sp. C2-2 [31] and *Paracoccus* sp. 32d [33].

In this study, we report on a gene of β-galactosidase from the marine bacterium *Alteromonas* sp. ML52, isolated from a deep-sea sample. This novel cold-adapted β-galactosidase belongs to the GH2 family and was overexpressed and characterized.

2. Results

2.1. Characterization and Identification of Strain ML52

Strain ML52 was isolated from deep-sea water in the Mariana Trench at a depth of 4000 m and found to produce intracellular β-galactosidase at 4 °C. Database searches showed that strain ML52 is related to the genus *Alteromonas*. As shown in the neighbor-joining tree (Figure 1) [34,35], strain ML52 formed a monophyletic cluster with *Alteromonas addita* R10SW13[T] (99.9% identity), *Alteromonas stellipolaris* LMG 21861[T] (99.8% identity) and *Alteromonas naphthalenivorans* SN2[T] (99.3% identity). Strain ML52 was able to produce β-galactosidase on 2216E X-Gal agar in the presence of either lactose or glucose (Figure 2), while the expression of β-galactosidase in *E. coli* containing a lac operon was repressed by glucose [36].

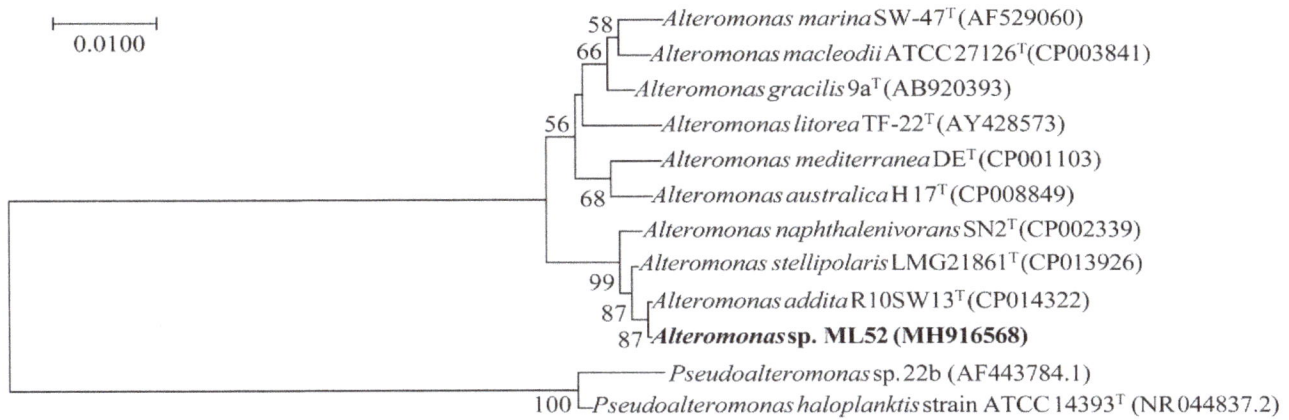

Figure 1. Neighbor-joining tree based on 16S rRNA gene sequences showing the phylogenetic position of strain ML52 and closely related *Alteromonas* and *Pseudoalteromonas* species. Bootstrap values (>50%) were calculated for 1000 replicates. The reference 16S rDNA sequences were collected from EzTaxon-e server (www.bacterio.net/) and the National Center for Biotechnology Information (NCBI) Database and aligned using the ClustalX 2.1 program (Conway Institute UCD Dublin, Dublin, Ireland). The phylogenetic tree was obtained using MEGA 7.0 software (Institute for Genomics and Evolutionary Medicine, Temple University, Tempe, AZ, USA).

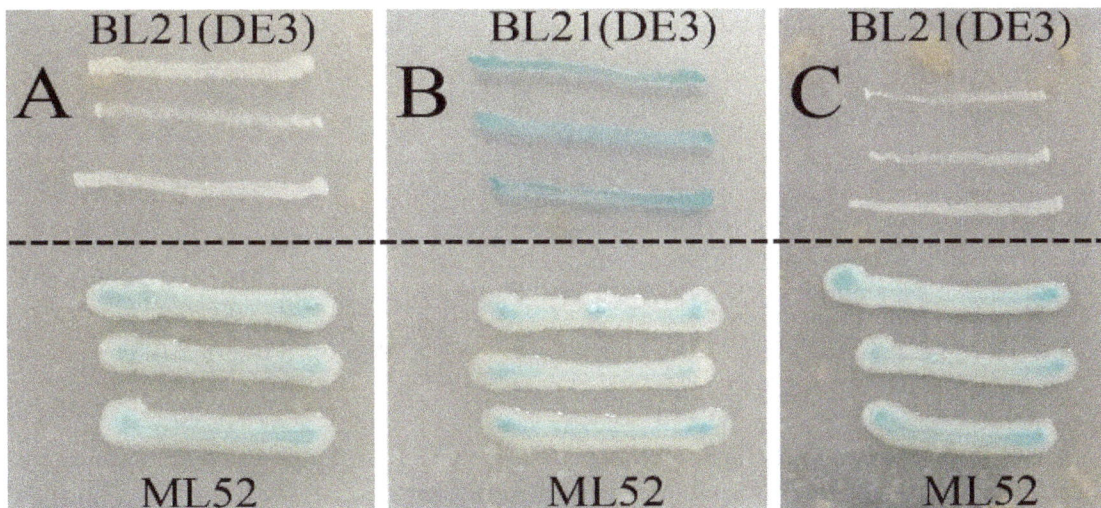

Figure 2. Effects of glucose and lactose on the expression of β-galactosidase of strain ML52. *E. coli* strain BL21(DE3) was used as a control. A, 2216E X-Gal agar with 2% glucose; B, 2216E X-Gal agar with 2% lactose; C, 2216E X-Gal agar with 2% glucose and lactose.

2.2. Molecular Cloning and Sequence Analysis

Deoxyribonucleic acid sequencing showed that *gal* consisted of an open reading frame of 3174 bp, encoding 1058 amino acid residues. The theoretical Mw and pI of the enzyme was 118,543 Da and 4.96, respectively. According to the BLAST results, *gal* had a highest identity of 99% to a putative β-galactosidase from *Alteromonas addita* (WP_062085674.1). At the time of writing, two characterized β-galactosidase genes from *Pseudoalteromonas* sp. 22b (AAR92204.1) and *Pseudoalteromonas haloplanktis* (CAA10470.1) exhibited the highest sequence identity (65%) to *gal*. Based on sequence comparisons, Gal was classified into glycoside hydrolase family 2. Amino acid sequence comparison of Gal and other characterized GH2 β-galactosidases are shown in Figure 3.

Figure 3. Sequence alignment of Gal and other GH2 β-galactosidase from different microorganisms. Identical residues are shaded in black and conserved residues are shaded in gray. The putative nucleophilic and catalytic amino acids are marked by a red asterisk and the regions relative to the formation of a tetramer are marked by blue boxes. Protein accession numbers and species are as follows: AAR92204.1, *Pseudoalteromonas* sp. 22b; CAA10470.1, *Pseudoalteromonas haloplanktis*; ABN72582.1, *E. coli* K-12; CAD29775.1, *Arthrobacter* sp. C2-2.

2.3. Expression and Purification of Recombinant Gal

The expression vector pET-24a was used for the expression of the *gal* gene from strain ML52 in *E. coli* BL21 (DE3). The recombinant enzyme with a 6-histidine tag was induced by IPTG and purified by Ni-NTA chromatography. After purification, approximately 55.2 mg of pure enzyme was obtained from 1L of induced culture. The apparent molecular weight of recombinant Gal was 126 kDa which was determined by SDS-PAGE (Figure 4), and which corresponded to the theoretical size. The relative molecular weight of recombinant Gal, which was determined by gel-filtration chromatography, was 485 kDa (results not shown). Hence recombinant Gal is probably a tetrameric protein, like *E. coli* β-galactosidase.

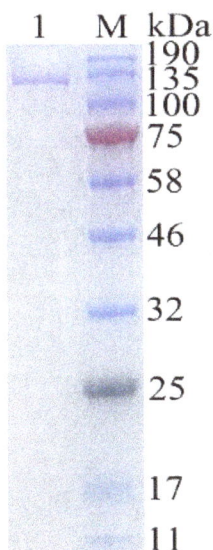

Figure 4. Twelve percent SDS-PAGE analysis of the purified β-galactosidase. Lane 1, purified β-galactosidase; lane M, protein marker. The gel was stained with 0.025% Coomassie blue R250.

2.4. Properties of Recombinant Gal

The optimum temperature for recombinant Gal was close to 35 °C and the enzyme had 19–30% activity at 4–10 °C (Figure 5A). Recombinant Gal maintained high activity over a pH range of 7 to 8.5 with maximum activity at pH 8. The enzyme was stable from a pH range between 6.5 and 9 (Figure 5B). Recombinant Gal was stable at temperatures between 4 and 20 °C, but its activity reduced rapidly at 30 °C and most activity was lost at 50 °C following half hour incubation (Figure 5C).

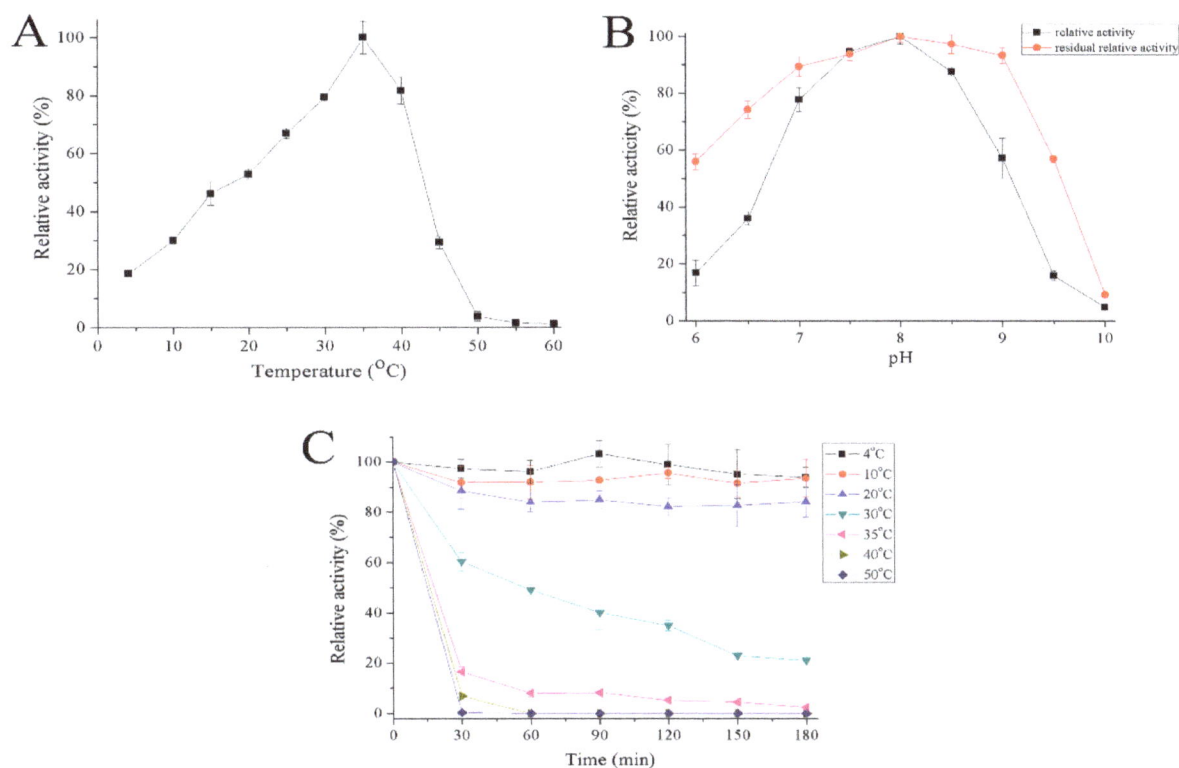

Figure 5. Effect of temperature on activity (**A**) and stability (**C**) of recombinant Gal and effect of pH (**B**) on activity of recombinant Gal. The optimum activity was set as 100% with specific activities of 434.3 U/mg, or 286 U/mg for the effect of temperature or pH, respectively.

Results outlining the effects of metal ions and chemicals on recombinant Gal are shown in Table 1. Reducing agents (DTT and 2-mercaptoethanol) and K^+ could stimulated the enzymatic activity of recombinant Gal. The addition of Ca^{2+} and urea slightly decreased enzyme activity whereas Mg^{2+}, Mn^{2+} and ionic detergent (SDS) had strong inhibitory effects. The presence of chelating agent (EDTA) and the ions Zn^{2+}, Ni^{2+}, and Cu^{2+} completely inhibited enzyme activity. Salt tolerance of the recombinant Gal was also investigated (Figure 6). Recombinant Gal maintained 53% residue activity in the presence of 1.5 M NaCl and 26% residue activity when NaCl concentration increased to 4 M.

Table 1. Effect of metal ions and chemicals on activity of recombinant Gal.

Ion/Reagent	Relative Activity (%)
None [1]	100 ± 1.3
K^+	123 ± 0.6
Ca^{2+}	74.9 ± 5.4
Mg^{2+}	32.4 ± 1.4
Mn^{2+}	21.3 ± 0.9
Ba^{2+}	48.2 ± 2.1
Zn^{2+}	0.1 ± 0.04
Ni^{2+}	0.2 ± 0.03
Cu^{2+}	0
EDTA	0.2 ± 0.02
Urea	89.9 ± 0.3
DTT	116.9 ± 7
2-Mercaptoethanol	108.9 ± 4
SDS	43.5 ± 1

[1] The activity of control (no additions) was set as 100% with a specific activity of 386.7 U/mg.

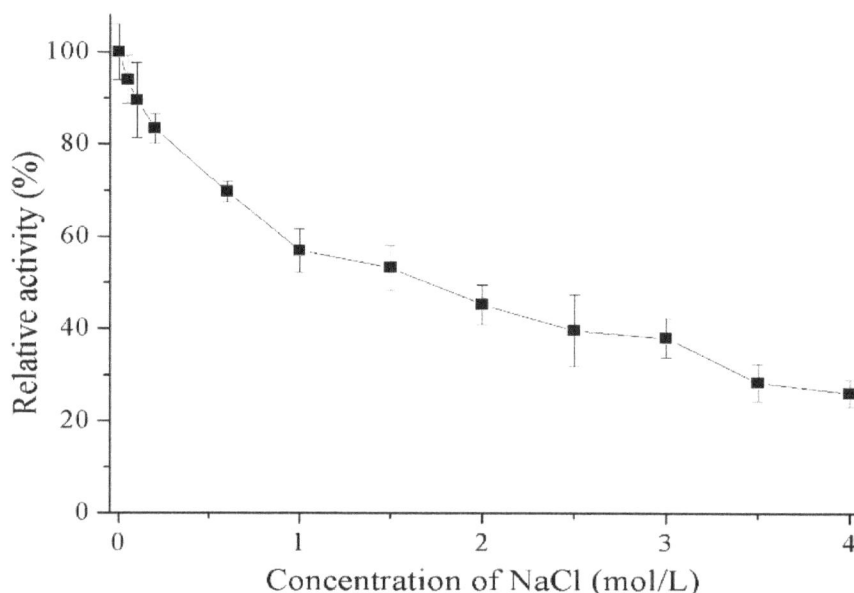

Figure 6. Effects of NaCl on the activity of recombinant Gal. Activity with 0 M NaCl was set as 100% with a specific activity of 445.3 U/mg.

The substrate specificity of recombinant Gal was investigated using seven chromogenic substrates (Table 2). The enzyme was specific to two β-D-galactopyranoside substrates, and showed no activity to other tested substrates. Kinetic parameters of recombinant Gal were determined at optimum temperature with ONPG (2-Nitrophenyl-β-D-galactopyranoside) and lactose as substrates, as shown in Table 3.

Table 2. Substrate specificity of recombinant Gal.

Substrate	Relative Activity (%)
2-Nitrophenyl-β-ᴅ-galactopyranoside (ONPG) [1]	100 ± 4.2
4-Nitrophenyl-β-ᴅ-galactopyranoside (PNPG)	12.8 ± 0.5
4-Nitrophenyl-α-ᴅ-galactopyranoside	0
2-Nitrophenyl-β-ᴅ-glucopyranoside	0
4-Nitrophenyl-β-ᴅ-glucopyranoside	0
4-Nitrophenyl-α-ᴅ-glucopyranoside	0
4-Nitrophenyl-β-ᴅ-xylopyranoside	0

[1] The activity of ONPG was set as 100% with a specific activity of 467.2 U/mg.

Table 3. Substrate specificity of recombinant Gal.

Substrate	V_{max} (U/mg)	K_m (mM)	k_{cat} (S^{-1})	k_{cat}/K_m (S^{-1} mM^{-1})
ONPG	464.7	0.14	3688.1	26343.6
Lactose	18.5	7.2	146.5	20.3

2.5. Hydrolysis of Lactose in Milk

The hydrolysis of milk lactose by recombinant Gal was determined at 4 °C, 10 °C and 25 °C (Figure 7 and Figure S1). The conversion rate of lactose in milk reached about 42% during the initial hour at 25 °C and 94% after 5 h incubation. At refrigerating temperatures of 4 °C and 10 °C, 58% and 64% of lactose was hydrolyzed after 8 h incubation, respectively. A lactose conversion rate of greater than 90% was reached after 24 h incubation and almost 100% after 48 h.

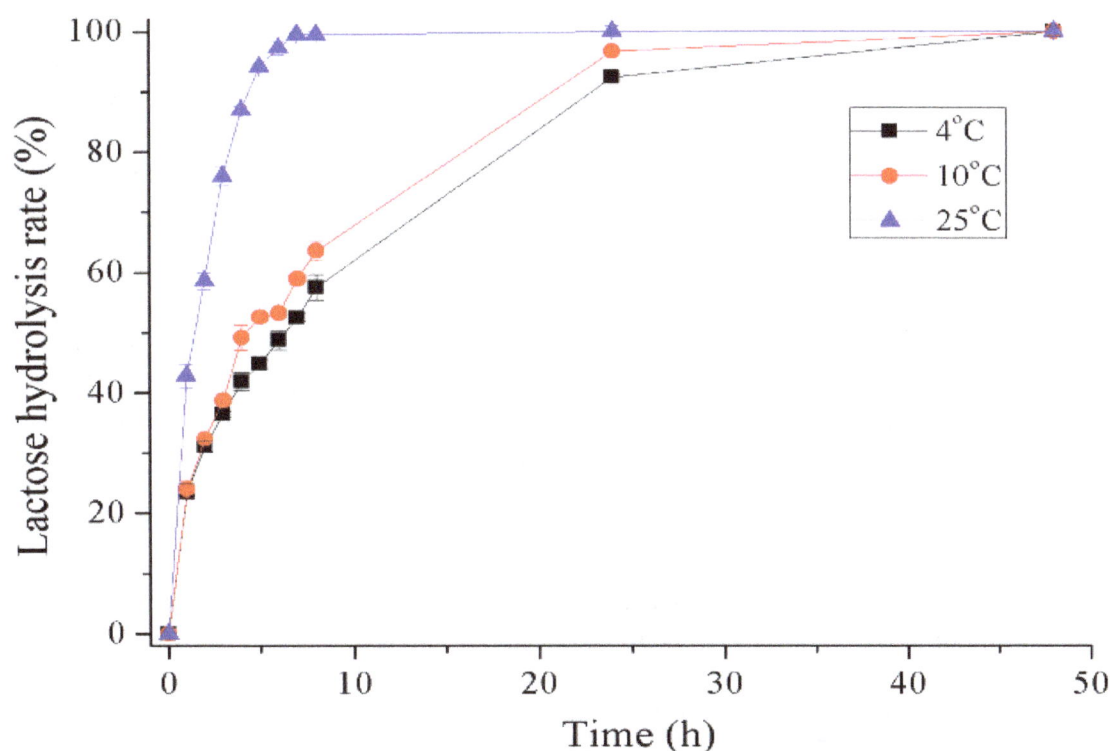

Figure 7. Hydrolysis of lactose in milk of recombinant Gal.

3. Discussion

A deep-sea bacterium *Alteromonas* sp. ML52 with β-galactosidase activity was obtained through plate screening. The β-galactosidase from strain ML52 was considered to be a constitutive enzyme because its expression was normal without lactose and not repressed by glucose. A novel GH2 β-galactosidase gene *gal* was cloned from strain ML52. Sequence alignment of Gal and other GH2 β-galactosidases (Figure 2) showed that Gal shares a conserved acid-base activity site and nucleophilic site typically found in this family of enzyme. Compared with a *E. coli* β-galactosidase, a mesophilic enzyme belonging to the GH2 family, the amino acid sequence composition of Gal showed significant decreases in Arg residues (4.5% versus 6.4%), Arg/Arg+Lys ratio (0.51 versus 0.77) and Pro residues (4.9% versus 6.1%) which are structural features of psychrophilic enzymes [30]. Skálová et al. [31]

compared two crystal structures of GH2 β-galactosidases, a tetrameric enzyme from *E. coli* and a hexameric enzyme from *Arthrobacter* sp. C2-2, and suggested that the outstanding loop region of domain 2, which participates in forming the contacts between subunits of the tetramer along with domain 3 in *E. coli* β-galactosidase; however, this kind of complementation does not occur in *Arthrobacter* sp. C2-2. The sequences relative to the formation of the tetramer of Gal in domain 2 and domain 3 were similar to that of the *E. coli* β-galactosidase, which are labeled in Figure 3. This theoretical prediction was tested by molecular weight determination. Thus, Gal is considered to be a homotetrameric protein.

Recombinant Gal has an optimum temperature of 35 °C and a low thermal stability over 30 °C. Other reported GH2 cold-adapted β-galactosidases possessed optimum temperatures between 10 and 40 °C (Table 4). Thus Gal, which was isolated from a deep-sea bacterium, is considered to be a cold-adapted enzyme. The optimum pH range for enzyme activity and stability was consistent with bacterial β-galactosidases which are normally active within a neutral pH range of 6 to 8. This feature indicates that Gal may be a suitable enzyme for lactose hydrolysis in milk (pH 6.5–6.8). The activating effect of K^+ and the inhibitory action of ions Zn^{2+}, Ni^{2+}, Cu^{2+} to the enzyme activity of recombinant Gal also occurred in a cold-adapted *Pseudoalteromonas* sp. 22b β-galactosidase [28], with the highest sequence identity to Gal. The chelating agent caused a complete inhibition of enzyme activity, indicating that the catalytic reaction of recombinant Gal may rely on the presence of metal ions. It was also noted that recombinant Gal preferred substrate ONPG to substrate PNPG. This phenomenon was also observed in cold-adapted β-galactosidases from *Paracoccus* sp. 32d [23] and *Arthrobacter psychrolactophilus* strain F2 [20], but many cold-adapted β-galactosidases, isolated from *Pseudoalteromonas* sp. 22b [28], *Arthrobacter* sp. 20B [37] and *Arthrobacter* sp. 32c [21], displayed higher levels of activity to PNPG. Remarkably, recombinant Gal showed significantly high affinity and reaction rate to the chromogenic substrate ONPG at optimum temperature compared to other GH2 cold-adapted β-galactosidases from *Arthrobacter psychrolactophilus* strain F2, *Arthrobacter* sp. and *Paracoccus* sp. 32d (Table 4). This property is similar to the *Pseudoalteromonas* sp. 22b β-galactosidase. For lactose, the natural substrate of β-galactosidase, recombinant Gal showed intermediate substrate affinity and catalytic efficiency at optimum temperature.

The major industry β-galactosidases were obtained from *Aspergillus* spp. and *Kluyveromyces* spp. [1,4,8]. The β-galactosidases from *Aspergillus* spp. have an optimum pH at acidic range (2.5–5.4) and are not suitable for hydrolysis of lactose in milk. The β-galactosidase from *Kluyveromyces lactis* is one of the most widely used commercial enzymes. The optimum pH (6.6–7) of *K. lactis* β-galactosidase is close to milk because the dairy environment is a natural habitat for this kind of yeast [4]. However, *K. lactis* β-galactosidase showed higher optimum temperature (40 °C) [38] and lower substrate affinity to both ONPG and lactose (K_m, 1.7 mM for ONPG and 17.3 mM for lactose) [8] compared with recombinant Gal.

Table 4. Biochemical characteristics of reported GH2 cold-adapted β-galactosidases.

Strain	Optimum		ONPG		Lactose		References
	Temperature (°C)	pH	K_m (mM)	k_{cat} (S^{-1})	K_m (mM)	k_{cat} (S^{-1})	
Alteromonas sp. ML52	35	8	0.14	3688.1	7.2	146.5	This work
Arthrobacter psychrolactophilus strain F2	10	8	2.7	12.7	42.1	3.02	[20]
Arthrobacter sp. 20B	25	6–8	-	-	-	-	[37]
Arthrobacter sp. C2-2	40	7.5	-	-	53.1	1106	[39]
Arthrobacter sp.	18	7	11.5	5.2	-	-	[17]
Paracoccus sp. 32d	40	7.5	1.17	71.81	2.94	43.23	[23]
Pseudoalteromonas haloplanktis TAE 79	-	8.5	-	203	2.4	33	[24]
Pseudoalteromonas sp. 22b	40	6–8	0.28	312	3.3	157	[28]

The results on lactose digestion in milk indicate that recombinant Gal could hydrolyze 100% of lactose in milk when incubated for 7 h at 25 °C and over 90% lactose was hydrolyzed after incubation for 24 h at 4 °C and 10 °C. No transglycosylation product was observed by HPLC during the hydrolysis process. For the *Pseudoalteromonas* sp. 22b cold-adapted β-galactosidase [28], 90% of lactose was hydrolyzed after 6 h at 30 °C and after 28 h at 15 °C. Another cold-adapted β-galactosidase from the *Arthrobacter psychrolactophilus* strain F2 [20] was able to hydrolyze approximately 70% of the lactose in milk at 10 °C after 24 h and displayed transglycosylation activity by forming a trisaccharide product during the reaction.

4. Materials and Methods

4.1. Isolation and Identification of Bacteria

Deep-sea water samples were collected from the Mariana Trench at a depth of 4000 m in September 2016. They were diluted and spread on marine agar 2216E (MA; BD Difco, Franklin Lakes, NJ, USA) containing 40 μg/mL X-Gal (5-bromo-4-chloro-3-indolyl-β-D-galactopyranoside, Sigma, St. Louis, MO, USA) and 2% lactose. After incubation at 4 °C for 2 weeks, the detectable blue colonies were repeatedly streaked on MA to obtain pure cultures. The 16S rRNA gene of strain ML52 was amplified by PCR from genomic DNA. To test the effects of glucose and lactose on the expression of β-galactosidase, strains were grown on MA containing 40 μg/mL X-gal and 2% glucose or lactose at 25 °C for 2 days.

4.2. Molecular Cloning and Sequence Analysis

Based on the DNA sequence of the predicted β-galactosidase gene (Accession number AMJ93096.1) from *Alteromonas addita* strain R10SW13 (Accession number CP014322.1) reported in the NCBI Database, two primers, forward primer-CG*GGATCC*ATGGCAAATGTTGCTCAA and reverse primer-CC*GCTCGAG*GCAATTTTCAGCACT (*Bam*HI and *Xho*I restriction sites are in italics) were designed. The PCR product was cloned into pMD20-T and sequenced by Tsingke, China. The plasmid pET-24a and amplified *gal* gene were then cleaved with the restriction endonucleases *Bam*HI and *Xho*I and the fragments were ligated with T4 DNA ligase (NEB), resulting in the plasmid pET-*gal*. The nucleotide sequence of *gal* was translated to an amino acid sequence using the ExPASY-Translate tool. A multiple alignment between Gal and other GH2 β-galactosidases from bacteria was constructed using ClustalW 2 (Conway Institute UCD Dublin, Dublin, Ireland) and GeneDoc 2.7 (Karl B. Nicholas et al., San Francisco, CA, USA). The theoretical molecular weight (Mw) and isoelectric point (pI) were calculated using the Compute pI/Mw tool (http://web.expasy.org/compute_pi/).

4.3. Expression and Purification

For expression of recombinant Gal, the plasmid pET-*gal* was transformed into *E. coli* BL21 (DE3) and cultivated at 37 °C in lysogeny broth (LB) containing 30 μg/mL kanamycin until the OD_{600} of the culture reached 0.6–0.8. Following this step, isopropyl-β-D-thiogalactopyranoside (IPTG) was added to the culture with a final concentration of 0.2 mM and the culture was further grown at 16 °C for 12 h. The cells were harvested by centrifugation at 8000 rpm for 10 min at 4 °C and suspended in a lysis solution (50 mM sodium phosphate buffer, pH 8). The mixture was then disrupted by sonication and the cell debris removed by centrifugation at 13,000 rpm for 15 min. The filtered supernatant was applied to a Ni^{2+}-chelating affinity column (His-Trap^{TM} HP, GE, Madison, WI, USA) which was previously equilibrated using an equilibration buffer (50 mM sodium phosphate buffer, 20 mM imidazole, 300 mM NaCl, pH 8) and subsequently eluted using a linear gradient of 20–250 mM imidazole in equilibration buffer. Enzyme purity was analyzed by SDS-PAGE (GenScript, Nanjing, China). Purified recombinant Gal was buffer-exchanged to 50 mM sodium phosphate buffer (pH 8) via centrifugal ultrafiltration (MW cut off 10 kDa, Millipore, Burlington, MA, USA). The protein concentration was determined using the BCA protein assay kit (Solarbio, Beijing, China) with bovine serum albumin as a standard.

4.4. Estimation of Molecular Weight

The purified enzyme was applied to a gel-filtration column (Superdex 200 HR 10/30, GE Healthcare, Madison, WI, USA) and eluted using a buffer containing 50 mM sodium phosphate (pH 7) and 150 mM NaCl. The standard proteins used were thyroglobumin (669 kDa), apoferritin (443 kDa), β-amylase (200 kDa), alcohol dehydrogenase (150 kDa), bovine serum albumin (66 kDa), and carbonic anhydrase (29 kDa) purchased from Sigma (St. Louis, MO, USA).

4.5. Beta-Galactosidase Activity Assay

Beta-galactosidase activity was determined in a 100 μL reaction mixture containing a final concentration of 0.4 ng/μL (0.83 nM) of purified recombinant Gal, 50 mM sodium phosphate buffer (pH 8) and 5 mM 2-nitrophenyl-β-D-galactopyranoside (ONPG). Each reaction was incubated for 5 min at 35 °C and quenched by the addition of 200 μL of 1 M Na_2CO_3. The absorbance of the released o-nitrophenol (ONP) was measured at 420 nm and quantified using an ONP standard curve. One unit (U) of enzyme activity was defined as the amount of enzyme required for the liberation of 1 μmol ONP per minute under the assay conditions. All assays were carried out in triplicate.

4.6. Effect of Temperature and pH

The optimum temperature of recombinant Gal was evaluated by incubating the reaction mixtures at different temperatures ranging from 4 °C to 60 °C. The optimum pH was determined by incubation in 10 mM Britton–Robinson buffers ranging from pH 6 to 10 (in increments of 0.5 pH units) at optimum temperature. The thermostability of recombinant Gal was determined by incubating enzymes at temperatures ranging from 4 °C to 50 °C for 180 min and the residual enzyme activity was determined every 30 min under the optimum conditions. The pH stability was determined after incubating the enzyme in Britton–Robinson buffers ranging from pH 6 to 10 (in increments of 0.5 pH units) at 4 °C for 3 h.

4.7. Effect of Metal Ions and Chemicals

The effect of metal ions and chemicals on recombinant Gal activity was determined after incubating the enzyme in water with 5 mM of KCl, $CaCl_2$, $MgCl_2$, $MnCl_2$, $BaCl_2$, $ZnCl_2$, $NiSO_4$, $CuCl_2$, EDTA, urea,

DTT, 2-mercaptoethanol or SDS at 4 °C for 1 h. The activity of the enzyme when incubated without any additional reagents was considered to be 100%. For salt stability, 0–4 M NaCl (final concentration) was added to the reaction mixtures and the activity determined using optimum conditions.

4.8. Determination of Substrate Specificity and Kinetic Parameters

The substrate specificity of the recombinant Gal was estimated using 5 mM (final concentration) ONPG, 4-Nitrophenyl-β-D-galactopyranoside (PNPG), 4-Nitrophenyl-α-D-galactopyranoside, 2-Nitrophenyl-β-D-glucopyranoside, 4-Nitrophenyl-β-D-glucopyranoside, 4-Nitrophenyl-α-D-glucopyranoside or 4-Nitrophenyl-β-D-xylopyranoside in 50 mM sodium phosphate buffer (pH 8) under the optimum conditions. For the activity of 4-nitrophenol-derived substrates, the absorbance of the released 4-nitrophenol (PNP) was measured at 405 nm and quantified using a PNP standard curve. Kinetic parameters of the recombinant Gal were determined at 35 °C and the reaction rate with ONPG (0.1–5 mM) and lactose (0.2–40 mM) as substrates was determined. The released glucose in the lactose hydrolysis reaction was measured using a commercial glucose oxidase-peroxidase assay kit (Shanghai Rongsheng Biotech Co., Ltd., Shanghai, China). One unit (U) of enzyme activity was defined as 1 μmol of glucose released per minute. The values of the kinetic constants were calculated using the Lineweaver-Burk method.

4.9. Hydrolysis of Lactose in Milk

The hydrolysis of lactose in milk was determined by incubating 100 μg purified Gal in 1 mL of commercial skim milk (Inner Mongolia Yili Industrial Group Co. Ltd., Hohhot, China) at 4 °C, 10 °C, or 25 °C for 48 h. The reaction was terminated by incubating the sample at 60 °C for 5 min and then adding an equal volume of 5% trichloroacetic acid (TCA) to the reaction mixture. The pH of the reaction mixture was adjusted to neutral pH using 1 M NaOH and centrifuged at 10,000 rpm for 10 min. The supernatant was analyzed by HPLC using the ZORBAX Carbohydrate Analysis Column (Agilent, Palo Alto, CA, USA), with 75% acetonitrile used as a mobile phase at a flow rate of 1 mL/min and a Shimadzu Refractive Index Detector (Kyoto, Japan).

4.10. Nucleotide Sequence Accession Numbers

The 16S rRNA gene sequence and β-galactosidase gene (*gal*) of strain ML52 were deposited in GenBank under the accession numbers MH916568 and MH925304, respectively.

5. Conclusions

In this study, a new GH2 β-galactosidase (Gal) was successfully cloned, purified and characterized from the deep-sea bacterium *Alteromonas* sp. ML52. Recombinant Gal is a cold-adapted enzyme and was able to efficiently hydrolyze lactose in milk at refrigerating temperature. These characteristics suggest that Gal could be a potential cold-active biocatalyst and usefully applied to the production of lactose-free milk in the dairy industry.

Author Contributions: Data curation, J.S., C.Y. and Z.Z.; Funding acquisition, J.S., W.W. and J.H.; Methodology, J.S., W.W. and J.H.; Project administration, J.S., C.Y., J.L. and F.D.; Writing-original draft, J.S.; Writing-review and editing, J.S. and J.H.

References

1. Oliveira, C.; Guimarães, P.M.; Domingues, L. Recombinant microbial systems for improved β-galactosidase production and biotechnological applications. *Biotechnol. Adv.* **2011**, *29*, 600–609. [CrossRef] [PubMed]
2. Ansari, S.A.; Satar, R. Recombinant β-galactosidases—Past, present and future: A mini review. *J. Mol. Catal. B-Enzym.* **2012**, *81*, 1–6. [CrossRef]

3.	Park, A.R.; Oh, D.K. Galacto-oligosaccharide production using microbial β-galactosidase: Current state and perspectives. *Appl. Microbiol. Biotechnol.* **2010**, *85*, 1279–1286. [CrossRef] [PubMed]
4.	Husain, Q. Beta galactosidases and their potential applications: A review. *Crit. Rev. Biotechnol.* **2010**, *30*, 41–62. [CrossRef] [PubMed]
5.	Chandrasekar, B.; van der Hoorn, R.A. Beta galactosidases in Arabidopsis and tomato—A mini review. *Biochem. Soc. Trans.* **2016**, *44*, 150–158. [CrossRef] [PubMed]
6.	Vasiljevic, T.; Jelen, P. Lactose hydrolysis in milk as affected by neutralizers used for the preparation of crude β-galactosidase extracts from *Lactobacillus bulgaricus* 11842. *Innov. Food Sci. Emerg. Technol.* **2002**, *3*, 175–184. [CrossRef]
7.	Kim, C.S.; Ji, E.S.; Oh, D.K. A new kinetic model of recombinant beta-galactosidase from *Kluyveromyces lactis* for both hydrolysis and transgalactosylation reactions. *Biochem. Biophys. Res. Commun.* **2004**, *316*, 738–743. [CrossRef] [PubMed]
8.	Panesar, P.S.; Panesar, R.; Singh, R.S.; Kennedy, J.F.; Kumar, H. Microbial production, immobilization and applications of beta-D-galactosidase. *J. Chem. Technol. Biotechnol.* **2006**, *81*, 530–543. [CrossRef]
9.	Horner, T.W.; Dunn, M.L.; Eggett, D.L.; Ogden, L.V. β-Galactosidase activity of commercial lactase samples in raw and pasteurized milk at refrigerated temperatures. *J. Dairy Sci.* **2011**, *94*, 3242–3249. [CrossRef] [PubMed]
10.	Rodriguez-Colinas, B.; Fernandez-Arrojo, L.; Santos-Moriano, P.; Ballesteros, A.; Plou, F. Continuous Packed Bed Reactor with Immobilized β-Galactosidase for Production of Galactooligosaccharides (GOS). *Catalysts* **2016**, *6*, 189. [CrossRef]
11.	Zhang, J.; Lu, L.; Lu, L.; Zhao, Y.; Kang, L.; Pang, X.; Liu, J.; Jiang, T.; Xiao, M.; Ma, B. Galactosylation of steroidal saponins by β-galactosidase from *Lactobacillus bulgaricus* L3. *Glycoconj. J.* **2016**, *33*, 53–62. [CrossRef] [PubMed]
12.	Maischberger, T.; Leitner, E.; Nitisinprasert, S.; Juajun, O.; Yamabhai, M.; Nguyen, T.H.; Haltrich, D. Beta-galactosidase from *Lactobacillus pentosus*: Purification, characterization and formation of galacto-oligosaccharides. *Biotechnol. J.* **2010**, *5*, 838–847. [CrossRef] [PubMed]
13.	Feller, G.; Gerday, C. Psychrophilic enzymes: Hot topics in cold adaptation. *Nat. Rev. Microbiol.* **2003**, *1*, 200–208. [CrossRef] [PubMed]
14.	Pakula, T.M.; Salonen, K.; Uusitalo, J.; Penttilä, M. The effect of specific growth rate on protein synthesis and secretion in the filamentous fungus *Trichoderma reesei*. *Microbiology* **2005**, *151*, 135–143. [CrossRef] [PubMed]
15.	Schmidt, M.; Stougaard, P. Identification, cloning and expression of a cold-active beta-galactosidase from a novel Arctic bacterium, *Alkalilactibacillus ikkense*. *Environ. Technol.* **2010**, *31*, 1107–1114. [CrossRef] [PubMed]
16.	Alikkunju, A.P.; Sainjan, N.; Silvester, R.; Joseph, A.; Rahiman, M.; Antony, A.C.; Kumaran, R.C.; Hatha, M. Screening and Characterization of Cold-Active β-Galactosidase Producing Psychrotrophic *Enterobacter ludwigii* from the Sediments of Arctic Fjord. *Appl. Biochem. Biotechnol.* **2016**, *180*, 477–490. [CrossRef] [PubMed]
17.	Coker, J.A.; Sheridan, P.P.; Loveland-Curtze, J.; Gutshall, K.R.; Auman, A.J.; Brenchley, J.E. Biochemical characterization of a beta-galactosidase with a low temperature optimum obtained from an Antarctic *Arthrobacter* isolate. *J. Bacteriol.* **2003**, *185*, 5473–5482. [CrossRef] [PubMed]
18.	Ghosh, M.; Pulicherla, K.K.; Rekha, V.P.; Raja, P.K.; Sambasiva Rao, K.R. Cold active β-galactosidase from *Thalassospira* sp. 3SC-21 to use in milk lactose hydrolysis: A novel source from deep waters of Bay-of-Bengal. *World J. Microbiol. Biotechnol.* **2012**, *28*, 2859–2869. [CrossRef] [PubMed]
19.	Fan, Y.; Hua, X.; Zhang, Y.; Feng, Y.; Shen, Q.; Dong, J.; Zhao, W.; Zhang, W.; Jin, Z.; Yang, R. Cloning, expression and structural stability of a cold-adapted β-galactosidase from *Rahnella* sp. R3. *Protein Expr. Purif.* **2015**, *115*, 158–164. [CrossRef] [PubMed]
20.	Nakagawa, T.; Ikehata, R.; Myoda, T.; Miyaji, T.; Tomizuka, N. Overexpression and functional analysis of cold-active beta-galactosidase from *Arthrobacter psychrolactophilus* strain F2. *Protein Expr. Purif.* **2007**, *54*, 295–299. [CrossRef] [PubMed]
21.	Hildebrandt, P.; Wanarska, M.; Kur, J. A new cold-adapted beta-D-galactosidase from the Antarctic *Arthrobacter* sp. 32c-gene cloning, overexpression, purification and properties. *BMC Microbiol.* **2009**, *9*, 151. [CrossRef] [PubMed]
22.	Wang, G.X.; Gao, Y.; Hu, B.; Lu, X.L.; Liu, X.Y.; Jiao, B.H. A novel cold-adapted β-galactosidase isolated from *Halomonas* sp. S62: Gene cloning, purification and enzymatic characterization. *World J. Microbiol. Biotechnol.* **2013**, *29*, 1473–1480. [CrossRef] [PubMed]

23. Wierzbicka-Woś, A.; Cieśliński, H.; Wanarska, M.; Kozłowska-Tylingo, K.; Hildebrandt, P.; Kur, J. A novel cold-active β-D-galactosidase from the *Paracoccus* sp. 32d—Gene cloning, purification and characterization. *Microb. Cell Fact.* **2011**, *10*, 108. [CrossRef]

24. Hoyoux, A.; Jennes, I.; Dubois, P.; Genicot, S.; Dubail, F.; François, J.M.; Baise, E.; Feller, G.; Gerday, C. Cold-adapted beta-galactosidase from the Antarctic psychrophile *Pseudoalteromonas haloplanktis*. *Appl. Environ. Microbiol.* **2001**, *67*, 1529–1535. [CrossRef] [PubMed]

25. Nakagawa, T.; Ikehata, R.; Uchino, M.; Miyaji, T.; Takano, K.; Tomizuka, N. Cold-active acid beta-galactosidase activity of isolated psychrophilic-basidiomycetous yeast *Guehomyces pullulans*. *Microbiol. Res.* **2006**, *161*, 75–79. [CrossRef] [PubMed]

26. Karan, R.; Capes, M.D.; DasSarma, P.; DasSarma, S. Cloning, overexpression, purification, and characterization of a polyextremophilic β-galactosidase from the Antarctic haloarchaeon *Halorubrum lacusprofundi*. *BMC Biotechnol.* **2013**, *13*, 3. [CrossRef] [PubMed]

27. Henrissat, B.; Davies, G. Structural and sequence-based classification of glycoside hydrolases. *Curr. Opin. Struct. Biol.* **1997**, *7*, 637–644. [CrossRef]

28. Cieśliński, H.; Kur, J.; Białkowska, A.; Baran, I.; Makowski, K.; Turkiewicz, M. Cloning, expression, and purification of a recombinant cold-adapted beta-galactosidase from antarctic bacterium *Pseudoalteromonas* sp. 22b. *Protein Expr. Purif.* **2005**, *39*, 27–34. [CrossRef] [PubMed]

29. Hu, J.M.; Li, H.; Cao, L.X.; Wu, P.C.; Zhang, C.T.; Sang, S.L.; Zhang, X.Y.; Chen, M.J.; Lu, J.Q.; Liu, Y.H. Molecular cloning and characterization of the gene encoding cold-active beta-galactosidase from a psychrotrophic and halotolerant *Planococcus* sp. L4. *J. Agric. Food Chem.* **2007**, *55*, 2217–2224. [CrossRef] [PubMed]

30. Uchil, P.D.; Nagarajan, A.; Kumar, P. β-Galactosidase. *Cold Spring Harb. Protoc.* **2017**, 774–779. [CrossRef] [PubMed]

31. Skálová, T.; Dohnálek, J.; Spiwok, V.; Lipovová, P.; Vondráčková, E.; Petroková, H.; Dusková, J.; Strnad, H.; Králová, B.; Hasek, J. Cold-active beta-galactosidase from *Arthrobacter* sp. C2-2 forms compact 660 kDa hexamers: Crystal structure at 1.9A resolution. *J. Mol. Biol.* **2005**, *353*, 282–294. [CrossRef] [PubMed]

32. Zhang, L.; Wang, K.; Mo, Z.; Liu, Y.; Hu, X. Crystallization and preliminary X-ray analysis of a cold-active β-galactosidase from the psychrotrophic and halotolerant *Planococcus* sp. L4. *Acta Crystallogr. Sect. F Struct. Biol. Cryst. Commun.* **2011**, *67*, 911–913. [CrossRef] [PubMed]

33. Rutkiewicz-Krotewicz, M.; Pietrzyk-Brzezinska, A.J.; Sekula, B.; Cieśliński, H.; Wierzbicka-Woś, A.; Kur, J.; Bujacz, A. Structural studies of a cold-adapted dimeric β-D-galactosidase from *Paracoccus* sp. 32d. *Acta Crystallogr. D Struct. Biol.* **2016**, *72*, 1049–1061. [CrossRef] [PubMed]

34. Saitou, N.; Nei, M. The neighbor-joining method: A new method for reconstructing phylogenetic trees. *Mol. Biol. Evol.* **1987**, *4*, 406–425. [CrossRef] [PubMed]

35. Kumar, S.; Stecher, G.; Tamura, K. MEGA7: Molecular Evolutionary Genetics Analysis Version 7.0 for Bigger Datasets. *Mol. Biol. Evol.* **2016**, *33*, 1870–1874. [CrossRef] [PubMed]

36. Arukha, A.P.; Mukhopadhyay, B.C.; Mitra, S.; Biswas, S.R. A constitutive unregulated expression of β-galactosidase in *Lactobacillus fermentum* M1. *Curr. Microbiol.* **2015**, *70*, 253–259. [CrossRef] [PubMed]

37. Białkowska, A.M.; Cieśliński, H.; Nowakowska, K.M.; Kur, J.; Turkiewicz, M. A new beta-galactosidase with a low temperature optimum isolated from the Antarctic *Arthrobacter* sp. 20B: Gene cloning, purification and characterization. *Arch. Microbiol.* **2009**, *191*, 825–835. [CrossRef] [PubMed]

38. Zhou, Q.Z.K.; Chen, X.D. Effects of temperature and pH on the catalytic activity of the immobilized β-galactosidase from *Kluyveromyces lactis*. *Biochem. Eng. J.* **2001**, *9*, 33–40. [CrossRef]

39. Karasová-Lipovová, P.; Strnad, H.; Spiwok, V.; Maláa, Š.; Králová, B.; Russell, N.J. The cloning, purification and characterisation of a cold-active β-galactosidase from the psychrotolerant Antarctic bacterium *Arthrobacter* sp. C2-2. *Enzym. Microb. Technol.* **2003**, *33*, 836–844. [CrossRef]

10

Microbial Degradation of Amino Acid-Containing Compounds using the Microcystin-Degrading Bacterial Strain B-9

Haiyan Jin [1,*], Yoshiko Hiraoka [2], Yurie Okuma [2], Elisabete Hiromi Hashimoto [2], Miki Kurita [2], Andrea Roxanne J. Anas [2], Hitoshi Uemura [3], Kiyomi Tsuji [3] and Ken-Ichi Harada [1,2,*]

[1] Graduate School of Environmental and Human Science, Meijo University, Tempaku, Nagoya 468-8503, Japan
[2] Faculty of Pharmacy, Meijo University, Tempaku, Nagoya 468-8503, Japan; y715o.c5@gmail.com (Y.H.); g0773315@ccalumni.meijo-u.ac.jp (Y.O.); elisabete.utfpr@gmail.com (E.H.H.); g0673319@ccalumni.meijo-u.ac.jp (M.K.); anasaroj@meijo-u.ac.jp (A.R.J.A.)
[3] Kanagawa Prefectural Institute of Public Health, Shimomachiya, Chigasaki, Kanagawa 253-0087, Japan; uemura.aklt@pref.kanagawa.jp (H.U.); tsuji.df7@pref.kanagawa.jp (K.T.)
* Correspondence: 163891501@ccalumni.meijo-u.ac.jp (H.J.); kiharada@meijo-u.ac.jp (K.-I.H.);

Abstract: Strain B-9, which has a 99% similarity to *Sphingosinicella microcystinivorans* strain Y2, is a Gram-negative bacterium with potential for use in the degradation of microcystin-related compounds and nodularin. We attempted to extend the application area of strain B-9 and applied it to mycotoxins produced by fungi. Among the tested mycotoxins, only ochratoxin A was completely hydrolyzed to provide the constituents ochratoxin α and L-phenylalanine, and levels of fumonisin B1 gradually decreased after 96 h. However, although drugs including antibiotics released into the aquatic environment were applied for microbial degradation using strain B-9, no degradation occurred. These results suggest that strain B-9 can only degrade amino acid-containing compounds. As expected, the tested compounds with amide and ester bonds, such as 3,4-dimethyl hippuric acid and 4-benzyl aspartate, were readily hydrolyzed by strain B-9, although the sulfonamides remained unchanged. The ester compounds were characteristically and rapidly hydrolyzed as soon as they came into contact with strain B-9. Furthermore, the degradation of amide and ester compounds with amino acids was not inhibited by the addition of ethylenediaminetetraacetic acid (EDTA), indicating that the responsible enzyme was not MlrC. These results suggest that strain B-9 possesses an additional hydrolytic enzyme that should be designated as MlrE, as well as an esterase.

Keywords: microcystin-degrading bacteria; mycotoxin; protease; esterase; inhibitor

1. Introduction

Microcystins (MCs) are typical compounds produced by cyanobacteria, such as *Microcystis*, *Anabaena*, and *Planktothrix* [1]. They are cyclic heptapeptides showing potent hepatotoxicity and tumor-promoting activity [1]. In the environment, there are many bacteria which work to degrade such hazardous and harmful compounds. The first MC-degrading bacterium was isolated and identified as a *Sphingomonas* strain (ACM-3962) in 1994 [2]. Similar bacteria capable of degrading MC were reported by Dziga et al. [3]. As per their review [3], many MC-degrading microorganisms have been found and identified, and the corresponding genetic aspects with respect to MC degradation have been studied. However, in related published papers, no substrates other than MCs have been applied [4]. The purpose of the present study is to elucidate the inherent function and role of MC-degrading microorganisms in the aquatic environment.

Strain B-9, isolated from Lake Tsukui, Japan, exhibits degradation activity against MCs [1]. This strain belongs to the genus *Sphingosinecella* sp., and, based on the 16S rDNA sequence (GenBank accession no. AB084247), has 99% similarity to *Sphingosinecella microcystinivorans* strain Y2, a type of MC-degrading bacteria [5]. The *Sphingomonas* sp. strain ACM-3962 [2] was the first strain reported to degrade MCs. The cloning and molecular characterization of four genes from strain ACM-3962 revealed the presence of three hydrolytic enzymes (MlrA, MlrB, and MlrC), together with a putative oligopeptide transporter (MlrD) [6,7]. The three hydrolytic enzymes were putatively characterized as metalloproteases (MlrA and MlrC) or serine proteases (MlrB). The microcystinase MlrA catalyzes the initial ring opening of microcystin-LR (MC-LR) at the (2*S*,3*S*,8*S*,9*S*)-3-amino-9-methoxy-2,6,8-trimethyl-10-phenyldeca-4(*E*),6(*E*)-dienoic acid (Adda)-Arg peptide bond to give linearized MC-LR (Adda-Glu-Mdha-Ala-Leu-β-MeAsp-Arg). This further degrades to Adda-Glu-Mdha-Ala by MlrB, and the third enzyme, MlrC, hydrolyzes the tetrapeptide into smaller peptides and amino acids [6–8]. In terms of advancements, recombinant MlrA and MlrC have been prepared, and the degradation scheme has been almost completely verified [9,10]. The use of typical protease inhibitors, such as ethylenediaminetetraacetic acid (EDTA) and 1,10-phenanthroline, results in the accumulation of linear MC-LR and the tetrapeptide, which allows for the classification of the enzymes MlrA and MlrC as metalloproteases, [6,7]. Meanwhile, phenylmethylsulfonyl fluoride (PMSF) characterizes MlrB as a possible serine protease [6,7].

We extended the area of application of strain B-9 for bioremediation and applied it to the secondary fungal metabolites of mycotoxins that may have mutagenic, carcinogenic, cytotoxic, and endocrine-disrupting effects. These substances frequently contaminate agricultural commodities despite efforts for prevention, so successful detoxification methods are needed. The application of microorganisms to degrade mycotoxins is a possible strategy that shows potential for example in food and feed processing [11]; in antibiotics used in human and veterinary medicine (which can enter the environment via wastewater treatment plant effluents, hospitals, and processing plant effluents); in the application of agricultural waste and biosolids to fields; and in the case of leakage from waste-storage containers and landfills [12,13]. Such antibiotic pollution may facilitate the development and spread of antibiotic resistance [4].

Strain B-9 degrades MC-LR, the most toxic of the MCs, within 24 h [14,15]. After the discovery of strain B-9, we advanced our research on the degradation of the following compounds: non-toxic cyanobacterial cyclic peptides that are structurally different from MCs [16]; representative cyclic peptides (antibiotics) produced by bacteria [17]; physiologically-active cardiovascular and neuropeptides [18]; and the glucagon/vasoactive intestinal polypeptide (VIP) family of peptides [19]. The aforementioned experiments [17–19] confirmed that strain B-9 could degrade the tested peptides completely. During the application of strain B-9 to remove mycotoxins and drugs released in the aquatic environment, we obtained interesting results concerning the function of this strain. The purpose of this study was to demonstrate the additional hydrolytic enzymes (such as protease and/or esterase) of strain B-9.

2. Results

2.1. Mycotoxin and Drugs with Amide, Ester, and Sulfonamide Bonds

As already mentioned [14,15], strain B-9 can degrade the phycotoxins microcystin and nodularin, as well as non-toxic cyclic peptides and linear peptides. In this study, we applied strain B-9 to mycotoxins. We monitored the degradation behavior of the five mycotoxins using HPLC and LC/MS. No degradation was observed in zearalenone, deoxynivalenol, or patulin, while the peak of ochratoxin A completely disappeared (Figure 1). The peak of fumonisin B1 reduced to a certain extent after 96 h (Figure S1). Figure 1 shows (a) HPLC chromatograms; (b) total ion chromatograms, and selected ion monitoring (SIM) (c) at m/z 166.1 and (d) at m/z 257.0 of the reaction products of ochratoxin A by microbial degradation using strain B-9 at 0 h. Although ochratoxin A appeared at 18.7 min in the HPLC chromatogram, such a peak was not observed in the data at 0 h (Figure 1A). It was found that the peak at 20.28 min was derived from strain B-9 itself by comparison of the chromatograms of ochratoxin A and a mixture of the ochratoxin A and strain B-9 broth. Consequently, the peaks at 3.66 and 13.78 min were derived from ochratoxin A, which corresponded to phenylalanine and ochratoxin α, respectively (Figure 2). After 96 h, the latter was still observed, whereas the presence of the former was significantly reduced (Figure 1B).

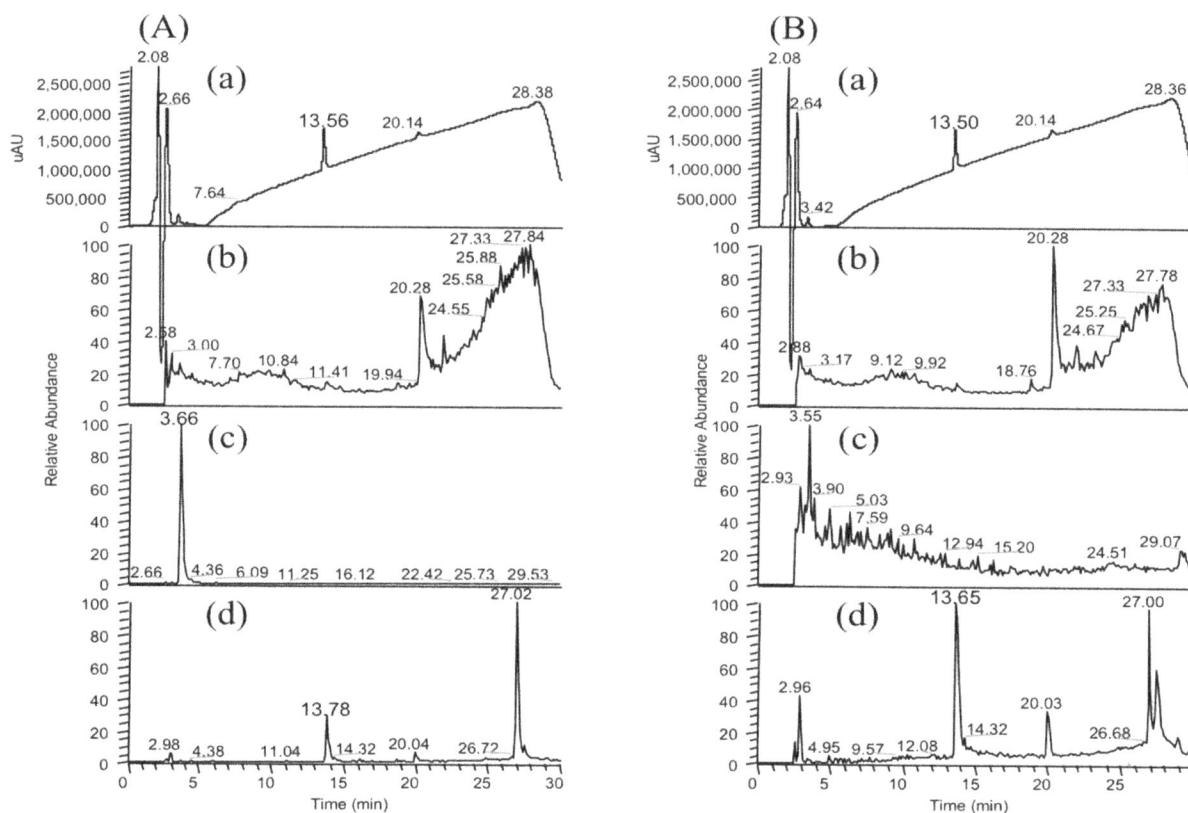

Figure 1. Degradation behavior of ochratoxin A at (**A**) 0 h and (**B**) 96 h after microbial degradation by strain B-9. (**a**) HPLC chromatograms; (**b**) total ion chromatograms, and selected ion monitoring (SIM) (**c**) at m/z 166.1 and (**d**) m/z 257.0 of the reaction products of ochratoxin A.

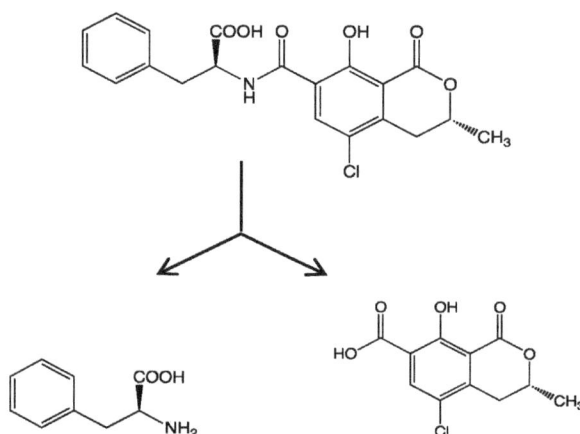

Figure 2. Microbial degradation of ochratoxin A (molecular weight (MW): 403.0) using strain B-9 to provide phenylalanine (MW: 165.1) and ochratoxin α (MW: 256.0).

We tried to degrade drugs including antibiotics released in the aquatic environment using strain B-9 and selected the following drugs with amide, ester, and sulfonamide bonds: oxytetracycline (OTC), sulphaminomethoxime, sulfadimethoxime, oseltamivir, crotamiton, *N,N*-diethyl-*m*-toluamide, and acetaminophen. Although these were treated in the same manner as the mycotoxins, no degradation was observed (data not shown). These results suggested that strain B-9 can degrade only amino acid-containing compounds.

2.2. Amino Acid-Containing Compounds (Amides, Esters, and Sulfonamides)

The following commercially available amino acid-containing compounds with different bonds were selected. Amides: 3,4-dimethylhippuric acid, D- and L-*N*-acetylphenylalanine, *N*-carboben zoxy-L-phenylalanine-L-phenylalanine, and L-leucine-2-naphthylamide; esters: L-serine benzyl ester, and 4-benzyl L-aspartate; and sulfonamides: *N*-(*p*-toluenesulfonyl)-L-phenylalanine, and *N*-(1-naphthalenesulfonyl)-L-phenylalanine (Figure 3). L-Leucine-2-naphthylamide was treated in the same manner as the mycotoxins. The degradation proceeded smoothly and the starting material peak disappeared within 3 h (Figure 4a). A new peak was formed in the HPLC chromatogram. LC/MS showed that the starting material peak appeared after 18.13 min and the new peaks at 2.63 and 9.41 min were detected at 3 h (Figure 5). These peaks were determined to be 2-naphthylamine and leucine by selected ion chromatograms (SIM) at m/z 144.1 and m/z 132.03, respectively. The results indicated that L-leucine-2-naphthylamide was subjected to microbial degradation using strain B-9 to provide 2-naphthyl amine and leucine. As shown in Figure 4, the remaining amide compounds were also degraded and characteristic degradation behavior was observed. In addition, 3,4-dimethylhippuric acid was similarly degraded in the case of L-leucine-2-naphthylamide (Figure 4b). While *N*-acetyl-L-phenylalanine disappeared within 24 h, the D-amino acid derivative continued to appear at 96 h (Figure 4c,d). In the case of *N*-carbobenzoxy-L-phenylalanine-L-phenylalanine, the starting material peak disappeared as soon as it came into contact with strain B-9 (Figure 4e).

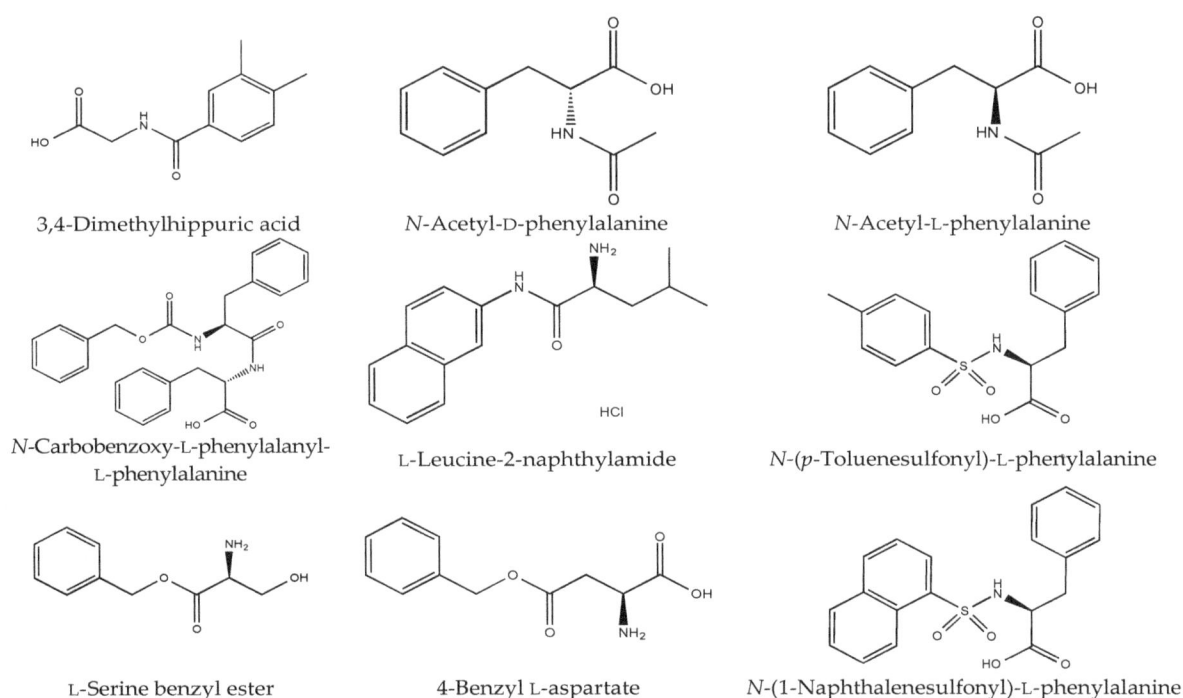

Figure 3. Structures of the tested amino acid-containing compounds. Amides: 3,4-dimethylhip puric acid, D- and L-*N*-acetylphenylalanines, *N*-carbobenzoxy-L-phenylalanine-L-phenylalanine, and L-leucine-2-naphthylamide; esters: serine benzyl ester, and 4-benzyl aspartate; and sulfonamides: *N*-(1-naphthalenesulfonyl)-phenylalanine, and *N*-(*p*-toluenesulfonyl)-L-phenylalanine.

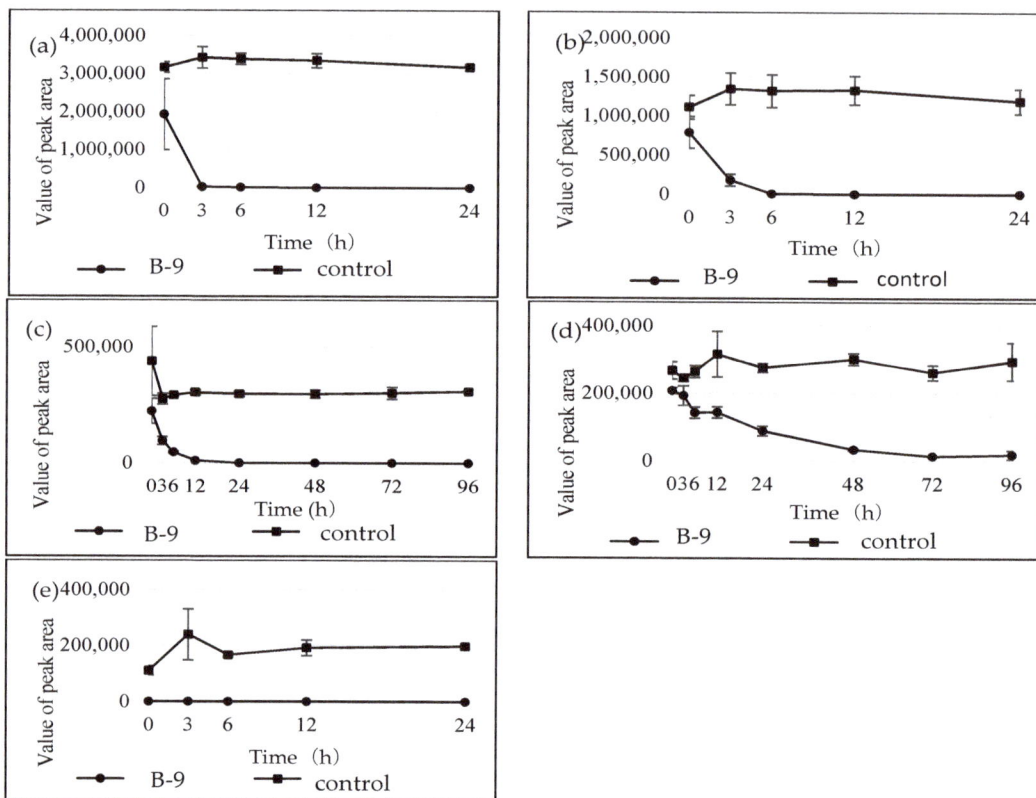

Figure 4. Time courses for the degradation of the tested compounds by strain B-9. (**a**) L-leucine-2-naphthylamide; (**b**) 3,4-dimethylhippuric acid; (**c**) N-acetyl-L-phenylalanine; (**d**) N-acetyl-D-phenyl alanine; and (**e**) N-carbobenzoxy-L-phenyl-L-phenylalanine.

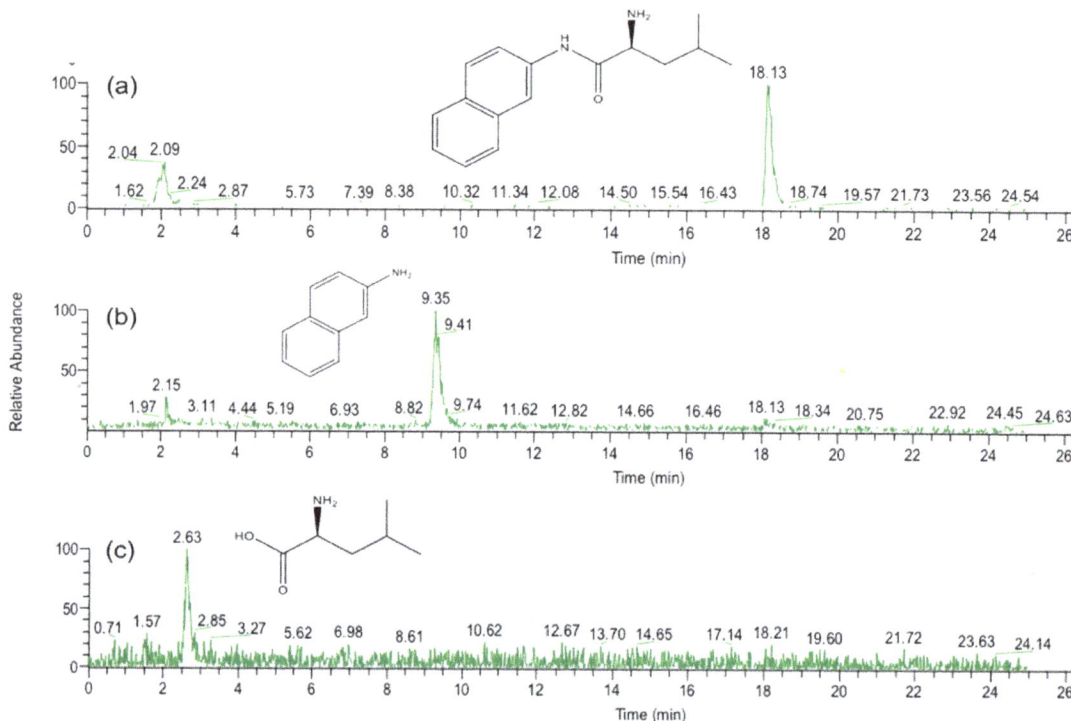

Figure 5. Selected ion chromatograms (SIMs) (**a**) at m/z 257.1 at 0 h; (**b**) at m/z 144.1 at 0 h; and (**c**) at m/z 132.03 at 3 h for reaction products of L-leucine-2-naphthylamide on microbial degradation using strain B-9.

There was a common degradation behavior of the tested ester compounds, in which the starting material peaks disappeared as soon as strain B-9 came into contact with the compounds, as shown in Figure 6a,b. The resulting benzyl alcohol continued to be detected by LC/MS during the experiment (data not shown). Figure 6c,d show the degradation behavior of the sulfonamide compounds. These compounds could not be degraded by strain B-9 during the experimental period.

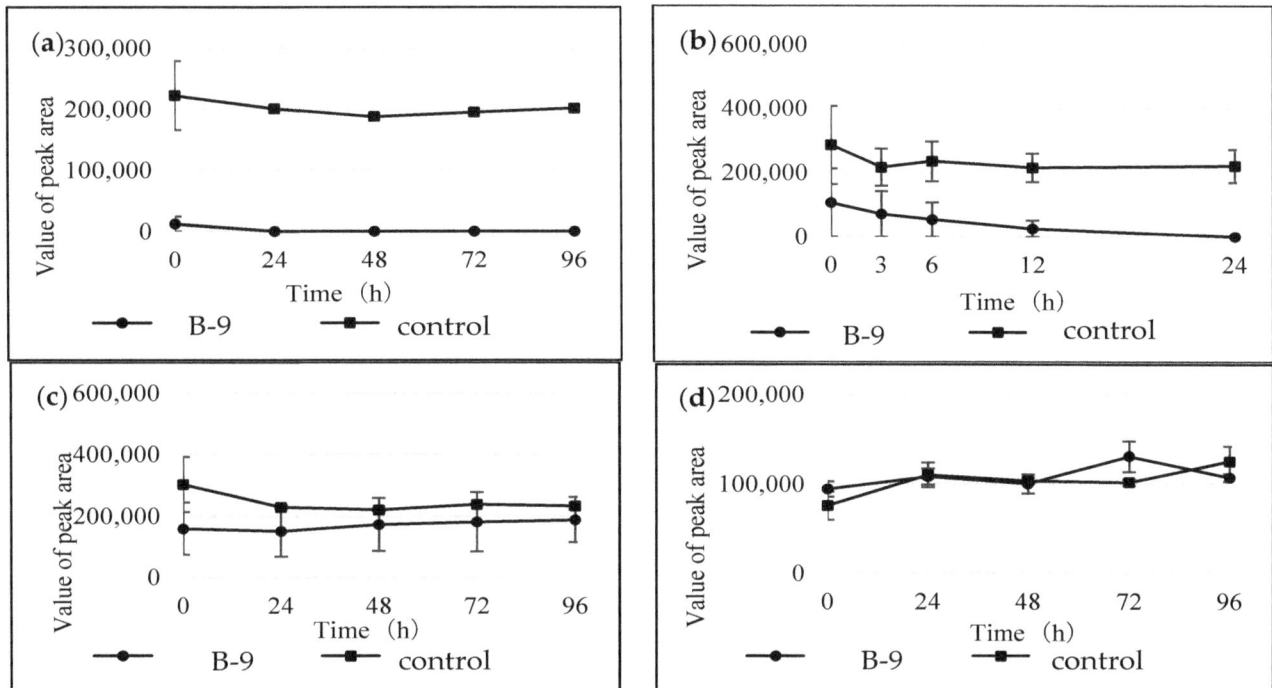

Figure 6. Time courses for the degradation of the tested compounds by strain B-9. (**a**) L-serine benzyl ester; (**b**) 4-benzyl L-aspartate; (**c**) N-(1-naphthalenesulfonyl)-L-phenylalanine; and (**d**) N-(p-toluenesulfonyl)-L-phenylalanine.

2.3. Inhibition of Hydrolysis of Amino Acid-Containing Compounds Using EDTA and PMSF

When inhibitors such as EDTA or PMSF were used at 10-mM concentrations to inhibit the degradation of MCs, the microbial degradation was effectively inhibited. Consequently, the concentration was set at 10 mM in this study. To check the inhibitory activity of the prepared solution, MC-LR was subjected to the microbial degradation in the presence or absence of the inhibitor. While MC-LR (22.7 min) and the resulting tetrapeptide (20.9 min) disappeared within 24 h in the HPLC chromatogram without the inhibitor (Figure S2A), the MC-LR disappeared and the tetrapeptide continued to appear even after 72 h in the HPLC chromatogram with the inhibitor (Figure S2B). These results were consistent with our previous findings that EDTA inhibits MlrC. Figure 7 shows the time course of the degradation using EDTA and the tested compounds. These were: (a) 3,4-dimethylhippuric acid; (b) N-carbobenzoxy-L-phenyl-L-phenylalanine; (c) L-leucine-2-naphthylamide; (d) 4-benzyl L-aspartate; and (e) L-serine benzyl ester. They showed a common degradation behavior in that the microbial degradation of the amide and ester compounds using strain B-9 was not inhibited by EDTA. These results indicated that the degradation mentioned above was not related to MlrC. In the case of PMSF, the following compounds tested positive: 3,4-dimethylhippuric acid; L-leucine-2-naphthylamide; and 4-benzyl-L-aspartate; while other compounds tested negative: N-carbobenzoxy-L-phenylalanine-L-phenylalanine and a-serine benzyl ester. No definitive conclusive information was obtained.

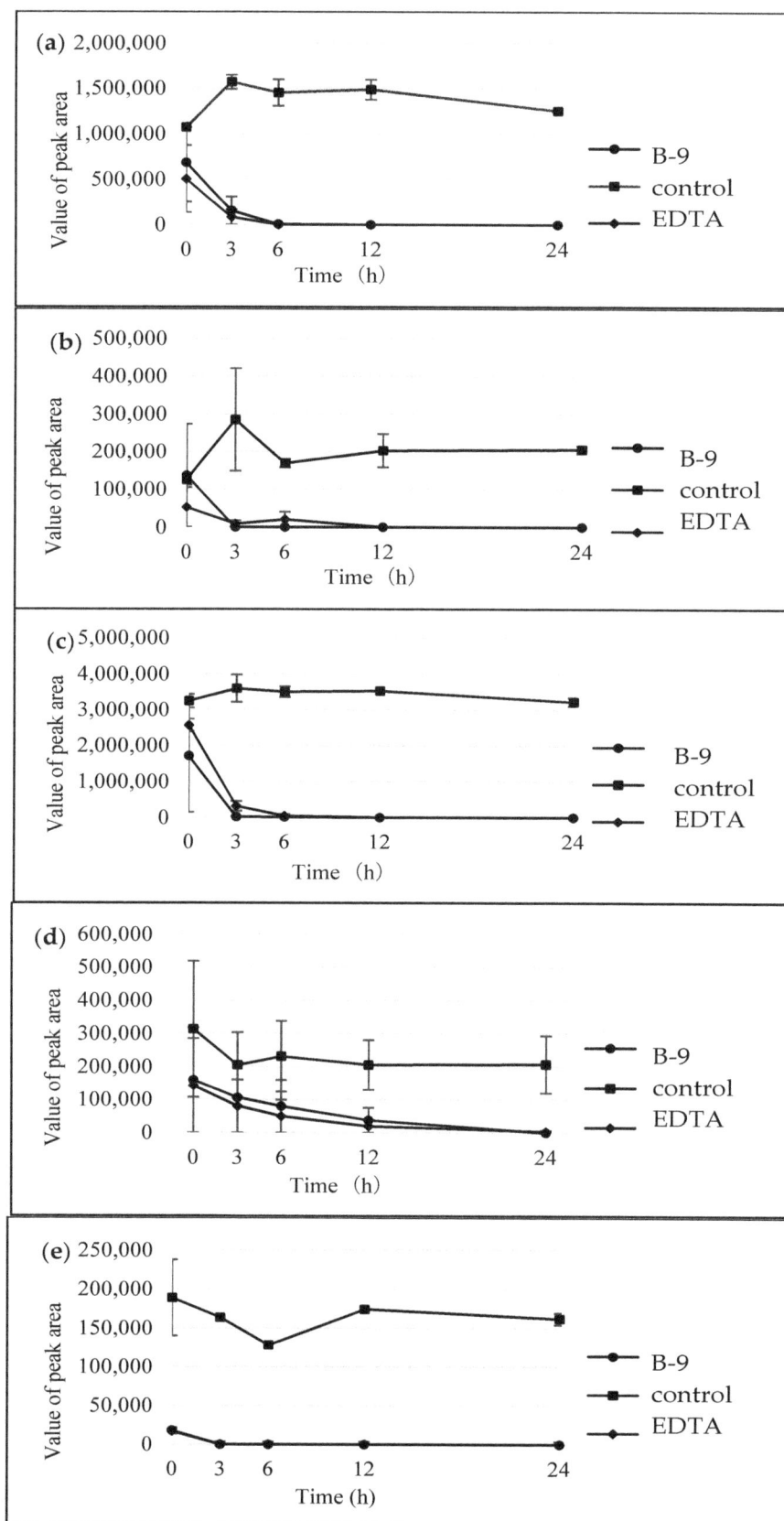

Figure 7. Time courses for the degradation of the tested compounds by B-9 in the presence of ethylenediaminetetraacetic acid (EDTA): (**a**) 3,4-dimethylhippuric acid; (**b**) *N*-carbobenzoxy-L-phenyl-L-phenylalanine; (**c**) L-leucine-2naphthyl amide; (**d**) 4-benzyl L-aspartate; and (**e**) L-serine benzyl ester.

3. Discussion

Strain B-9, which has 99% similarity to *Sphingosinicella microcystinivorans* strain Y2 [5], is Gram-negative and has, in several studies, shown promise for the degradation of MC-related compounds and nodularin [14,15]. Subsequently, we applied strain B-9 to other types of substrates, such as cyanobacterial peptides including depsipeptides [16], and bacterial cyclic peptides including depsipeptides [16], which are structurally different from the MCs and nodularin. Based on these results, the hydrolytic behavior using this strain is suggested as follows: (1) the reaction essentially occurs at a peptide bond in a cyclic peptide moiety to give a linearized peptide, which is more quickly hydrolyzed compared to their original ones; (2) strain B-9 primarily hydrolyzes an ester bond in a depsipeptide, in which the resulting peptides are further hydrolyzed; (3) a cyclic peptide is hydrolyzed at the acyclic part, and no further reaction occurs; and (4) the resulting linearized peptide is more quickly hydrolyzed compared to the cyclic one. In some cases, it is hard to detect the degraded peptides or amino acids due to rapid hydrolysis [16].

To confirm these observations and to further investigate the hydrolytic activities of the strain, we extended our study to physiologically active peptides such as neuropeptides and cardiovascular peptides [18]. The tested peptides were classified into two groups: (1) linear peptides, and (2) cyclic peptides with a loop formed by disulfide bond formation. The linearized peptides degraded faster than the loop-containing peptides because the loop formed by the disulfide bond was regarded as one of the degradation-resistant factors. Hydrolysis of the peptides occurred through the cleavage of various peptide bonds, and strain B-9 may bear similarities to the mammalian neutral endopeptidase (NEP) 24.11, a 94-kDa zinc metalloendoprotease widely distributed in humans and involved in the processing of peptide hormones due to its broad selectivity [20]. In a separate study, we observed the degradation behavior of the linear peptides—the glucagon/VIP family peptides (3200–5000 Da)—by strain B-9 in the absence or presence of two protease inhibitors, EDTA and PMSF. Consequently, we confirmed that one of the B-9 proteases (presumed to be MlrB), which is not inhibited by EDTA, cleaved bioactive peptides in the manner of an endopeptidase similar to NEP. Another protease, which is not inhibited by PMSF, corresponded to MlrC and cleaved the resulting middle-sized peptides to smaller peptides or amino acids [19].

In the present study, we attempted to extend the applications of strain B-9, applying it to mycotoxins produced by fungi. Among the tested mycotoxins, only ochratoxin A was completely hydrolyzed to provide the constituents ochratoxin α and L-phenylalanine (Figure 2), and fumonisin B1 levels gradually decreased to a certain extent after 96 h due to the formation of a new peak at 12.7 min (Figure S1). However, although drugs including antibiotics released into the aquatic environment were applied for microbial degradation using strain B-9, no degradation occurred. These results suggest that strain B-9 can only degrade amino acid-containing compounds. As expected, the tested compounds with amide and ester bonds were readily hydrolyzed by strain B-9, although the sulfonamides were not degraded (Figures 4 and 6). In particular, the ester compounds were characteristically and rapidly hydrolyzed as soon as they came into contact with strain B-9 (Figure 6). Furthermore, the degradations of the amide and ester compounds containing amino acids were not inhibited by the addition of EDTA, suggesting that the responsible enzyme is not MlrC.

It is understood that MlrC is found in the final stage in the microbial degradation of MC, catalyzing the degradation from linearized MC and tetrapeptide to smaller peptides and amino acids. Indeed, Dziga et al. reported the role of MlrC in MC degradation, in which linearized MC and tetrapeptides could be degraded to provide Adda by the cleavage of a peptide bond between Adda and Glu by a recombinant MlrC [14]. These results suggest that strain B-9 possesses an additional hydrolytic enzyme that should be designated as MlrE. Furthermore, the results of the present study suggest that strain B-9 possesses an esterase. As mentioned above, strain B-9 degraded depsipeptides such as aeruginosins and mikamycin A, in which the cleavage at the ester bond was predominant over that of other peptide bonds [16]. However, there may be a possibility that known proteases are responsible for the ester bond cleavage.

Since their discovery in 1994, it has been believed that MC-degrading microorganisms are only responsible for MC degradation [2]. Although many reports on MC-degrading microorganisms have appeared since then [3], few papers have described their substantial function and role in the aquatic environment. As reported by our group, one of the MC-degrading microorganisms, strain B-9, is applicable to structurally diverse peptide compounds, suggesting a different function. We should investigate the detailed function of each hydrolytic and transporter enzyme, as well as a system composed of these enzymes to understand their inherent roles under aquatic conditions.

4. Materials and Methods

4.1. Chemicals

As protease inhibitors, EDTA-2Na (purity: >99.5%) and PMSF (purity: >98.5%), were purchased from Dojindo Laboratories (Kumamoto, Japan) and Sigma-Aldrich Japan (Tokyo, Japan), respectively. Acetonitrile (ACN, LC/MS grade, purity: 99.8%), methanol (MeOH, LC/MS grade, purity: 99.7%), ethanol (EtOH, special grade, purity: 99.5%), formic acid (FA, LC/MS grade, purity: 99.5%), acetic acid (AcOH, LC/MS grade, purity: 99.5%), trifluoroacetic acid (TFA, special grade, purity: 98.0%), ammonium carbonate (special grade), and 28% ammonia solution (NH_4OH, special grade) were purchased from Wako Pure Chemical Industries, Ltd. (Osaka, Japan). Water used for the preparation of all the solutions was purified using a Milli-Q apparatus (Millipore, Billerica, MA, USA); LC/MS analysis used ultrapure water from Wako. The mycotoxins (ochratoxin A, fumonisin B1, zearalenone, deoxynivalenol, patulin) were purchased from Sigma (St. Louis, MO, USA). Drugs with amide, ester, and sulfonamide bonds and amino acid-containing compounds were obtained from the following companies: Aldrich and Sigma Japan (Tokyo, Japan), Nacalai Tesque (Kyoto, Japan), Tokyo Chemical Industry (Tokyo, Japan), and Wako Pure Chemical Industries, Ltd. (Osaka, Japan).

4.2. MC-Degrading Bacterium

Bacterial strain B-9, isolated from the surface water of Lake Tsukui, Kanagawa, Japan, was previously reported to degrade various MCs and nodularin [15]. This bacterium was inoculated into a flask containing 100 mL of Sakurai medium composed of peptone, yeast extract, and glucose, and incubated at 27 °C at 200 revolutions per minute (rpm) for 3 days.

4.3. Degradation of Tested Compounds

Two milligrams of the tested compounds was dissolved in 1 mL of EtOH and 50 μL of the solution was evaporated to dryness. One milliliter of the preincubated cell broth of strain B-9 (containing approximately 3×10^6 colony forming units (CFU) mL^{-1}) was added to the residue. The resulting solution was mixed, and then incubated at 27 °C for 5, 15, 30, 60, and 120 min. After incubation, 50 μL of each of these mixtures was added to 50 μL of MeOH containing 0.2% FA and filtered using an Ultrafree-MC membrane centrifuge-filtration unit (hydrophilic polytetrafluoroethylene (PTFA), 0.20 μm, Millipore, Bedford, MA, USA) to stop the degradation and to eliminate proteins. Each supernatant was then analyzed by HPLC and LC/MS.

4.4. Enzyme Inhibition

Enzyme inhibitors were prepared as follows: EDTA was prepared as a 200-mM stock solution in water and was used at an assay concentration of 10 mM. PMSF was prepared as a 200-mM stock solution in EtOH and was used at an assay concentration of 10 mM. The cell broth and required inhibitor were preincubated at 27 °C for 30 min.

4.5. High-Performance Liquid Chromatography

The degradation process was monitored by HPLC-photo diode array (PDA) at 220 or 254 nm. The system consisted of a pair of LC 10AD VP pumps, a DGU 12A degasser, a CTO 6A column oven, an SPD 10A VP photodiode array detector, and an SCL 10A VP system controller (Shimadzu, Kyoto, Japan). Five microliters of the filtered sample were loaded onto a TSK-gel Super ODS column (2.0 μm, 2.0 × 100 mm, TOSOH, Tokyo, Japan) at 40 °C. The mobile phase was 0.1% formic acid in water (A) and 0.1% formic acid in methanol (B). The gradient conditions were initially 40–90% B for 20 min, and the flow rate was 200 μL/min.

4.6. Liquid Chromatography/Ion Trap Mass Spectrometry

The sample, column, mobile phase, and gradient conditions were the same as those used for the HPLC analysis (12). The LC separation was performed using the Agilent 1100 HPLC system (Agilent Technologies, Palo Alto, CA, USA). Five microliters of the sample was filtrated using an Ultrafree-MC membrane centrifuge filtration unit (hydrophilic PTFE, 0.20 μm, Millipore, Bedford, MA, USA) and loaded onto a TSK-gel Super ODS column (2.0 μm, 2.0 × 100 mm, TOSOH, Tokyo, Japan) at 40 °C. The mobile phase was 0.1% formic acid in water (A) and 0.1% formic acid in acetonitrile (B). The flow rate was 200 μL/min with UV detection at 254 nm. The gradient conditions were initially 10–90% B for 40 min. The entire eluate was directed into the mass spectrometer, where it was diverted to waste 2.5 min after injection to avoid any introduction of salts into the ion source. The MS analysis was accomplished using a Finnigan LCQ Deca XP plus ITMS (Thermo Fischer Scientific, San Jose, CA, USA) equipped with an electrospray ionization (ESI) interface. The ESI conditions in the positive ion mode were as follows: capillary temperature 300 °C, sheath gas flow rate 35 (arbitrary unit), ESI source voltage 5000 V, capillary voltage 43 V, and tube lens offset 15 V. Various scan ranges were used according to the molecular weights of the tested compounds.

Supplementary Materials: Figure S1: Total ion chromatograms (a) and (b) and selected ion monitoring (SIM) at *m/z* 722.4 (c) and *m/z* 564.3 (d) of fumonisin B1 and a reaction product by microbial degradation using B-9 at 0 h (A) and 96 h (B), respectively, Figure S2: (A) HPLC chromatograms of MC-LR by B-9 without EDTA after 0, 6 and 24 h. (B) HPLC chromatograms of MC-LR by B-9 with EDTA after 0, 6 and 72 h.

Acknowledgments: K.-I.H. and H.J. gratefully acknowledge Tatsuko Sakai at the Analytical Services Center, Faculty of Pharmacy, Meijo University, for assistance and support.

Author Contributions: K.-I.H. and H.J. conceptualized the research. K.T., H.U. and Y.O. performed the degradation experiments of drugs including antibiotics with the supervision of K.-I.H.; E.H.H. and M.K. performed the experiments of mycotoxin degradation with the guidance of K.-I.H.; Y.H. and H.J. performed degradation of amide and esters with the guidance of K.-I.H.; M.K., Y.H. and H.J. ran the LC/MS and analyzed the LC/MS data under the supervision of A.R.J.A.; H.J. and K.-I.H. wrote the manuscript. All co-authors agreed to the contents of the paper.

References

1. Tsuji, K.; Asakawa, M.; Anzai, Y.; Sumino, T.; Harada, K.-I. Degradation of microcystins using immobilized microorganism isolated in an eutrophic lake. *Chemosphere* **2006**, *65*, 117–124. [CrossRef] [PubMed]
2. Jones, G.J.; Bourne, D.G.; Blakeley, R.L.; Doelle, H. Degradation of the cyanobacterial hepatotoxin microcystin by aquatic bacteria. *Nat. Toxins* **1994**, *2*, 228–235. [CrossRef] [PubMed]
3. Dziga, D.; Wasylewski, M.; Wladyka, B.; Nybom, S.; Meriluoto, J. Microbial degradation of microcystins. *Chem. Res. Toxicol.* **2013**, *26*, 841–852. [CrossRef] [PubMed]
4. Martinez, J.L. Antibiotics and antibiotic resistance genes in natural environments. *Science* **2008**, *321*, 365–367. [CrossRef] [PubMed]
5. Maruyama, T.; Park, H.D.; Ozawa, K.; Tanaka, Y.; Sumino, T.; Hamana, K.; Hiraishi, A.; Kato, K. *Sphingosinicella microcystinivorans* gen. nov., sp. nov., a microcystin-degrading bacterium. *Int. J. Syst. Evol. Microbiol.* **2006**, *56*, 85–89. [CrossRef] [PubMed]
6. Bourne, D.G.; Jones, G.J.; Blakeley, R.L.; Jones, A.; Negri, A.P.; Riddles, P. Enzymatic pathway for the bacterial degradation of the cyanobacterial cyclic peptide toxin microcystin, LR. *Appl. Environ. Microbiol.* **1996**, *62*, 4086–4094. [PubMed]

7. Bourne, D.G.; Riddles, P.; Jones, G.J.; Smith, W.; Blakeley, R.L. Characterisation of a gene cluster involved in bacterial degradation of the cyanobacterial toxin microcystin, LR. *Environ. Toxicol.* **2001**, *16*, 523–534. [CrossRef] [PubMed]

8. Hashimoto, E.H.; Kato, H.; Kawasaki, Y.; Nozawa, Y.; Tsuji, K.; Hirooka, E.Y.; Harada, K.-I. Further investigation of microbial degradation of microcystin using advanced Marfey's method. *Chem. Res. Toxicol.* **2009**, *22*, 391–398. [CrossRef] [PubMed]

9. Dziga, D.; Wasylewski, M.; Zielinska, G.; Meriluoto, J.; Wasylewski, M. Heterologous expression and characterization of microcystinase. *Toxicon* **2012**, *59*, 578–586. [CrossRef] [PubMed]

10. Dziga, D.; Wasylewski, M.; Szetela, A.; Bochenska, O.; Wladyka, B. Verification of the role of MlrC in microcystin biodegradation by studies using a heterologously expressed enzyme. *Chem. Res. Toxicol.* **2012**, *25*, 1192–1194. [CrossRef] [PubMed]

11. Cserháti, M.; Kriszt, B.; Krifaton, C.; Szoboszlay, S.; Háhn, J.; Tóth, S.; Nagy, I.; Kukolya, J. Mycotoxin-degradation profile of *Rhodococcus* strains. *Int. J. Food Microbiol.* **2013**, *166*, 176–185. [CrossRef] [PubMed]

12. Kümmerer, K. The presence of pharmaceuticals in the environment due to human use—Present knowledge and future challenges. *J. Environ. Manag.* **2009**, *90*, 2354–2366. [CrossRef] [PubMed]

13. Sarmah, A.K.; Meyer, M.T.; Boxall, A.B.A. A global perspective on the use, sales, exposure pathways occurrence, fate and effects of veterinary antibiotics (VAs) in the environment. *Chemosphere* **2006**, *65*, 725–759. [CrossRef] [PubMed]

14. Harada, K.-I.; Imanishi, S.; Kato, H.; Mizuno, M.; Ito, E.; Tsuji, K. Isolation of Adda from microcystin-LR by microbial degradation. *Toxicon* **2004**, *44*, 107–109. [CrossRef] [PubMed]

15. Imanishi, S.; Kato, H.; Mizuno, M.; Tsuji, K.; Harada, K.-I. Bacterial degradation of microcystins and nodularin. *Chem. Res. Toxicol.* **2005**, *18*, 591–598. [CrossRef] [PubMed]

16. Kato, H.; Imanishi, S.Y.; Tsuji, K.; Harada, K.-I. Microbial degradation of cyanobacterial cyclic peptides. *Water Res.* **2007**, *41*, 1754–1762. [CrossRef] [PubMed]

17. Kato, H.; Tsuji, K.; Harada, K.-I. Microbial degradation of cyclic peptides produced by bacteria. *J. Antibiot.* **2009**, *62*, 181–190. [CrossRef] [PubMed]

18. Kondo, F.; Okada, S.; Miyachi, A.; Kurita, M.; Tsuji, K.; Harada, K.-I. Microbial degradation of physiologically active peptides by strain B-9. *Anal. Bioanal. Chem.* **2012**, *403*, 1783–1791. [CrossRef] [PubMed]

19. Miyachi, A.; Kondo, F.; Kurita, M.; Tsuji, K.; Harada, K.-I. Microbial Degradation of linear peptides by strain B-9 of *Sphingosinicella* and its application in peptide quantification using liquid chromatography-mass spectrometry. *J. Biosci. Bioeng.* **2015**, *119*, 724–728. [CrossRef] [PubMed]

20. Stephenson, S.L.; Kenny, A.J. The hydrolysis of α-human atrial natriuretic peptide by pig kidney microvillar membranes is initiated by endopeptidase-24.11. *Biochem. J.* **1987**, *243*, 183–187. [CrossRef] [PubMed]

11

Biochemical Characterization and Degradation Pattern of a Unique pH-Stable PolyM-Specific Alginate Lyase from Newly Isolated *Serratia marcescens* NJ-07

Benwei Zhu [†,*], **Fu Hu** [†], **Heng Yuan, Yun Sun and Zhong Yao** *

College of Food Science and Light Industry, Nanjing Tech University, Nanjing 211816, China; hufu@njtech.edu.cn (F.H.); yuanheng17@njtech.edu.cn (H.Y.); sunyun_food@njtech.edu.cn (Y.S.)
* Correspondence: zhubenwei@njtech.edu.cn (B.Z.); yaozhong@njtech.edu.cn (Z.Y.);

† These authors contributed equally to this work.

Abstract: Enzymatic preparation of alginate oligosaccharides with versatile bioactivities by alginate lyases has attracted increasing attention due to its featured characteristics, such as wild condition and specific products. In this study, AlgNJ-07, a novel polyM-specific alginate lyase with high specific activity and pH stability, has been purified from the newly isolated marine bacterium *Serratia marcescens* NJ-07. It has a molecular weight of approximately 25 kDa and exhibits the maximal activity of 2742.5 U/mg towards sodium alginate under 40 °C at pH 9.0. Additionally, AlgNJ-07 could retain more than 95% of its activity at pH range of 8.0–10.0, indicating it possesses excellent pH-stability. Moreover, it shows high activity and affinity towards polyM block and no activity to polyG block, which suggests that it is a strict polyM-specific alginate lyase. The degradation pattern of AlgNJ-07 has also been explored. The activity of AlgNJ-07 could be activated by NaCl with a low concentration (100–300 mM). It can be observed that AlgNJ-07 can recognize the trisaccharide as the minimal substrate and hydrolyze the trisaccharide into monosaccharide and disaccharide. The TLC and ESI-MS analysis indicate that it can hydrolyze substrates in a unique endolytic manner, producing not only oligosaccharides with Dp of 2–5 but also a large fraction of monosaccharide. Therefore, it may be a potent tool to produce alginate oligosaccharides with lower Dps (degree of polymerization).

Keywords: *Serratia marcescens*; polyM-specific; alginate lyase; oligosaccharides

1. Introduction

Alginate is the major component of cell wall of brown algae [1]. It is a linear anionic polysaccharide and consists of α-L-guluronate (G) and its C5 epimer β-D-mannuronate (M), which are linked by α-1, 4-glycosidic bonds [2]. The two monomeric units are arranged into three groups: poly-α-L-guluronate (polyG), poly-β-D-mannuronate (polyM), and the heteropolymer (polyMG) [3]. Due to its high viscosity, gelling properties, and versatile activities, alginate has been widely applied in food, chemical, and pharmaceutical industries [4–6]. However, the applications of this polysaccharide are still limited by its high molecular weight and poor solubility [7]. The alginate oligosaccharide, as the degradation product of alginate, retains various specific physiological functions and activities of polysaccharide but possesses good bioavailability [8]. For instance, Pack et al. found that alginate oligosaccharide (AOS) can reduce plasma LDL-cholesterol levels by regulating the expression of LDLR [9]. Iwamoto et al. studied the effect of AOS with different structures on the induction of cytokine production from RAW264.7 cells and found that G8 and M7 showed the most potent activity [10]. Yamamoto et al. reported that mannuronate oligomers (M3–M7) could induce the production and secretion of multiple cytokines, such as tumor necrosis factor- α (TNF-α), granulocyte colony-stimulating factor (GCSF), and monocyte chemoattractant protein-1 (MCP-1) [11].

Alginate lyase, a member of polysaccharide lyase, can catalyze the alginate by the β-elimination, producing unsaturated oligosaccharides with double bonds between C4 and C5 [12]. Until now, a number of alginate lyases have been identified, gene-cloned, purified, and characterized from various sources, such as marine and terrestrial bacteria, marine mollusks, and algae [13–18]. According to the substrate specificities, alginate lyases can be classified into three types: polyM-specific lyases (EC 4.2.2.3), polyG-specific lyases (EC 4.2.2.11), and bifunctional lyases (EC 4.2.2.-) [19]. Additionally, the alginate lyases are generally organized into seven polysaccharide lyase (PL) families according to the sequence similarity, namely PL-5, -6, -7, -14, -15, -17, and -18 families [20]. Moreover, in terms of the mode of action, alginate lyases can be grouped into endolytic and exolytic alginate lyases [21]. Endolytic enzymes can cleave glycosidic bonds inside alginate polymer and release unsaturated oligosaccharides as main products [22], while exolytic ones can further degrade oligosaccharides into monomers [23]. Now alginate lyases, especially endolytic enzymes, have been widely used to produce alginate oligosaccharides for food and nutraceutical industries [24,25]. Moreover, the enzymes can also be used to elucidate the fine structures of alginate and prepare protoplast of brown algae [26–28]. Furthermore, alginate lyases also show great potential in the treatment of cystic fibrosis by degrading the polysaccharide biofilm of pathogen bacterium [29]. So far, many alginate lyases originating from marine microorganisms have been well characterized. However, few of these enzymes have been commercially used in the food and nutraceutical industries due to the poor substrate specificity and low activity [30–36]. Thus, to explore novel enzymes with high activity and high substrate specificity will be of great importance for both research and commercial purposes.

In this work, a new alginate lyase with high substrate specificity and pH stability has been identified and characterized from *Serratia marcescens* NJ-07. To evaluate the enzyme for potential use in the food and nutraceutical industries, the kinetics and analysis of degrading products has also been characterized, which suggests that it would be a potential candidate for expanding applications of alginate lyases.

2. Results and Discussions

2.1. Screening and Identification of Strain NJ-07

The strain was isolated from rotten red algae from the Yellow Sea. The 16S rRNA sequence of the strain was sequenced and submitted to GeneBank (accession number MH119760). According to the phylogenetic analysis of 16S rRNA sequence (Figure 1), the strain was assigned to the genus Serratia and named *Serratia marcescens* NJ-07.

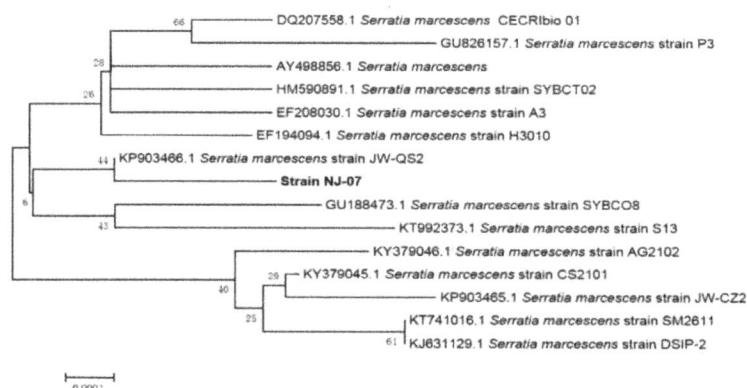

Figure 1. The phylogenetic analysis of strain NJ-07 and other similar strains. The phylogenetic tree was constructed by MEGA 6.0 on the basis of the 16S rRNA gene sequences of strain AlgNJ-07 and other known Serratia species.

2.2. Purification of Alginate Lyase

The strain NJ-07 was cultured in optimized liquid medium for 40 h until alginate lyase reached the highest activity. The supernatant containing alginate lyase was subjected to further purification by anion exchange chromatography with Source 15Q. After purification, the alginate lyase was purified 7.43-fold with a yield of 68.1%. The final specific activity of the purified alginate lyase was 2742.5 U/mg towards sodium alginate. The result of SDS-PAGE showed a single protein band with a molecular weight of 25 kDa (Figure 2), which was designated as AlgNJ-07. The alginate lyases are grouped into three types based on their molecular weights: small alginate lyases (25–30 kDa), medium-sized alginate lyases (around 40 kDa), and large alginate lyases (>60 kDa). As a result, the AlgNJ-07 belongs to the small ones. Similarly, the AlyA from *Azotobacter chroococcum* 4A1M has a small molecular weight of 24 kDa [31]. While the AlyA from *Pseudomonas* sp. E03, AlyA from *Pseudomonas aeruginosa*, and AlyA from *Pseudomonas* sp. strain KS-408 possess the medium-sized molecular weights of 40.4 kDa, 43 kDa, and 44.5 kDa, respectively [33,34,36]. The ALYII from *Pseudomonas* sp. OS-ALG-9 has a large molecular weight of 79 kDa [32].

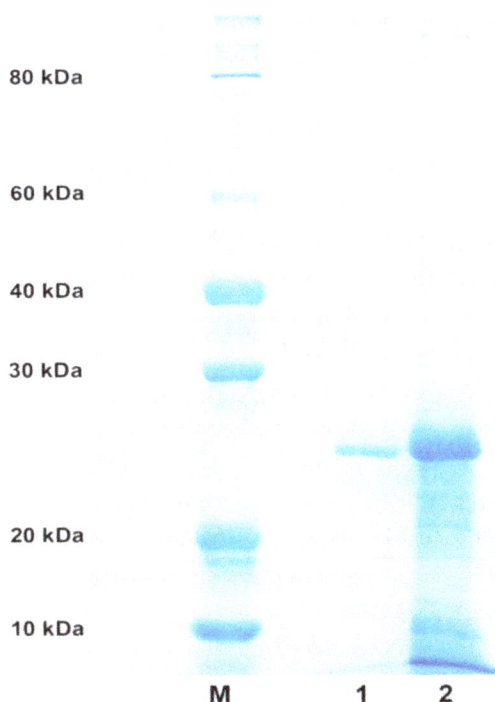

Figure 2. The SDS-PAGE analysis of purified alginate lyase AlgNJ-07. Lane M: the protein molecular weight standard; lane 1: the purified AlgNJ-07; lane 2: the crude enzyme from supernatant.

2.3. Substrate Specifity and Enzymatic Kinetics of the Enzyme

Seven kinds of polysaccharide substrates were used to investigate the substrate specificity of the enzyme (Table 1). The alginate lyase showed higher activity towards sodium alginate and polyM, but no activity towards polyG. Additionally, the AlgNJ-07 displayed no activity towards pullulan, pectin, xylan, and heparin. Therefore, the AlgNJ-07 is a novel polyM-specific alginate lyase. Until now, hundreds of alginate lyases have been identified and characterized. However, only a few enzymes exhibited the polyM-specific activity, such as AlgA from *Pseudomonas* sp. E03 [34], ALYII from *Pseudomonas* sp. OS-ALG-9 [32], the AlyA from *Azotobacter chroococcum* 4A1M [31], AlgL from *Pseudomonas aeruginosa* [30], AlyA from *Pseudomonas aeruginosa* [36], AlyA from *Pseudomonas* sp. strain KS-408, and AlyM from unknown marine bacterium [33,35]. They all displayed preference to polyM substrate and very low activity toward polyG substrate. However, compared with these characterized enzymes, AlgNJ-07 showed no activity toward polyG, indicating it is a novel alginate lyase with strict polyM-specific substrate specificity.

Table 1. The substrate specificity of AlgNJ-07 towards various substrates.

Substrate	Activity (U/mg)
Sodium alginate	2742.5
PolyM	3842.3
PolyG	N.D. *
Pullulan	N.D.
Pectin	N.D.
Xylan	N.D.
Heparin	N.D.

* No activity detected.

The kinetics of AlgNJ-07 towards sodium alginate and polyM were calculated according to the hyperbolic regression analysis. As shown in Table 2, the K_m values of AlgNJ-07 with sodium alginate and polyM as substrates were 0.53 mM and 0.27 mM. The results showed that AlgNJ-07 had a much lower K_m values towards polyM than sodium alginate, indicating that it showed higher affinity towards polyM than that to sodium alginate. The k_{cat}/K_m values of AlgNJ-07 towards polyM ($115 \, \text{mM}^{-1}\cdot\text{s}^{-1}$) was higher than alginate ($64 \, \text{mM}^{-1}\cdot\text{s}^{-1}$), which indicates that the enzyme possesses higher catalytic efficiency towards M block than to MG block. The polyM-specific alginate lyase AlgL from *Pseudomonas aeruginosa* showed different K_m and k_{cat} values towards polyM substrates with various Dps and it exhibited different affinity and catalytic efficiency towards those substrates. The variation in k_{cat}/K_m with substrate length suggests that AlgL operates in a processive manner [30].

Table 2. The kinetics parameters of AlgNJ-07.

Substrate	Sodium Alginate	polyM
K_m (mM)	0.53	0.27
V_{max} (nmol/s)	74	67
k_{cat} (s^{-1})	34	31
k_{cat}/K_m (s^{-1}/mM)	64	115

2.4. Biochemical Characterization of AlgNJ-07

The enzyme showed maximum activity at 40 °C (Figure 3A) and was stable below 40 °C (Figure 3B). It possessed approximately 50% activity after incubation at 40 °C for 30 min and was gradually inactivated as the temperature increased. The thermal degeneration curve of AlgNJ-07 was shown in Figure 4. The enzyme could retain more than 70% of its total activity after being incubated at 40 °C for 60 min, which indicates it possesses better thermal stability. The optimal temperature for polyM-specific alginate lyase from *Pseudomonas* sp. strain KS-408 was 37 °C [33]. The AlgA from *Pseudomonas* sp. E03 and ALYII from *Pseudomonas* sp. OS-ALG-9 both exhibited their maximal activity at 30 °C [33,34]. While the AlyA from *Azotobacter chroococcum* 4A1M showed the highest activity at 60 °C, which shows potential in industrial applications [31].

The optimal pH for the enzyme activity was 9.0 (Figure 3C) and retained more than 80% activity at a broad pH range from pH 8.0 to 10.0 (Figure 3D) after incubation for 24 h. However, this enzyme was mostly stable at pH 9.0 and retained more than 80% activity at a broad pH range from 7.0 to 10.0. Interestingly, it could retain about 40% of its activity at pH 11.0. Thus, AlgNJ-07 was an alkaline-stable lyase and it could retain stability in a broader pH range. While most of the other characterized polyM-specific alginate lyases exhibited their maximal activity around neutral pH. For instance, the AlgA from *Pseudomonas* sp. E03 possessed its optimal pH of 8.0 [34], the AlyA from

Pseudomonas sp. strain KS-408 displayed its maximal activity at pH of 9.0 [34]. While the AlyA from *Azotobacter chroococcum* 4A1M had a lower optimal pH of 6.0 [31].

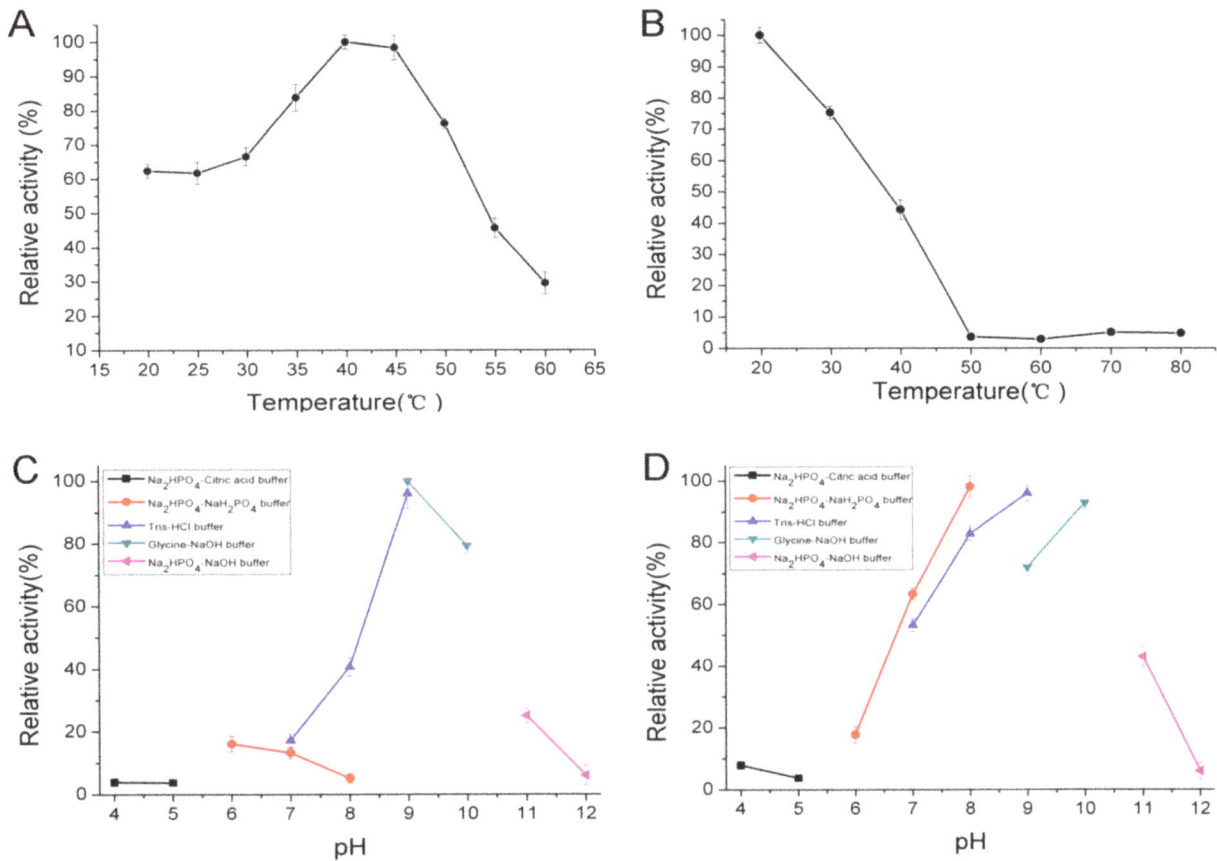

Figure 3. The biochemical characteristics of AlgNJ-07. (**A**) The optimal temperature of AlgNJ-07. (**B**) The thermal stability of AlgNJ-07. (**C**) The optimal pH of AlgNJ-07. (**D**) The pH stability of AlgNJ-07. Each value represents the mean of three replicates ± standard deviation.

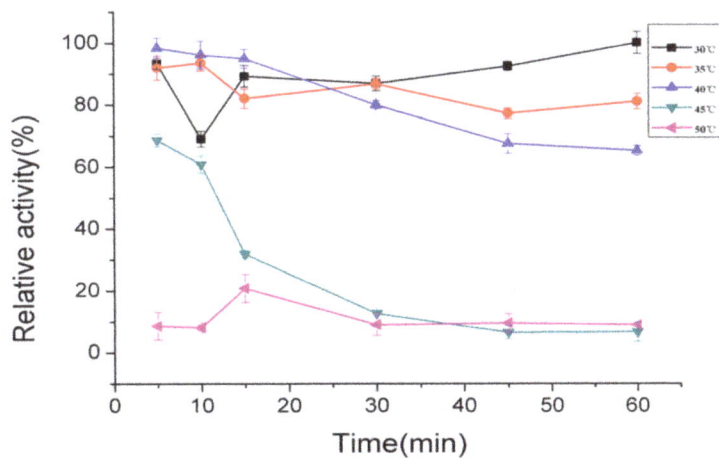

Figure 4. The thermal degeneration curve of AlgNJ-07. The maximal activity of the treated enzyme was regarded as 100% and the other relative activity was determined.

The effects of metal ions on the activity of AlgNJ-07 are shown in Table 3. It was observed that Na^+ could enhance the activity of the enzyme, while some divalent ions such as Zn^{2+}, Cu^{2+}, Mn^{2+}, and Co^{2+} inhibited the activity. Interestingly, the reported activators such as Mg^{2+} and Ca^{2+} displayed slight inhibitory effects on activity of AlgNJ-07. While Ca^{2+} can activate the activities of the AlyA from *Pseudomonas* sp. strain KS-408 [33], the AlyA from *Pseudomonas* sp. E03 [34], the AlyA from *Azotobacter chroococcum* 4A1M [31], and ALYII from *Pseudomonas* sp. OS-ALG-9 [32] could enhance the substrate-binding ability of the enzyme.

Table 3. The effect of metal ions on activity of AlgNJ-07.

Reagent	Relative Activity (%)
Control	100 ± 0.5
K^+ (100 mM)	87 ± 0.5
K^+ (300 mM)	94 ± 0.3
K^+ (500 mM)	92 ± 2.2
Na^+ (100 mM)	106 ± 0.6
Na^+ (300 mM)	120 ± 1.1
Na^+ (500 mM)	103 ± 2.6
Zn^{2+}	1 ± 0.3
Cu^{2+}	5 ± 0.5
Mn^{2+}	4 ± 0.1
Co^{2+}	25 ± 0.3
Ca^{2+}	90 ± 0.3
Fe^{3+}	18 ± 0.1
Mg^{2+}	98 ± 0.5
Ni^{2+}	72 ± 1.2

To determine the number of substrate binding subsites in the active tunnel of AlgNJ-07, we compared the degrading capability of AlgNJ-07 to oligosaccharide substrates with different Dps. As shown in Figure 5, purified disaccharide cannot be further degraded by the enzyme even under more focused conditions (high enzyme concentration and prolonged incubation time). The trisaccharide was the shortest chain that can be recognized and cleaved by AlgNJ-07, producing monosaccharide and disaccharide. The result indicated that trisaccharide was the shortest substrate for AlgNJ-07.

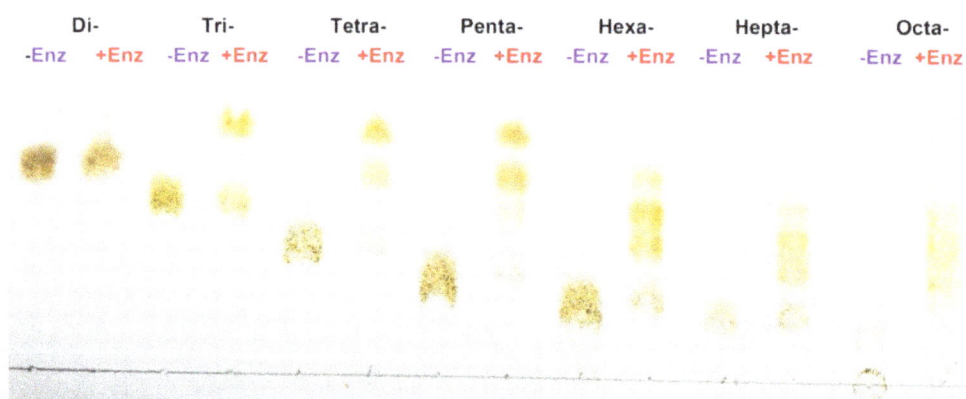

Figure 5. TLC analysis of hydrolysis products of oligosaccharides with Dps (2–8) for determination of substrate binding sites of AlgNJ-07 (−Enz: enzyme free; +Enz: AlgNJ-07 added).

The degradation products of sodium alginate and polyM by AlgNJ-07 were analyzed by TLC plate (Figure 6). As the proceeding of hydrolysis, oligosaccharides with high Dp (6–8) appeared. After incubation for 48 h, dimers, trimers, and tetramers turned out to be the main hydrolysis products for sodium alginate and polyM. Interestingly, the enzyme could release monosaccharide with processing of the hydrolysis. The distributions of the degradation products for the above two kinds of substrates were similar, and the results indicate that AlgNJ-07 can hydrolyze the substrates in a unique endolytic manner.

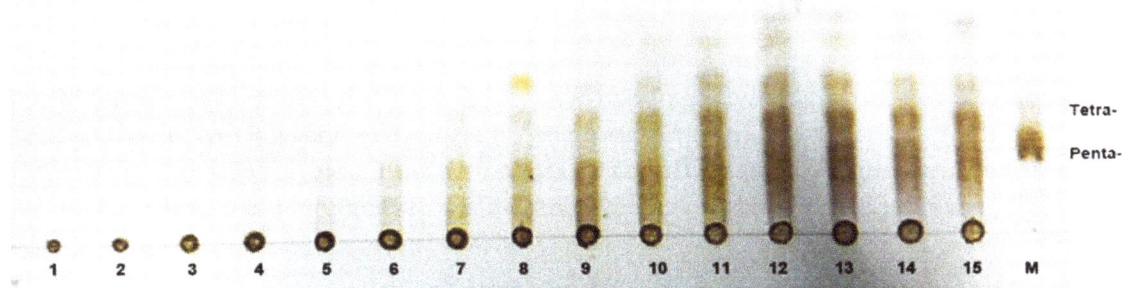

Figure 6. TLC analysis of the AlgNJ-07 hydrolysis products for different times. Lane 1–15, the samples taken by 0 min, 1 min, 3 min, 5 min, 10 min, 15 min, 30 min, 45 min, 60 min, 2 h, 4 h, 12 h, 24 h, 36 h, and 48 h. Lane M, the oligosaccharide standards of tetramer and pentamer.

In order to further determine the composition of the degradation products, the hydrolysates (1 mL) were then loaded onto a carbograph column to remove salts after removing other proteins, followed by being concentrated, dried, and re-dissolved in 1 mL methanol with the final concentration of 1 mg/mL. The degradation products were then analyzed by ESI-MS. As shown in Figure 7, monosaccharides, disaccharides, and trisaccharides account for a major fraction of the hydrolysates of two kinds of substrates. This result indicate that AlgNJ-07 may be a potential tool for the enzymatic hydrolysis of sodium alginate to produce oligosaccharides with lower Dps. The distribution of degradation products of other polyM-specific enzymes is similar, such as AlgA from *Pseudomonas* sp. E03 [34] and AlyA from *Pseudomonas* sp. strain KS-408 [33], which mainly produced oligosaccharides with Dp of 2–5 in an endolytic manner. However, the AlgL from *Pseudomonas aeruginosa* generated dimeric and trimeric products, and the rapid-mixing chemical quench studies indicate that AlgL can operate as an exopolysaccharide lyase [30]. None of those enzymes could produce monosaccharide during the hydrolytic procedure, which indicates that the AlgNJ-07 possesses a unique manner for releasing products.

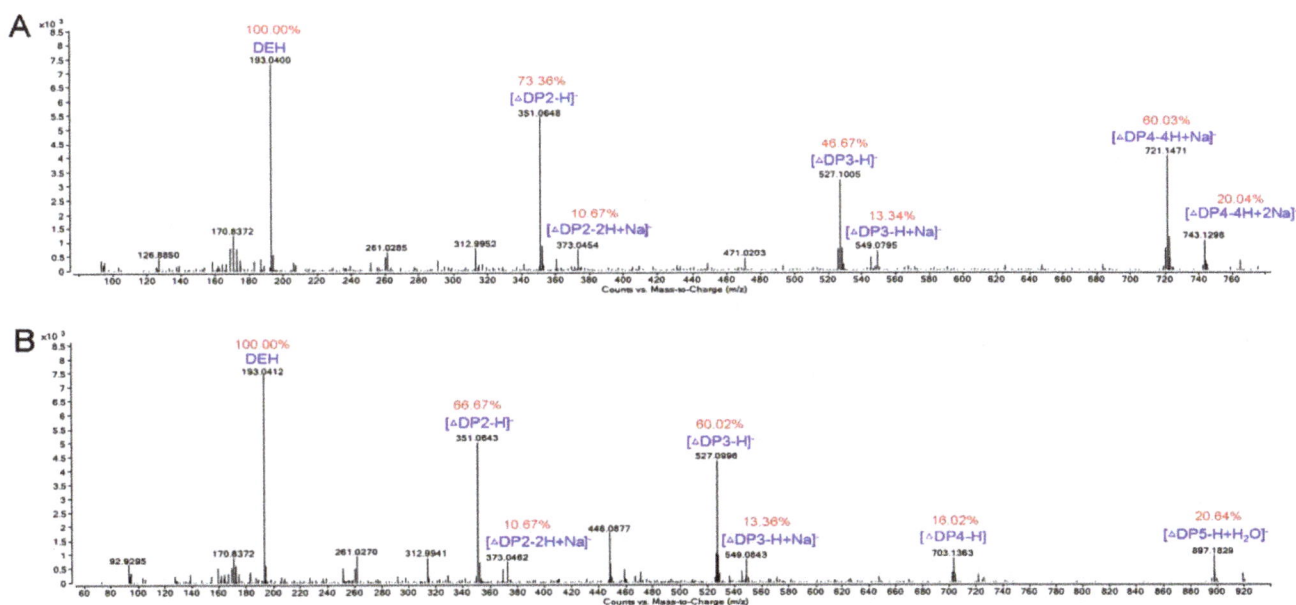

Figure 7. ESI-MS analysis of the degradation products of AlgNJ-07 with (**A**) alginate and (**B**) the polyM as substrate. The data highlighted in red represent the relative abundance of peaks.

3. Materials and Methods

3.1. Materials

Sodium alginate derived from brown seaweed was purchased from Sigma (St. Louis, MO, USA). PolyM (purity: about 99%) and polyG (purity: about 99%) were purchased from Qingdao BZ Oligo Biotech Co., Ltd. (Qingdao, China). The SOURCETM 15Q 4.6/100 PE column was purchased from GE HealthCare Bio-Sciences (Uppsala, Sweden). Other chemicals and reagents used in this study were of analytical grade.

3.2. Screening and Identification of Strain NJ-07

The samples were collected from the coast of the Yellow Sea, washed by sterilized sea water and then spread on sodium alginate-agar plates. The plates were incubated at 30 °C for 36 h and the positive colonies showing clear zones were picked out from the selection plates. The re-screening process was conducted as follows. Strains with clear hydrolytic zones were selected and incubated aerobically in a fermentation medium (modified marine broth 2216 medium containing 5 g/L $(NH_4)_2SO_4$, 19.45 g/L NaCl, 12.6 g/L $MgCl_2 \cdot 6H_2O$, 6.64 g/L $MgSO_4 \cdot 7H_2O$, 0.55 g/L KCl, 0.16 g/L $NaHCO_3$, 1 g/L ferric citrate, and 10 g/L sodium alginate) at 30 °C and 200 rpm. Furthermore, the activity of alginate lyase was determined by 3,5-dinitrosalicylic acid (DNS) colorimetry [37]. Among the isolates, the most active strain NJ-07 was selected for further studies. To identify the NJ-07 strain, the 16S rRNA gene of the strain was amplified through PCR by using universal primers. The purified PCR fragment was sequenced and compared with reported 16S rRNA sequences in GenBank by using BLAST. A phylogenetic tree was constructed using CLUSTAL X and MEGA 6.0 through neighbor-joining method [38].

3.3. Production and Purification of the Alginate Lyase

The strain NJ-07 was propagated in a fermentation medium with shaking for 40 h at 30 °C. The culture medium was centrifuged ($10,000 \times g$, 60 min) to completely remove the sludge and the cell-free supernatant was fractionated at 30% and 80% ammonium sulfate saturation. The precipitated protein with 30% ammonium sulfate saturation was discarded, and the precipitated protein with 80% ammonium sulfate saturation was suspended in distilled water and dialyzed in a dialysis bag (MWCO: 8000–14,000 Da) against the distilled water and freeze-dried successively. Protein contents were determined by the Bradford method [39]. The obtained enzyme powder was dissolved in 5 mL Tris-HCl buffer (pH 9.0) with 4% as the final concentration, then the enzyme solution was applied to a SOURCETM 15Q 4.6/100 PE column equilibrated with a linear gradient of 0–0.5 M NaCl in an equilibrating buffer under a flow rate of 1 mL/min. The eluents were monitored continuously at 280 nm for protein and fractions were assayed for activity against sodium alginate. Fractions were collected and monitored for the presence of alginate lyase. The purity of the fractions was assessed by SDS-PAGE. Pure fractions with activity were stored at −80 °C.

3.4. Enzyme Activity Assay

The purified enzyme (0.1 mL) was mixed with 0.9 mL Tris-HCl (20 mM, pH 8.0, 1% sodium alginate) and incubated at 40 °C for 10 min. The reaction was stopped by heating in boiling water for 10 min. The enzyme activity was then assayed by measuring the increased absorbance at 235 nm due to the formation of double bonds between C4 and C5 at the nonreducing terminus by β-elimination. One unit was defined as the amount of enzyme required to increase the absorbance at 235 nm by 0.01 per min [40].

3.5. Substrate Specificity and Kinetic Measurement of Alginate Lyase

The purified enzyme was reacted with 1% of sodium alginate, polyM, polyG, pectin, xylan, and heparin. The assays of enzyme activity for sodium alginate, polyM, and polyG were defined as described previously, whereas the assays for pectin, xylan, and heparin were determined by using the DNS method. The kinetic parameters of the purified enzyme toward sodium alginate and polyM were determined by measuring the enzyme activity with substrates at different concentrations (0.1–8.0 mg/mL). As sodium alginate is a polymer consisting of random combinations of mannuronic acid and guluronic acid residues. Since they both have the same molecular weight (MW), substrate molarity was calculated using the MW of 176 g/mol for each monomer of uronic acid in the polymer. The concentrations of the product were determined by monitoring the increase in absorbance at 235 nm using the extinction coefficient of 6150 M^{-1} cm^{-1}. Velocity (V) at the tested substrate concentration was calculated as follows: V (mol/s) = (milliAU/min × min/60 s × AU/1000 milliAU × 1 cm)/(6150 M^{-1} cm^{-1}) × (2 × 10^{-4} L). The K_m and V_{max} values were calculated by hyperbolic regression analysis as described previously [41]. Additionally, the turnover number (k_{cat}) of the enzyme was calculated by the ration of V_{max} versus enzyme concentration ([E]).

3.6. Biochemical Characterization of AlgNJ-07

The effects of pH on the enzyme activity were evaluated by incubating the purified enzyme in buffers with different pHs (4.0–12.0) at 40 °C under the assay conditions described previously. The pH stability depended on the residual activity after the enzyme was incubated in buffers with different pH (4.0–12.0) for 24 h and then residual activity was determined at 40 °C under the assay conditions. Meanwhile, the effects of temperatures (20–60 °C) on the purified enzyme were investigated at pH 9.0. The thermal stability of the enzyme was determined at pH 9.0 under the assay conditions described previously after incubating the purified enzyme at 30–50 °C for 30 min. The buffers with different pHs used were phosphate-citrate (pH 4.0–5.0), NaH_2PO_4-Na_2HPO_4 (pH 6.0–8.0), Tris–HCl (pH 7.0–9.0), glycine-NaOH (pH 9.0–10.0), and Na_2HPO_4–NaOH (pH 11.0–12.0). In addition, the thermally-induced denaturation was also investigated by incubating the enzyme at 30–50 °C for 0–60 min.

The influence of metal ions on the activity of the enzyme was performed by incubating the purified enzyme at 4 °C for 24 h in the presence of various metal compounds at a concentration of 1 mM. Then, the activity was measured under standard test conditions. The reaction mixture without any metal ions was used as a control.

3.7. Substrate Binding Subsites of AlgNJ-07

To determine the smallest substrate and the number of substrate binding subsites in its catalytic tunnel of AlgNJ-07, hydrolysis reactions were carried out using oligosaccharides with different Dps (Dp 2–8) at a concentration of 10 mg/mL in 10 μL reaction mixture (pH 9.0). The reaction mixtures were incubated at 40 °C with AlgNJ-07 for 24 h. The hydrolysates were loaded onto a carbograph column (Alltech, Grace Davison Discovery Sciences, Carnforth, UK) to remove salts after removing proteins, and then concentrated, dried, and re-dissolved in 1 mL methanol. The degradation products were analyzed by TLC with the solvent system (1-butanol/formic acid/water 4:6:1) and visualized by heating TLC plate at 130 °C for 5 min after spraying with 10% (v/v) sulfuric acid in ethanol.

3.8. TLC and ESI-MS Analysis of the Degradation Products of AlgNJ-07

To investigate the degradation pattern of AlgNJ-07, the reaction mixtures (800 μL) containing 1 μg purified enzyme and 2 mg substrates (sodium alginate and polyM) were incubated at 30 °C for 0–48 h. The hydrolysis products were analyzed by TLC as above. To further determine the composition of the products, ESI-MS was used. The supernatants (2 μL) were loop-injected to an LTQ XL linear ion trap mass spectrometer (Thermo Fisher Scientific, Waltham, MA, USA) after centrifugation. Samples were introduced by direct infusion into the electrospray ionization source (ESI) and mass spectra (MS) were collected. To help elucidate the structure of the ESI-MS peaks, the MS spectra were collected concurrently by isolating specific m/z anions, and the oligosaccharides were detected in a negative-ion mode using the following settings: ion source voltage, 4.5 kV; capillary temperature, 275–300 °C; tube lens, 250 V; sheath gas, 30 arbitrary units (AU); and scanning the mass range, 150–2000 m/z.

4. Conclusions

An alginate lyase-producing bacterium was isolated and identified as *Serratia marcescens* NJ-07. The alginate lyase AlgNJ-07 was purified by anion-exchange chromatography. It had a molecular weight of approximately 25 kDa and exhibited the maximal activity of 2742.52 U/mg under 40 °C at pH 9.0. Additionally, AlgNJ-07 could retain more than 95% of its activity at pH range of 8.0–10.0, which indicates it possesses excellent pH-stability. It showed high activity and affinity toward polyM block and no activity on polyG block, suggesting it is a strict polyM-specific alginate lyase. TLC and ESI-MS analysis indicated that it can hydrolyze substrates in a unique endolytic manner and produce oligosaccharides with Dp of 2–5 and a large fraction of monosaccharides. Therefore, it may be a potent tool to produce alginate oligosaccharides with lower Dps.

Acknowledgments: The authors gratefully acknowledge the financial support of the National Natural Science Foundation of China (No. 31601410).

Author Contributions: B.Z. and F.H. conceived and designed the experiments; B.Z. and F.H. performed the experiments; Y.S., H.Y. and Z.Y. analyzed the data; B.Z. wrote the paper. All authors reviewed the manuscript.

References

1. Gacesa, P. Enzymic degradation of alginates. *Int. J. Biochem.* **1992**, *24*, 545–552. [CrossRef]
2. Pawar, S.N.; Edgar, K.J. Alginate derivatization: A review of chemistry, properties and applications. *Biomaterials* **2012**, *33*, 3279–3305. [CrossRef] [PubMed]
3. Mabeau, S.; Kloareg, B. Isolation and analysis of the cell walls of brown algae: *Fucus spiralis, F. Ceranoides, F. Vesiculosus, F. Serratus*, bifurcaria bifurcata and laminaria digitata. *J. Exp. Bot.* **1987**, *38*, 1573–1580. [CrossRef]
4. Fujihara, M.; Nagumo, T. An influence of the structure of alginate on the chemotactic activity of macrophages and the antitumor activity. *Carbohydr. Res.* **1993**, *243*, 211–216. [CrossRef]
5. Otterlei, M.; Ostgaard, K.; Skjak-Braek, G.; Smidsrod, O.; Soon-Shiong, P.; Espevik, T. Induction of cytokine production from human monocytes stimulated with alginate. *J. Immunother.* **1991**, *10*, 286–291. [CrossRef] [PubMed]
6. Bergero, M.F.; Liffourrena, A.S.; Opizzo, B.A.; Fochesatto, A.S.; Lucchesi, G.I. Immobilization of a microbial consortium on ca-alginate enhances degradation of cationic surfactants in flasks and bioreactor. *Int. Biodeterior. Biodegrad.* **2017**, *117*, 39–44. [CrossRef]
7. Yang, J.S.; Xie, Y.J.; He, W. Research progress on chemical modification of alginate: A review. *Carbohydr. Polym.* **2011**, *84*, 33–39. [CrossRef]
8. Tai, H.B.; Tang, L.W.; Chen, D.D.; Irbis, C.; Bioconvertion, L.O. Progresses on preparation of alginate oligosaccharide. *Life Sci. Res.* **2015**, *19*, 75–79.
9. Do, J.R.; Back, S.Y.; Kim, H.K.; Lim, S.D.; Jung, S.K. Effects of aginate oligosaccharide on lipid metabolism in mouse fed a high cholesterol diet. *J. Korean Soc. Food Sci. Nutr.* **2014**, *43*, 491–497.
10. Iwamoto, M.; Kurachi, M.; Nakashima, T.; Kim, D.; Yamaguchi, K.; Oda, T.; Iwamoto, Y.; Muramatsu, T. Structure-activity relationship of alginate oligosaccharides in the induction of cytokine production from raw264.7 cells. *FEBS Lett.* **2005**, *579*, 4423–4429. [CrossRef] [PubMed]
11. Yamamoto, Y.; Kurachi, M.; Yamaguchi, K.; Oda, T. Induction of multiple cytokine secretion from raw264.7 cells by alginate oligosaccharides. *J. Agric. Chem. Soc. Jpn.* **2007**, *71*, 238–241. [CrossRef]
12. Wong, T.Y.; And, L.A.P.; Schiller, N.L. Alginate lyase: Review of major sources and enzyme characteristics, structure-function analysis, biological roles, and applications. *Annu. Rev. Microbiol.* **2000**, *54*, 289–340. [CrossRef] [PubMed]

13. Hashimoto, W.; Miyake, O.; Momma, K.; Kawai, S.; Murata, K. Molecular identification of oligoalginate lyase of sphingomonas sp. Strain a1 as one of the enzymes required for complete depolymerization of alginate. *J. Bacteriol.* **2000**, *182*, 4572–4577. [CrossRef] [PubMed]

14. Zhu, B.; Tan, H.; Qin, Y.; Xu, Q.; Du, Y.; Yin, H. Characterization of a new endo-type alginate lyase from *Vibrio* sp. W13. *Int. J. Biol. Macromol.* **2015**, *75*, 330–337. [CrossRef] [PubMed]

15. Rahman, M.M.; Inoue, A.; Tanaka, H.; Ojima, T. Isolation and characterization of two alginate lyase isozymes, akaly28 and akaly33, from the common sea hare aplysia kurodai. *Comp. Biochem. Phys.* **2010**, *157*, 317–325. [CrossRef] [PubMed]

16. Li, J.W.; Dong, S.; Song, J.; Li, C.B.; Chen, X.L.; Xie, B.B.; Zhang, Y.Z. Purification and characterization of a bifunctional alginate lyase from *Pseudoalteromonas* sp. Sm0524. *Mar. Drugs* **2011**, *9*, 109–123. [CrossRef] [PubMed]

17. Zhu, X.; Li, X.; Shi, H.; Zhou, J.; Tan, Z.; Yuan, M.; Yao, P.; Liu, X. Characterization of a novel alginate lyase from marine *Bacteriumvibrio furnissii* H1. *Mar. Drugs* **2018**, *16*, 30. [CrossRef] [PubMed]

18. Chen, P.; Zhu, Y.; Men, Y.; Zeng, Y.; Sun, Y. Purification and characterization of a novel alginate lyase from the marine *Bacterium bacillus* sp. Alg07. *Mar. Drugs* **2018**, *16*, 86. [CrossRef] [PubMed]

19. Zhu, B.; Yin, H. Alginate lyase: Review of major sources and classification, properties, structure-function analysis and applications. *Bioengineered* **2015**, *6*, 125–131. [CrossRef] [PubMed]

20. Henrissat, B. A classification of glycosyl hydrolases based on amino acid sequence similarities. *Biochem. J.* **1991**, *280 Pt 2*, 309–316. [CrossRef] [PubMed]

21. Kim, H.S.; Lee, C.G.; Lee, E.Y. Alginate lyase: Structure, property, and application. *Biotechnol. Bioprocess Eng.* **2011**, *16*, 843. [CrossRef]

22. Kim, H.T.; Ko, H.J.; Kim, N.; Kim, D.; Lee, D.; Choi, I.G.; Woo, H.C.; Kim, M.D.; Kim, K.H. Characterization of a recombinant endo-type alginate lyase (alg7d) from *Saccharophagus degradans*. *Biotechnol. Lett.* **2012**, *34*, 1087–1092. [CrossRef] [PubMed]

23. Heetaek, K.; Jaehyuk, C.; Wang, D.M.; Jieun, L.; Heechul, W.; Ingeol, C.; Kyoungheon, K. Depolymerization of alginate into a monomeric sugar acid using alg17c, an exo-oligoalginate lyase cloned from *Saccharophagus degradans* 2–40. *Appl. Microbiol. Biotechnol.* **2012**, *93*, 2233–2239.

24. Zhang, Z.; Yu, G.; Guan, H.; Zhao, X.; Du, Y.; Jiang, X. Preparation and structure elucidation of alginate oligosaccharides degraded by alginate lyase from vibro sp. 510. *Carbohydr. Res.* **2004**, *339*, 1475–1481. [CrossRef] [PubMed]

25. Zhu, B.; Chen, M.; Yin, H.; Du, Y.; Ning, L. Enzymatic hydrolysis of alginate to produce oligosaccharides by a new purified endo-type alginate lyase. *Mar. Drugs* **2016**, *14*, 108. [CrossRef] [PubMed]

26. Min, K.H.; Sasaki, S.F.; Kashiwabara, Y.; Umekawa, M.; Nisizawa, K. Fine structure of smg alginate fragment in the light of its degradation by alginate lyases of pseudomonas sp. *J. Biochem.* **1977**, *81*, 555–562. [CrossRef] [PubMed]

27. Boyen, C.; Kloareg, B.; Polne-Fuller, M.; Gibor, A. Preparation of alginate lyases from marine molluscs for protoplast isolation in brown algae. *Phycologia.* **1990**, *29*, 173–181. [CrossRef]

28. Inoue, A.; Kagaya, M.; Ojima, T. Preparation of protoplasts from laminaria japonica using native and recombinant abalone alginate lyases. *J. Appl. Phycol.* **2008**, *20*, 633–640. [CrossRef]

29. Alkawash, M.A.; Soothill, J.S.; Schiller, N.L. Alginate lyase enhances antibiotic killing of mucoid *Pseudomonas aeruginosa* in biofilms. *APMIS* **2006**, *114*, 131–138. [CrossRef] [PubMed]

30. Farrell, E.K.; Tipton, P.A. Functional characterization of algl, an alginate lyase from *Pseudomonas aeruginosa*. *Biochemistry* **2012**, *51*, 10259–10266. [CrossRef] [PubMed]

31. Haraguchi, K.; Kodama, T. Purification and propertes of poly(β-D-mannuronate) lyase from azotobacter chroococcum. *Appl. Microbiol. Biotechnol.* **1996**, *44*, 576–581. [CrossRef]

32. Kraiwattanapong, J.; Motomura, K.; Ooi, T.; Kinoshita, S. Characterization of alginate lyase (alyii) from *Pseudomonas* sp. Os-alg-9 expressed in recombinant escherichia coli. *World J. Microb. Biotechnol.* **1999**, *15*, 105–109. [CrossRef]

33. Kam, N.; Park, Y.J.; Lee, E.Y.; Kim, H.S. Molecular identification of a polym-specific alginate lyase from *Pseudomonas* sp. Strain ks-408 for degradation of glycosidic linkages between two mannuronates or mannuronate and guluronate in alginate. *Can. J. Microbiol.* **2011**, *57*, 1032–1041. [CrossRef] [PubMed]

34. Zhu, B.W.; Huang, L.S.X.; Tan, H.D.; Qin, Y.Q.; Du, Y.G.; Yin, H. Characterization of a new endo-type polym-specific alginate lyase from *Pseudomonas* sp. *Biotechnol. Lett.* **2015**, *37*, 409–415. [CrossRef] [PubMed]

35. Romeo, T.; Iii, J.F.P. Purification and structural properties of an extracellular (1–4)-.beta.-D-mannuronan-specific alginate lyase from a marine bacterium. *Biochemistry* **1986**, *25*, 8385–8391. [CrossRef]

36. Eftekhar, F.; Schiller, N.L. Partial purification and characterization of a mannuronan-specific alginate lyase from *Pseudomonas aeruginosa*. *Curr. Microbiol.* **1994**, *29*, 37–42. [CrossRef]

37. Miller, G.L. Use of dinitrosalicylic acid reagent for determination of reducing sugar. *Anal. Biochem.* **1959**, *31*, 426–428. [CrossRef]

38. Saitou, N.; Nei, M. The neighbor-joining method: A new method for reconstructing phylogenetic trees. *Mol. Biol. Evol.* **1987**, *4*, 406–425. [PubMed]

39. Kruger, N.J. The bradford method for protein quantitation. *Methods Mol. Biol.* **1988**, *32*, 9–15.

40. Zhu, B.; Ni, F.; Sun, Y.; Yao, Z. Expression and characterization of a new heat-stable endo-type alginate lyase from deep-sea bacterium *Flammeovirga* sp. Nj-04. *Extremophiles* **2017**, *21*, 1027–1036. [CrossRef] [PubMed]

41. Swift, S.M.; Hudgens, J.W.; Heselpoth, R.D.; Bales, P.M.; Nelson, D.C. Characterization of algmsp, an alginate lyase from *Microbulbifer* sp. 6532a. *PLoS ONE* **2014**, *9*, e112939. [CrossRef] [PubMed]

12

Biochemical Characterization and Elucidation of Action Pattern of a Novel Polysaccharide Lyase 6 Family Alginate Lyase from Marine Bacterium *Flammeovirga* sp. NJ-04

Qian Li, Fu Hu, Benwei Zhu *, Yun Sun and Zhong Yao

College of Food Science and Light Industry, Nanjing Tech University, Nanjing 211816, China; njlq@njtech.edu.cn (Q.L.); hufu@njtech.edu.cn (F.H.); sunyun_food@njtech.edu.cn (Y.S.); yaozhong@njtech.edu.cn (Z.Y.)
* Correspondence: zhubenwei@njtech.edu.cn.

Abstract: Alginate lyases have been widely used to prepare alginate oligosaccharides in food, agricultural, and medical industries. Therefore, discovering and characterizing novel alginate lyases with excellent properties has drawn increasing attention. Herein, a novel alginate lyase FsAlyPL6 of Polysaccharide Lyase (PL) 6 family is identified and biochemically characterized from *Flammeovirga* sp. NJ-04. It shows highest activity at 45 °C and could retain 50% of activity after being incubated at 45 °C for 1 h. The Thin-Layer Chromatography (TLC) and Electrospray Ionization Mass Spectrometry (ESI-MS) analysis indicates that FsAlyPL6 endolytically degrades alginate polysaccharide into oligosaccharides ranging from monosaccharides to pentasaccharides. In addition, the action pattern of the enzyme is also elucidated and the result suggests that FsAlyPL6 could recognize tetrasaccharide as the minimal substrate and cleave the glycosidic bonds between the subsites of −1 and +3. The research provides extended insights into the substrate recognition and degradation pattern of PL6 alginate lyases, which may further expand the application of alginate lyases.

Keywords: alginate lyase; polysaccharide lyase of family 6; characterization; degradation pattern

1. Introduction

Alginate is a linear acidic polysaccharide that constitutes the cell wall of brown algae [1]. It consists of two uronic acids, namely the β-ᴅ-mannuronate (M) and the α-ʟ-guluronate (G), which are randomly arranged into different blocks [2]. The alginate has been widely used in food, agricultural and medical industries due to its favorable properties and versatile activities. However, the applications of alginate have been greatly limited by its disadvantages such as high molecular weight, low solubility, and poor bioavailability. In addition, the alginate molecule could not get into the circulation system due to its huge molecular structure. Therefore, it could not exhibit its physiological activities. Alginate oligosaccharides, as the degrading products of alginate, are smaller with excellent solubility and bioavailability than the polysaccharides. In addition, the physiological effects, such as anticoagulant, antioxidant, and antineoplastic activities, can also be retained after degradation. Therefore, they have been widely used as anticoagulants, plant growth accelerators and tumor inhibitors in food, agricultural, and medical fields [3–5]. Therefore, it holds great promise to degrade the alginate to prepare functional alginate oligosaccharides [6].

Alginate lyases could degrade alginate to oligosaccharides by β-elimination mechanism and therefore they belong to the Polysaccharides Lyase (PL) family [7]. Recently, alginate lyases have drawn increasing attention for preparing alginate oligosaccharides with the advantages such as high efficiency and specificity and mild degrading conditions [8]. Up to now, numerous alginate lyases have been isolated, identified, and characterized [9]. Unfortunately, only a few show high activity and thermal stability, which are essential properties for industrial applications [10,11]. Previously, two alginate lyases with excellent characteristics have been identified from the *Flammeovirga* sp. NJ-04. In this study, a novel alginate lyase of PL 6 family has been cloned and characterized from the strain. The biochemical properties and degrading pattern of the enzyme have been investigated and this research would further expand the applications of alginate lyases in related fields.

2. Results and Discussion

2.1. Sequence Analysis of FsAlyPL6

The gene of FsAlyPL6 was cloned and analyzed from *Flammeovirga* sp. NJ-04. The open reading frame (ORF) consisted of 2238 bps and encoded a putative alginate lyase of 745 amino acid residues with a theoretical molecular mass of 83.09 kDa. According to the conserved domain analysis, the FsAlyPL6 contained an N-terminal catalytic domain (Met^1-Asn^{366}) and a C-terminal domain (Gln^{367}-Lys^{745}). Based on the sequence alignments shown in Figure 1, FsAlyPL6 shared the highest identity (45%) with AlyGC (BAEM00000000.1) from *Glaciecola chathamensis* S18K6T, which indicated FsAlyPL6 is a new member of family PL6. In addition, FsAlyPL6 contained three conserved regions "NG(G/A)E", "KS", and "R(H/S)G" (marked in Figure 1), which are involved in substrate binding and catalytic activity [12]. The alginate lyases of PL6 family can be divided into three subfamilies, namely subfamilies 1, 2, and 3. In order to confirm the subfamilies of FsAlyPL6, the phylogenetic tree was used to compare the sequence homology with alginate lyases from diverse subfamilies. As is shown in Figure 2, FsAlyPL6 clustered with representative enzymes of subfamily 1, which indicated FsAlyPL6 is a new member of the subfamily 1 alginate lyase.

2.2. Expression and Purification of FsAlyPL6

The gene of FsAlyPL6 was ligated into pET-21a(+) and then the recombinant plasmid was transformed into *E. coli BL21* (DE3) for heterologously expression. The recombinant FsAlyPL6 was then purified by Ni-NTA sepharose affinity chromatography and analyzed by SDS-PAGE (Figure 3). A clear band (about 80 kDa) can be observed in gel, which was consistent with the theoretical molecular mass of 83.09 kDa. Three kinds of substrates (sodium alginate, polyM, and polyG) were employed to determine the substrate specificity of FsAlyPL6. As shown in Table 1, FsAlyPL6 exhibited higher activity towards sodium alginate (483.95 U/mg) and it showed lower activity towards to polyM (221.5 U/mg). However, it showed the lowest activity towards to polyG (19.35 U/mg). Accordingly, FsAlyPL6 is a polyMG-preferred lyase like most of PL6 family alginate lyases with the exceptions of Patl3640 from *Pseudoalteromonas atlantica* T6c and Pedsa0631 from *Pedobacter saltans* [13]. Both of them preferred polyG to polyMG blocks. In addition, TsAly6A from *Thalassomonas* sp. LD5 [14], OalS6 from *Shewanella* sp. Kz7 [15], OalC6 from *Cellulophaga* sp. SY116 [16], and AlyF from *Vibrio* sp. OU02 [17] are all characterized as polyG-preferred alginate lyases.

The kinetic parameters of FsAlyPL6 towards sodium alginate, polyM, and polyG were calculated based on the hyper regression analysis. As shown in Table 1, the K_m values of FsAlyPL6 towards sodium alginate, polyM, and polyG were 0.50 mg/mL, 1.52 mg/mL, and 1.62 mg/mL, respectively. FsAlyPL6 had a lower K_m value towards sodium alginate than to polyM and polyG. Accordingly, FsAlyPL6 exhibited higher affinity towards MG-block than to M-block and G-block. The k_{cat} values of FsAlyPL6 towards sodium alginate, polyM and polyG were 33.98 s^{-1}, 17.66 s^{-1}, and 4.98 s^{-1}, respectively. It indicated that FsAlyPL6 had higher catalytic efficiency towards MG-block than to the other two blocks.

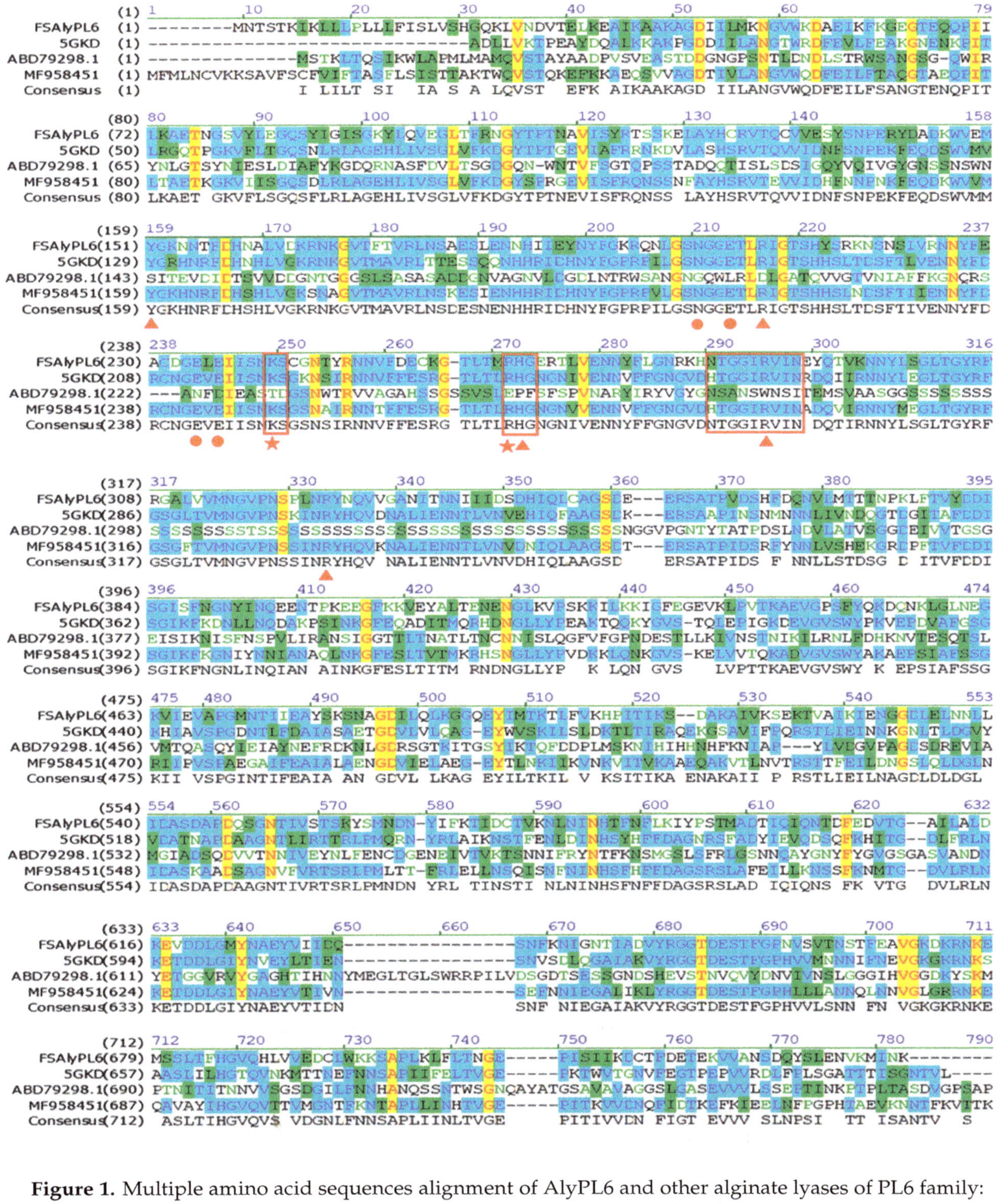

Figure 1. Multiple amino acid sequences alignment of AlyPL6 and other alginate lyases of PL6 family: AlyGC (BAEM00000000.1) from *Glaciecola chathamensis* S18K6T, polysaccharide lyase (ABD79298.1) from *Saccharophagus degradans* 2–40, and TsAly6A (MF958451) from *Thalassomonas* sp. LD5. Three boxes enclose conserved regions. Residues in FsAlyPL6, which are responsible for the enzymatic activity Ca^{2+} binding and catalysis, are marked in triangle, dots, and stars, respectively.

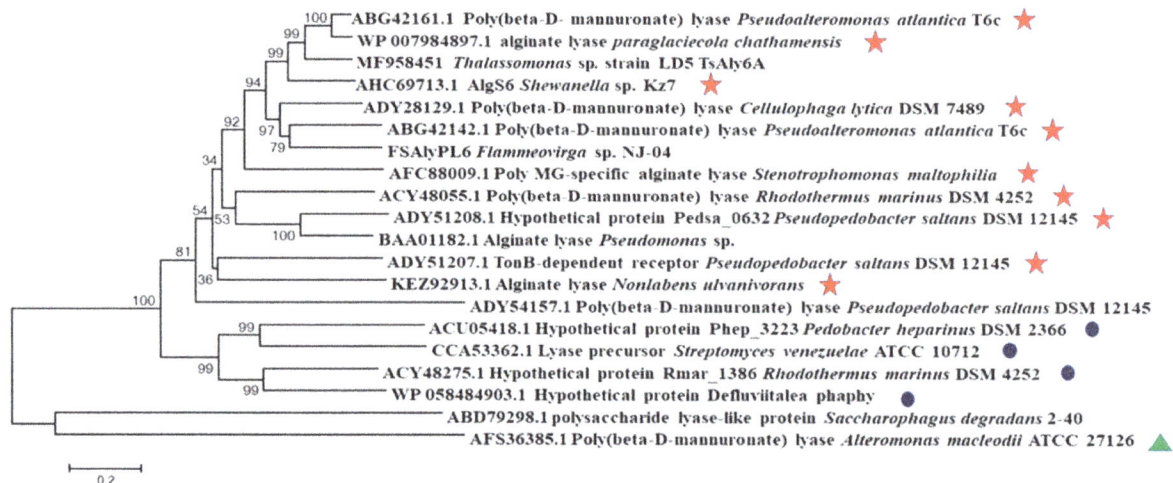

Figure 2. Phylogenetic analysis of FsAlyPL6 with other alginate lyases of PL6 family based on amino acid sequence comparisons. The species names are indicated along with accession numbers of corresponding alginate lyase sequences. Bootstrap values of 1000 trials are presented in the branching points. The subfamilies 1, 2, and 3 are marked with stars, dots, and triangle, respectively.

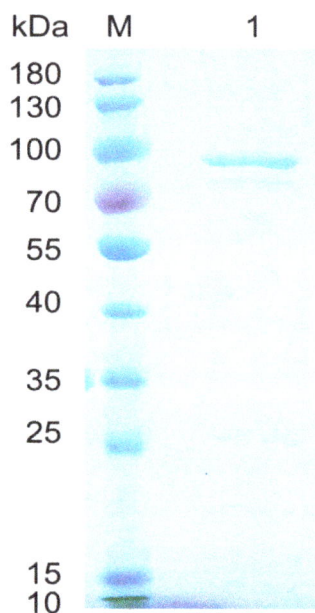

Figure 3. Sodium dodecyl sulfate polyacrylamide gel electrophoresis (SDS-PAGE) analysis of purified FsAlyPL6. Lane M protein: restrained marker (Thermo Scientific, Waltham, MA, USA); lane 1: purified FsAlyPL6.

Table 1. Specificity and kinetics of FsAlyPL6.

Substrate	Sodium Alginate	PolyM	PolyG
Activity (U/mg)	483.95	221.5	19.35
K_m (mg/mL)	0.50	1.52	1.62
V_{max} (nmol/s)	1.36	0.71	0.20
k_{cat} (s^{-1})	33.98	17.66	4.98
k_{cat}/K_m (mL·s^{-1}·mg^{-1})	62.91	11.58	3.08

2.3. Biochemical Characterization of FsAlyPL6

The optimal temperature of FsAlyPL6 is 45 °C and it retains more than 90% of maximal activity after being incubated at 45 °C for 1 h (Figure 4A). Compared with other PL6 family alginate lyases,

FsAlyPL6 exhibits preferable thermal characteristics than most PL6 family alginate lyases. For example, AlyF of *Vibrio* OU02 showed the maximal activity at 30 °C [17] and AlyGC from *G. chathamensis* S18K6T has an optimal temperature of 30 °C [12]. OalC6 of *Cellulophaga* sp. SY116 exhibits highest activity at 40 °C and retains about 80% of highest activity after being incubated at 40 °C for 1 h [16]. In addition, FsAlyPL6 retains 95% activity after being incubated at 35 °C for 60 min and inactivated gradually with temperature increased (Figure 4B). This remarkable characteristic indicated FsAlyPL6 possesses great potential in industrial applications for preparation alginate oligosaccharides. The optimal pH of FsAlyPL6 is 9.0 and it retains about 90% activity incubated at pH 9.0–10.0 for 12 h (Figure 4C,D), which indicated FsAlyPL6 is an alkaline-stable lyase. To the best of our knowledge, few alginate lyases of PL6 family are alkaline-stable lyases, and most of them exhibit the maximal activities around neutral pH values such as OalC6 of *Cellulophaga* sp. SY116 has an optimal pH of 6.6 and it retains only 60% of its maximal activity after being incubated at pH 6.0 for 6 h [16]. The OalS6 from *Shewanella* sp. Kz7 exhibits maximal activity at pH 7.2 and retains 80% after being hatched at pH 6.0–8.0 for 24 h [15]. The influences of metal ions on enzyme activity were also investigated. As shown in Table 2, like TsAly6A from *Thalassomonas* sp. LD5 [14], the activity of FsAlyPL6 can be activated by Ca^{2+} and Mg^{2+}. FsAlyPL6 is inhibited by various divalent metal ions such as Cu^{2+}, Zn^{2+} and Ni^{2+}, which is similar to OalS6 from *Shewanella* sp. Kz7 [15].

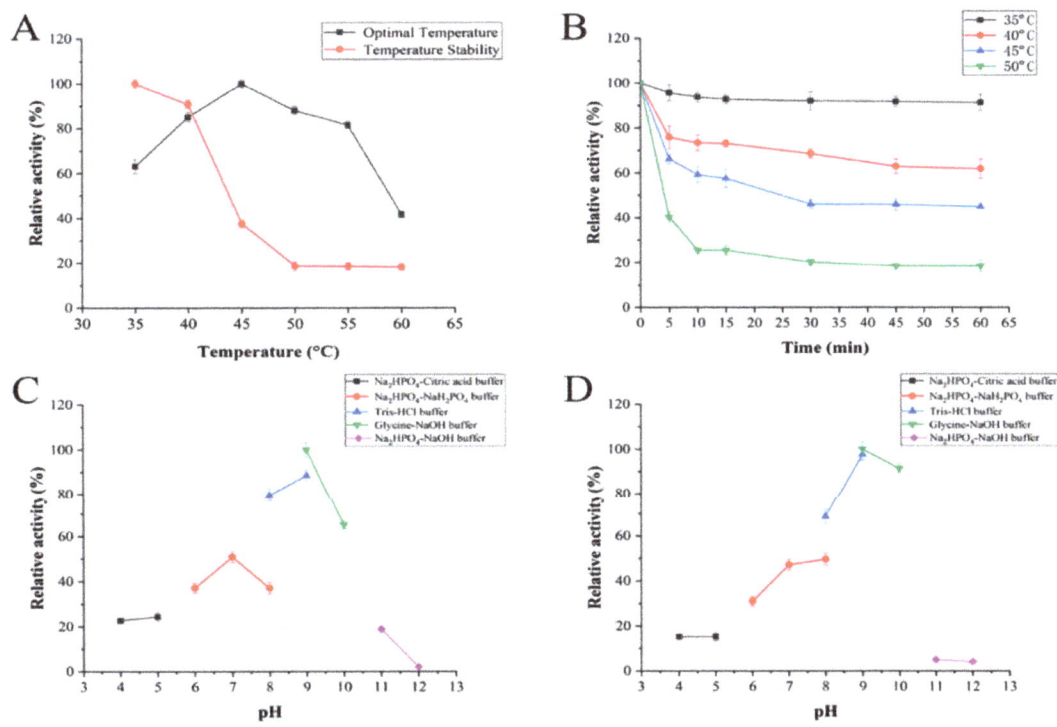

Figure 4. Biochemical characterization of FsAlyPL6: (**A**) The optimal temperature and thermal stability of FsAlyPL6; (**B**) the thermal-induced denaturation of FsAlyPL6; (**C**) the optimal pH of the FsAlyPL6; (**D**) the pH stability of FsAlyPL6.

Table 2. *Cont.*

Reagent	Relative Activity (%)
Control	100.00 ± 2.97
K^+	93.26 ± 2.23
Na^+	118.57 ± 1.08
Ca^{2+}	104.33 ± 1.12
Mg^{2+}	102.31 ± 2.78
Co^{2+}	22.14 ± 1.32

Table 2. Effects of metal ions on activity of FsAlyPL6.

Reagent	Relative Activity (%)
Zn^{2+}	24.88 ± 3.57
Cu^{2+}	15.28 ± 1.20
Ni^{2+}	50.19 ± 3.93
Mn^{2+}	6.46 ± 0.60
Fe^{3+}	26.55 ± 1.21

2.4. Action Pattern and Substrate Docking of FsAlyPL6 Product Analysis

To elucidate the action mode of FsAlyPL6, the degradation products of three substrates for different times (0–48 h) were analyzed by TLC (Figure 5). As the degrading process continues, three kinds of substrates are degraded into oligosaccharides with lower degrees of polymerization (DPs) (2–5) and monosaccharide, which indicated that FsAlyPL6 can cleave the glycosidic bonds within the substrates in an endolytic manner. The ESI-MS results indicated that degradation products of FsAlyPL6 towards the three different substrates include monosaccharide, and oligosaccharides with different DPs (2–5) can be detected (Figure 6A–C). Most of PL6 family enzymes are endo-type alginate lyases, which produce oligosaccharides with DPs (2–4). However, the Patl3640 from *Pseudoalteromonas atlantica* T6c [13], Pedsa0631 from *Pedobacter saltans* [13], OalS6 from *Shewanella* sp. Kz7 [15], and OalC6 from *Cellulophaga* sp. SY116 degrade the substrates into monosaccharides in an exolytic manner [16].

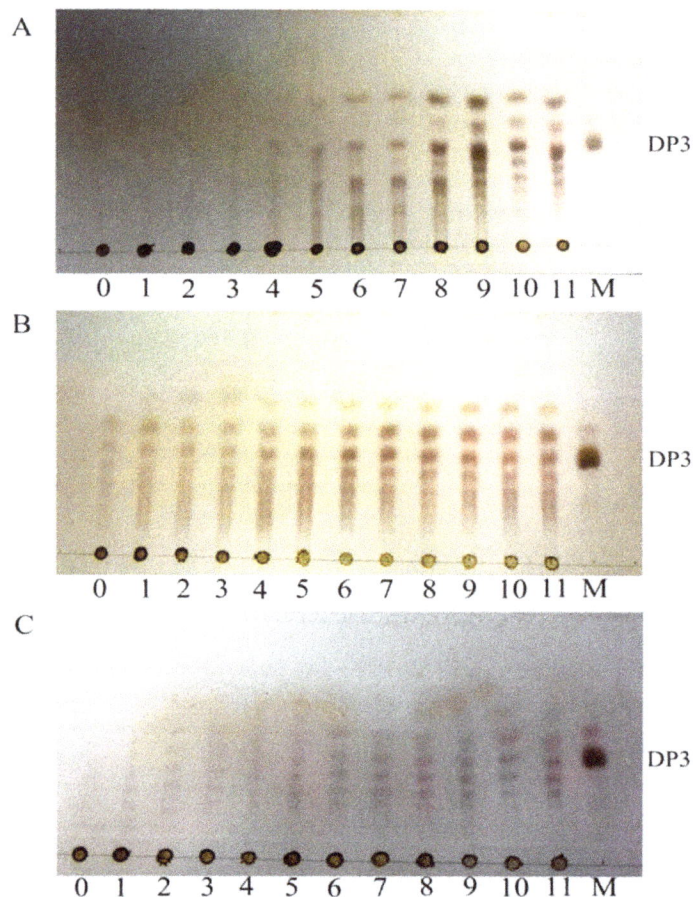

Figure 5. TLC analysis of degrading products of FsAlyPL6 towards alginate (**A**), polyM (**B**), and polyG (**C**). Lane M, the oligosaccharide standard; lanes 0–11, the samples taken by 0 min, 5 min, 10 min, 15 min, 30 min, 60 min, 2 h, 6 h, 12 h, 24 h, and 48 h, respectively.

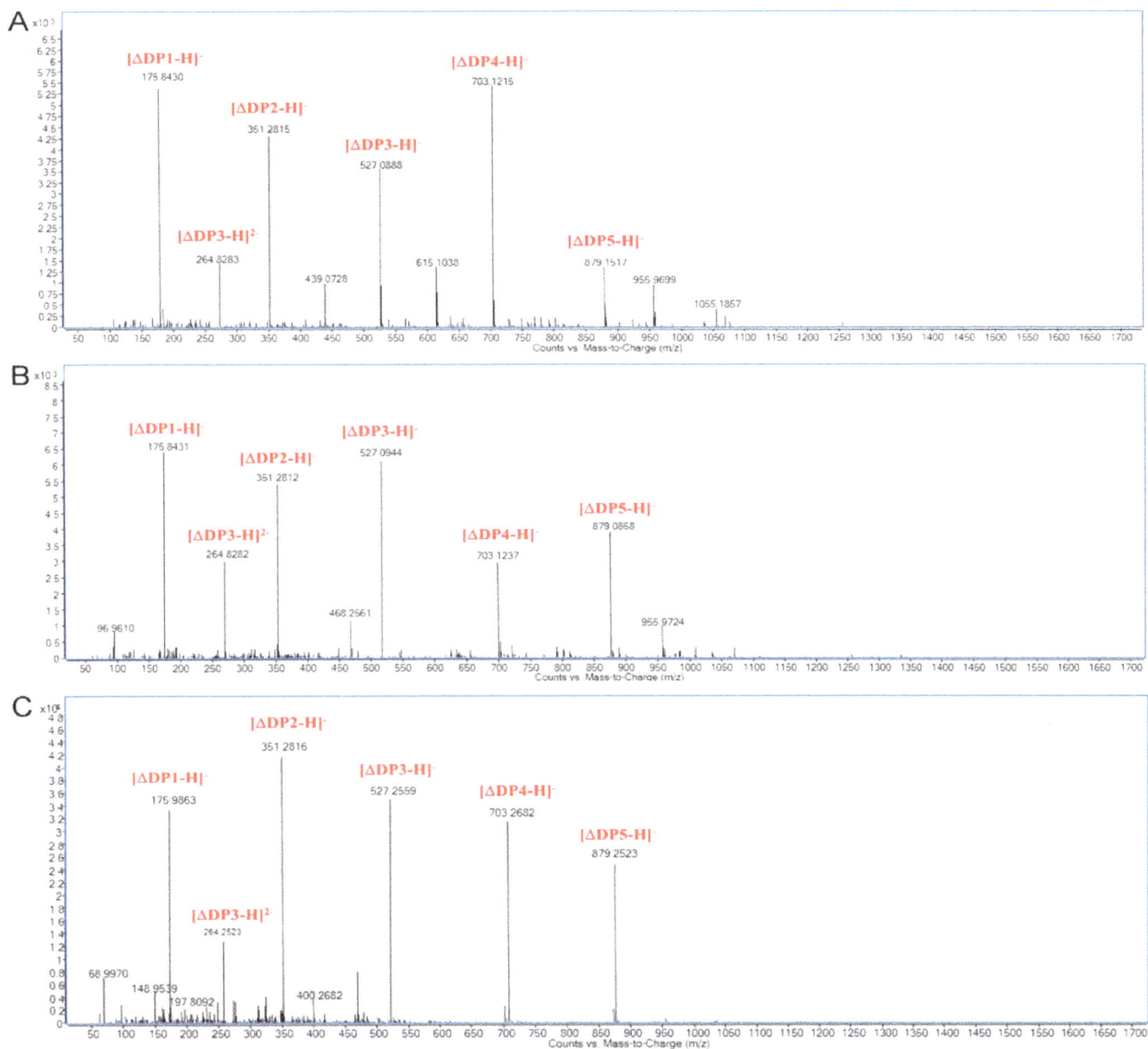

Figure 6. ESI-MS analysis of products of FsAlyPL6 towards alginate (**A**), polyM (**B**), and polyG (**C**).

The three-dimensional model of the FsAlyPL6 was constructed by PHYRE2 and the tetrasaccharide (MMMM) was docked into the FsAlyPL6. Because the sequence similarity between FsAlyPL6 and AlyGC was high (45%), the protein model was successfully constructed with 100% confidence. As shown in Figure 7A, the overall structure of FsAlyPL6 was predicted to fold into a "twin tower-like" structure (Figure 7A), which is similar to the structure of AlyGC (Figure 7B). However, AlyGC is an exo-type alginate lyase and FsAlyPL6 degrade alginate into oligosaccharide in an endolytic manner. The key residues for substrate recognition were identified by the sequence alignment and protein–substrate interactions. As shown in Figure 7C, the residues R_{239}, R_{263}, K_{218}, E_{213}, and Y_{332} are were highly conserved and involved in the interaction between the protein and substrates in subsites -1, $+1$, $+2$ and $+3$, respectively (Figure 8A,B). Based on the docking and β-elimination mechanism, the residues K_{218} and R_{239} acted as the Brønsted base and Brønsted acid, respectively, in the cleavage reaction of FsAlyPL6 on alginate, which is consistent with the residues of AlyGC (Figure 8B).

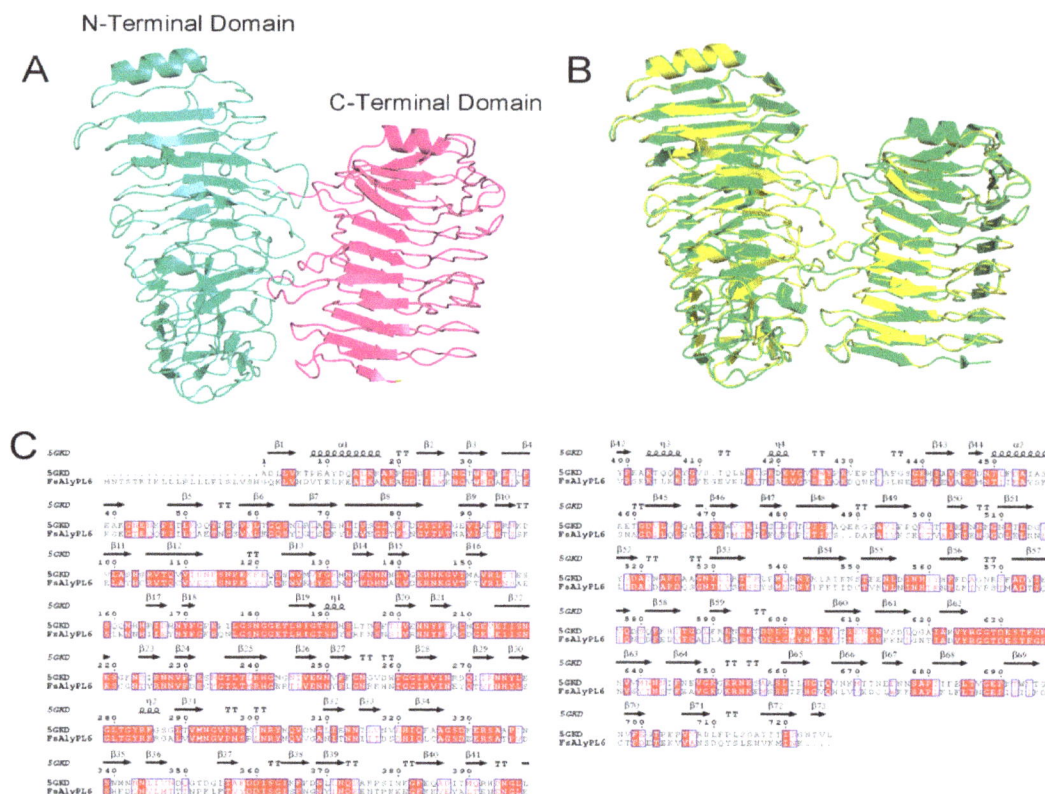

Figure 7. (**A**) Overall structure of FsAlyPL6; (**B**) the structural comparison of FsAlyPL6 (green) and AlyGC (yellow); (**C**) sequence alignments of FsAlyPL6 and AlyGC.

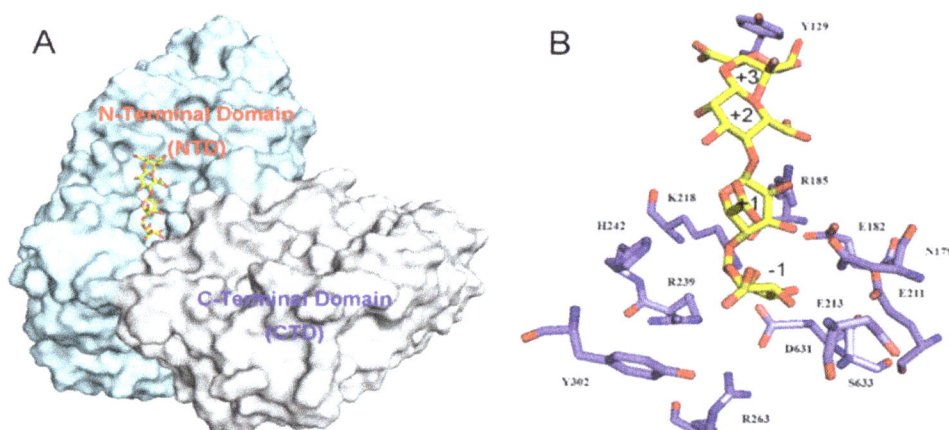

Figure 8. (**A**) Stereo view of the alginate tetrasaccharide (MMMM) bound to the tunnel-shaped active site of FsAlyPL6. (**B**) The presentation of catalytic residues responsible for binding and catalyzing the substrate.

3. Materials and Methods

3.1. Materials and Strains

Sodium alginate (M/G ratio: 77/23) was purchased from Sigma-Aldrich (St. Louis, MO, USA). PolyG and polyM (purity: about 95%; M/G ratio: 3/97 and 97/3, respectively) were purchased from Qingdao BZ Oligo Biotech Co., Ltd. (Qingdao, China). *Flammeovirga* sp. NJ-04 was isolated from the South China Sea and conserved in our laboratory. It was cultured at 35 °C in 2216E medium (Difoc). *Escherichia coli* DH5α and *E. coli* BL21 (DE3) were used for plasmid construction and as the hosts for gene expression, respectively. These strains were cultured at 37 °C in Luria-Bertani (LB) broth or on LB broth agar plates (LB broth was supplemented with 1.5% agar and contained 100 µg/mL ampicillin).

3.2. Cloning and Sequence Analysis of Alginate Lyase

As previously reported, a gene cluster for degrading alginate has been identified within the genome of the strain *Flammeovirga* sp. NJ-04 [10]. According to the sequence of the putative alginate lyase gene sequence (WP_044204792.1), a pair of special primers was designed as described in Supplementary Materials. For gene expression, the alginate lyase gene *FsAlyPL6* was subcloned and then ligated into pET-21a(+) expression vector. The theoretical molecular (Mw) and isoelectric point (pI) were calculated using Compute pI/Mw tool (https://web.expasy.org/compute_pi/). Molecular Evolutionary Genetics Analysis (MEGA) Program version 6.0 (Center for Evolutionary Medicine and Informatics, The Biodesign Institute, Tempe, AZ, USA) was applied to construct a phylogenetic tree through a neighbor-joining method based on alginate lyase protein sequences of PL6 family. The Vector NTI (Invitrogen, Thermo Scientific, Waltham, MA, USA) was used to obtain multiple sequence alignment. The homology modeling and docking was built by Protein Homology/analogY Recognition Engine V 2.0 (Structural Bioinformatics Group, Imperial College, London, Britain).

3.3. Hereologous Expression and Purification of the Recombinant Enzyme

The recombinant plasmid pET-21a(+)–*FsAlyPL6* was transformed into *E. coli* BL21 (DE3). It was then cultured in an LB medium (containing100 µg/mL of ampicillin) at 37 °C by shaking at 200 rpm for 5 h, followed by being induced with 0.1 mM IPTG at 25 °C for 36 h when OD_{600} reached 0.6. The purification of FsAlyPL6 was performed as follows. The cells were harvested by centrifugation and then sonicated in lysis buffer (50 mM Tris-HCl with 300 mM NaCl, pH 8.0). The cell homogenate that contained recombinant protein were purified by using a His-trap column (GE Healthcare, Uppsala, Sweden). SDS on 12% (*w/v*) resolving gel was applied to detect the purity of the recombinant protein.

3.4. Substrate Specificity and Enzymatic Kinetics

The reaction was performed using 20 µL FsAlyPL6 (4 µg) mixed with 180 µL 0.8% alginate, polyM, and polyG respectively. The enzyme activity was measured using the ultraviolet absorption method [11]. One unit was defined as the amounts of enzyme required to increase absorbance at 235 nm (extinction coefficient: $6150 \, M^{-1} \cdot cm^{-1}$) by 0.1 per min. The kinetic parameters of the FsAlyPL6 towards alginate, polyM, and polyG were investigated by measuring the enzyme activity with these substrates at different concentrations (0.4–10 mg/mL). Velocity (V), K_m, and V_{max} values were calculated as previously reported [10]. The radio of V_{max} versus enzyme concentration ([E]) was used to calculate the turnover number (k_{cat}) of the enzyme.

3.5. Biochemical Characterization of the Recombinant Enzyme FsAlyPL6

The effects of temperature on the enzyme activity were determined by testing the activity at different temperatures (35 °C to 60 °C). The thermal stability was characterized by measuring the residual activity after the purified FsAlyPL6 was incubated at 35–60 °C for 1 h. Furthermore, the thermally induced denaturation was also determined by measuring the residual activity after incubating the enzyme at 35–50 °C for 0–60 min. To investigated the optimal pH of the FsAlyPL6, 1% alginate mixed with different buffers at 45 °C (50 mM phosphate–citrate (pH 4.0–5.0), 50 mM NaH_2PO_4–Na_2HPO_4 (pH 6.0–8.0), 50 mM Tris–HCl (pH 7.0–9.0), and glycine–NaOH (pH 9.0–12.0)) were used as the substrates and the purified enzyme incubated with these substrates under standard conditions. Moreover, the pH stability was evaluated based on the residual activity after being incubated with indifferent buffers (pH 4.0–12.0) for 20 h. The effects of metal ions on the enzymatic activity were performed by incubating the FsAlyPL6 with substrates that contained various metal compounds with a final concentration of 1 mM. The reaction performed under standard tested conditions and the substrates blend without any metal ion was taken as the control.

3.6. Action Pattern and Degradation Product Analysis

In order to elucidate the action pattern of the FsAlyPL6, the thin-layer chromatography (TLC) was applied to analyze the degrading products of FsAlyPL6 towards sodium alginate, polyM and polyG. The reaction and treatment of the samples were performed as previously reported [10]. In order to investigate the composition of the degrading products, ESI-MS was employed as follows: The supernatant (2 μL) was loop-injected to an LTQ XL linear ion trap mass spectrometer (Thermo Fisher Scientific, Waltham, MA, USA) after centrifugation. The oligosaccharides were detected in a negative-ion mode using the following settings: ion source voltage, 4.5 kV; capillary temperature, 275–300 °C; tube lens, 250 V; sheath gas, 30 arbitrary units (AU); and scanning the mass range, 150–2000 *m/z*.

3.7. Molecular Modeling and Docking Analysis

Protein Homology/analogY Recognition Engine V 2.0 was applied to construct the three-dimensional structure of FsAlyPL6 according to the known structure of alginate lyase AlyGC from *Glaciecola chathamensis* S18K6T (PDB: 5GKD) with a sequence identity of 45%. The molecular docking of the FsAlyPL6 and MMMM was performed using Molecular Operating Environment (MOE, Chemical Computing Group Inc., Montreal, QC, Canada). The ligand-binding sites were defined using the bound ligand in the homology models. PyMOL (http://www.pymol.org) was used to visualize and analyze the modeled structure and to construct graphical presentations and illustrative figures.

4. Conclusions

In this study, we reported a new PL family alginate lyase FsAlyPL6 from the marine *Flammeovirga* sp. NJ-04. It preferred to degrade the polyMG block and showed highest activity at 45 °C and could retain 50% of activity after being incubated at 45 °C for 1 h. The FsAlyPL6 endolytically degraded alginate polysaccharide and released oligosaccharides with DPs of 1–5. In addition, it could recognize tetrasaccharide as the minimal substrate and cleave the glycosidic bonds between the subsites of −1 and +3 to release oligosaccharides. The research provides extended insights into the degradation pattern of PL6 alginate lyases and further expands the application of alginate lyases.

Author Contributions: Q.L. and F.H. conceived and designed the experiments; B.Z., Q.L., and F.H. performed the experiments; Y.S., Y.S., and Z.Y. analyzed the data; B.Z. wrote the paper. All authors reviewed the manuscript.

Acknowledgments: The work was supported by the National Natural Science Foundation of China (grant numbers: 31601410 and 21776137).

References

1.   Gacesa, P. Enzymic degradation of alginates. *Int. J. Biochem.* **1992**, *24*, 545–552. [CrossRef]
2.   Lee, K.Y.; Mooney, D.J. Alginate: Properties and biomedical applications. *Prog. Polym. Sci.* **2012**, *37*, 106–126. [CrossRef] [PubMed]
3.   An, Q.D.; Zhang, G.H.; Zhang, Z.C.; Zheng, G.S.; Luan, L.; Murata, Y.; Li, X. Alginate-deriving oligosaccharide production by alginase from newly isolated Flavobacterium sp. LXA and its potential application in protection against pathogens. *J. Appl. Microbiol.* **2010**, *106*, 161–170. [CrossRef] [PubMed]
4.   Iwamoto, M.; Kurachi, M.; Nakashima, T.; Kim, D.; Yamaguchi, K.; Oda, T.; Iwamoto, Y.; Muramatsu, T. Structure-activity relationship of alginate oligosaccharides in the induction of cytokine production from RAW264.7 cells. *FEBS Lett.* **2005**, *579*, 4423–4429. [CrossRef] [PubMed]

5.   Tusi, S.K.; Khalaj, L.; Ashabi, G.; Kiaei, M.; Khodagholi, F. Alginate oligosaccharide protects against endoplasmic reticulum- and mitochondrial-mediated apoptotic cell death and oxidative stress. *Biomaterials* **2011**, *32*, 5438–5458. [CrossRef] [PubMed]

6.   Falkeborg, M.; Cheong, L.Z.; Gianfico, C.; Sztukiel, K.M.; Kristensen, K.; Glasius, M.; Xu, X.; Guo, Z. Alginate oligosaccharides: Enzymatic preparation and antioxidant property evaluation. *Food Chem.* **2014**, *164*, 185–194. [CrossRef] [PubMed]

7.   Wong, T.Y.; And, L.; Schiller, N.L. Alginate Lyase: Review of Major Sources and Enzyme Characteristics, Structure-Function Analysis, Biological Roles, and Applications. *Annu. Rev. Microbiol.* **2000**, *54*, 289–340. [CrossRef] [PubMed]

8.   Zhu, B.; Chen, M.; Yin, H.; Du, Y.; Ning, L. Enzymatic Hydrolysis of Alginate to Produce Oligosaccharides by a New Purified Endo-Type Alginate Lyase. *Mar. Drugs* **2016**, *14*, 108. [CrossRef] [PubMed]

9.   Zhu, B.; Yin, H. Alginate lyase: Review of major sources and classification, properties, structure-function analysis and applications. *Bioengineered* **2015**, *6*, 125–131. [CrossRef] [PubMed]

10. Zhu, B.; Ni, F.; Sun, Y.; Yao, Z. Expression and characterization of a new heat-stable endo-type alginate lyase from deep-sea bacterium Flammeovirga sp. NJ-04. *Extremophiles* **2017**, *21*, 1027–1036. [CrossRef] [PubMed]

11. Inoue, A.; Anraku, M.; Nakagawa, S.; Ojima, T. Discovery of a Novel Alginate Lyase from Nitratiruptor sp. SB155-2 Thriving at Deep-sea Hydrothermal Vents and Identification of the Residues Responsible for Its Heat Stability. *J. Biol. Chem.* **2016**, *291*, 15551–15563. [CrossRef] [PubMed]

12. Xu, F.; Dong, F.; Wang, P.; Cao, H.Y.; Li, C.Y.; Li, P.Y.; Pang, X.H.; Zhang, Y.Z.; Chen, X.L. Novel Molecular Insights into the Catalytic Mechanism of Marine Bacterial Alginate Lyase AlyGC from Polysaccharide Lyase Family 6. *J. Biol. Chem.* **2017**, *292*, 4457–4468. [CrossRef] [PubMed]

13. Mathieu, S.; Henrissat, B.; Labre, F.; Skjak-Braek, G.; Helbert, W. Functional Exploration of the Polysaccharide Lyase Family PL6. *PLoS ONE* **2016**, *11*, e0159415. [CrossRef] [PubMed]

14. Gao, S.; Zhang, Z.L.; Li, S.Y.; Su, H.; Tang, L.Y.; Tan, Y.L.; Yu, W.G.; Han, F. Characterization of a new endo-type polysaccharide lyase (PL) family 6 alginate lyase with cold-adapted and metal ions-resisted property. *Int. J. Biol. Macromol.* **2018**, *120*, 729–735. [CrossRef] [PubMed]

15. Li, S.Y.; Wang, L.N.; Han, F.; Gong, Q.H.; Yu, W.G. Cloning and characterization of the first polysaccharide lyase family 6 oligoalginate lyase from marine *Shewanella* sp Kz7. *J. Biochem.* **2016**, *159*, 77–86. [CrossRef] [PubMed]

16. Li, S.Y.; Wang, L.N.; Chen, X.H.; Zhao, W.W.; Sun, M.; Han, Y.T. Cloning, Expression, and Biochemical Characterization of Two New Oligoalginate Lyases with Synergistic Degradation Capability. *Mar. Biotechnol.* **2018**, *20*, 75–86. [CrossRef] [PubMed]

17. Lyu, Q.; Zhang, K.; Shi, Y.; Li, W.; Diao, X.; Liu, W. Structural insights into a novel Ca^{2+}-independent PL-6 alginate lyase from *Vibrio* OU02 identify the possible subsites responsible for product distribution. *BBA Gen. Subjects* **2019**, *1863*, 1167–1176. [CrossRef] [PubMed]

13

Characteristics of a Novel Manganese Superoxide Dismutase of a Hadal Sea Cucumber (*Paelopatides* sp.) from the Mariana Trench

Yanan Li [1,2], **Xue Kong** [1,2] **and Haibin Zhang** [1,*]

[1] Institute of Deep-Sea Science and Engineering, Chinese Academy of Sciences, Sanya 572000, China; liyn@idsse.ac.cn (Y.L.); kongx@sidsse.ac.cn (X.K.)

[2] College of Earth and Planetary Sciences, University of Chinese Academy of Sciences, Beijing 100039, China

* Correspondence: hzhang@idsse.ac.cn.

Abstract: A novel, cold-adapted, and acid-base stable manganese superoxide dismutase (Ps-Mn-SOD) was cloned from hadal sea cucumber *Paelopatides* sp. The dimeric recombinant enzyme exhibited approximately 60 kDa in molecular weight, expressed activity from 0 °C to 70 °C with an optimal temperature of 0 °C, and resisted wide pH values from 2.2–13.0 with optimal activity (> 70%) at pH 5.0–12.0. The Km and Vmax of Ps-Mn-SOD were 0.0329 ± 0.0040 mM and 9112 ± 248 U/mg, respectively. At tested conditions, Ps-Mn-SOD was relatively stable in divalent metal ion and other chemicals, such as β-mercaptoethanol, dithiothreitol, Tween 20, Triton X-100, and Chaps. Furthermore, the enzyme showed striking stability in 5 M urea or 4 M guanidine hydrochloride, resisted digestion by proteases, and tolerated a high hydrostatic pressure of 100 MPa. The resistance of Ps-Mn-SOD against low temperature, extreme acidity and alkalinity, chemicals, proteases, and high pressure make it a potential candidate in biopharmaceutical and nutraceutical fields.

Keywords: expression; purification; deep-sea enzyme; pCold vector

1. Introduction

Reactive oxygen species (ROS) are necessary for various physiological functions, such as signaling pathways and immune responses; the mass accumulation of ROS will damage bio-macromolecules, leading to cell death and various diseases [1,2]. Superoxide dismutases (SODs, EC 1.15.1.1) are one of the most important antioxidant enzymes that clear ROS by converting them into oxygen and hydrogen peroxide. According to the different metal cofactors, several types, such as Cu,Zn-SOD, Mn-SOD, Fe-SOD, cambialistic SOD (activated with either Fe or Mn), Ni-SOD, and Fe,Zn-SOD, have been reported in many species [3–7].

Studies have shown that SODs are related to immune reactions in invertebrates, as exemplified by bacterial and viral invasion [4,8], environmental pollution [9,10], and temperature stimulation [11]. Recently, Xie et al. indicates that antioxidant is related to the deep-sea environmental adaptability [12]. On the other hand, point mutations and activity loss of SODs lead to serious diseases and death in vertebrates. For example, the mice model of mitochondria SOD-deficiency is characterized by neurodegeneration, myocardial injury, and perinatal death [13,14]. A strong link is observed between Alzheimer's disease, tumor, amyotrophic lateral sclerosis, and SODs [15,16]. Hence, the physiological significance of SODs allows their application in the therapeutic and nutraceutical fields. To date, SODs have been reported to exhibit positive effects on inflammatory diseases, arthritis tumor, and promotion [17–19]. An orally effective form of SOD (glisodin) has been developed by Isocell Pharma, and it showed cosmetic and health benefits in human subjects [20,21]. Producing SOD using engineered bacteria is one of the most promising methods to obtain high yield and inexpensive SODs for application. Therefore, the development of SODs with remarkable characteristics is particularly urgent.

Sea cucumbers are highly important commercial sea foods owing to their high nutritional value, and they are distributed from shallow water to the deep sea [22]. Although deep sea is an extremely low-temperature and high hydrostatic-pressured environment for most living organisms, holothurians dominate benthic megafaunal communities in hadal trenches and form "the kingdom of Holothuroidea" when food is abundant [23]. Extreme environments, such as the deep sea, are ideal for the development of new enzymes; numerous novel enzymes with unique activities, such as proteases and lipases, have been identified from the deep sea [24,25]. Considering the promising applications of SODs in therapeutic and nutraceutical fields, relationship with the adaptability of the deep-sea environment and limited studies in extreme organisms, especially in hadal sea cucumbers, we report a novel manganese superoxide dismutase from hadal sea cucumber *Paelopatides* sp. (Ps-Mn-SOD), which inhabits a depth of 6500 m in the Mariana Trench, analyzed its biochemical characteristics, and evaluated its stability for potential use in the food and preliminarily nutraceutical fields.

2. Results

2.1. Sequence Characteristics

The ORF of Ps-Mn-SOD is 768 bp long, encoding 255 amino acids. A signal peptide was detected at the N-terminal of deduced amino acid sequence. The N- and C-terminal domains spanned from Lys-34 to Ser-127 and Pro-137 to Leu-242, respectively. Four conserved amino acid residues, namely, His-63, His-119, Asp-209, His-213 are responsible for manganese coordination. A conserved residue of Tyr-35 is responsible for the second coordination sphere of the metal [26]. A highly conserved Mn-SOD signature sequence with the pattern D-x-[WF]-E-H-[STA]-[FY] existed in Ps-Mn-SOD (DVWEHAYY). The predicted secondary structure contained 13 α-helices and 4 β-strands. The deduced theoretical isoelectric point was 5.05, and the molecular weight was 29.29 kDa. The instability index of 36.97 classified the protein as stable. The 3D model of Ps-Mn-SOD was predicted using the x-ray template of *Bacillus subtilis*, which shared 45.27% sequence identity (PDB ID: 2RCV) [27]. This model shows that Ps-Mn-SOD is presented as a homodimer, and each subunit embraces one manganese ion. The global and per-residue model qualities were assessed using the QMEAN scoring function [28]. GMQE and QMEAN4 Z-scores reached 0.64 and −2.63, respectively, suggesting the accuracy of predicted 3D model of Ps-Mn-SOD. Figure 1 and Supplementary Figure S1 provide the related structural information of Ps-Mn-SOD.

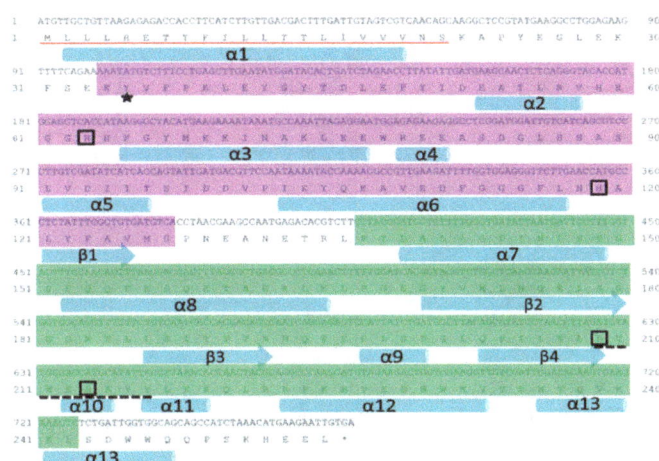

Figure 1. Nucleotide and corresponding amino acid sequences of Ps-Mn-SOD. The signal peptide is drawn with a red line. The signature sequence DVWEHAYY is underlined with dotted line. N- and C-terminal domains are marked with purple and green shades, respectively. Four conserved amino acid residues for manganese coordination are boxed. Asterisk points to the highly conserved Tyr-35 residue. Cylinders and arrows represent helices and strands, respectively.

2.2. Homology and Phylogenetic Analysis

Multiple alignment and pairwise homology analysis between Ps-Mn-SOD and other invertebrates were performed, and the results are shown in Figure 2 and Supplementary Table S1. Multiple alignment of Ps-Mn-SOD with other invertebrates indicated that four amino acids were responsible for manganese binding, and the signature sequences are highly conserved in different Mn-SOD sources and were also identified in Ps-Mn-SOD (Figure 2). The highest similarity and identity were shared with *Apostichopus japonicus* (83.9% and 78.0%), followed by *Capitella teleta* (66.9% and 47.9%), *Exaiptasia pallida* (66.3% and 47.7%), *Strongylocentrotus purpuratus* (65.1% and 47.0%), *Mizuhopecten yessoensis* (64.4% and 46.7%), and *Stylophora pistillata* (63.1% and 45.8%). To determine the type of SOD present, we performed phylogenetic analysis based on the amino acid sequences of the determined SOD types in Genebank (Figure 3). The results showed that the present SOD clustered with *A. japonicus* and evidently a Mn-SOD type with high bootstrap values.

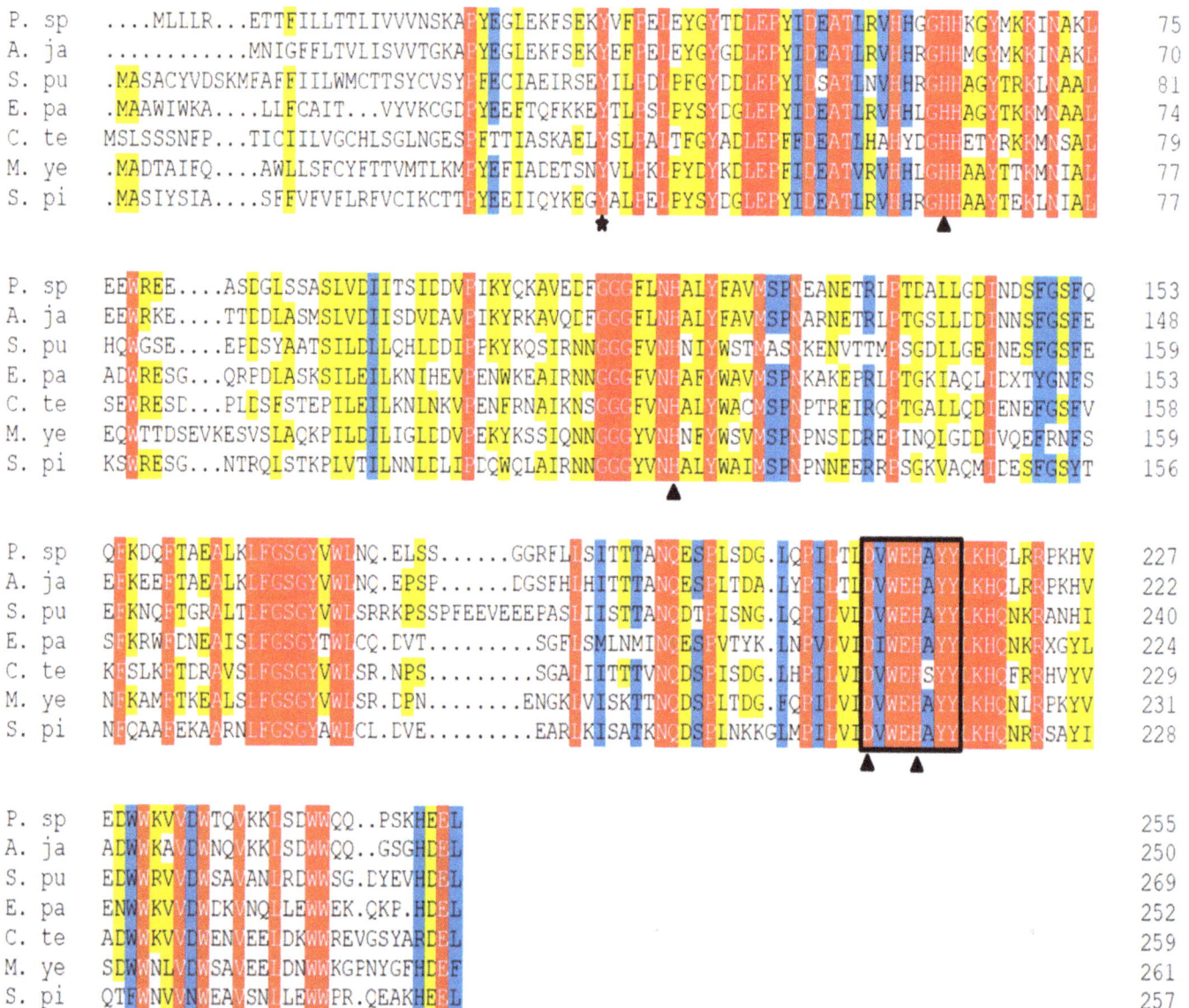

```
P. sp    ....MLLLR...ETTFILLTTLIVVVNSKAFYEGLEKFSEKYVFFEIEYGYTDLEFYIDEATLRVHHGGHHKGYMKKINAKL   75
A. ja    ...........MNIGFFLTVLISVVTGKAFYEGLEKFSEKYEFFEIEYGYGDLEFYIDEATLRVHRGHHMGYMKKINAKL   70
S. pu    .MASACYVDSKMFAFFIILWMCTTSYCVSYFFECIAEIRSEYILFDIFFGYDDLEFYIDSATLNVHHRGHHAGYTRKLNAAL   81
E. pa    .MAAWIWKA....LLFCAIT...VYVKCGDFYEEFTQFKKEYTLFSLFYSYDGLEFYIDEATLRVHHLGHHAGYTKKMNAAL   74
C. te    MSLSSSNFP...TICIIILVGCHLSGLNGESFTTIASKAELYSLFAITFGYADLEFFFDEATLHAHYDGHHETYRKKMNSAL   79
M. ye    .MADTAIFQ....AWLLSFCYFTTVMTLKMFYEFIADETSNYVLFKIFYDYKDLEFFIDEATVRVHHLGHHAAYTTKMNIAL   77
S. pi    .MASIYSIA....SFFVFVFLRFVCIKCTTFYEEIIQYKEGYALFEIFYSYDGLEFYIDEATLRVHRGHHAAYTEKLNIAL   77

P. sp    EEWREE....ASDGISSASLVDIITSIDDVFIKYQKAVEDFGGGFLNHALYFAVMSPNEANETRLFTDALLGDINDSFGSFQ   153
A. ja    EEWRKE....TTDDLASMSLVDIISDVDAVFIKYRKAVQDFGGGFLNHALYFAVMSPNARNETRLFTGSLLDDINNSFGSFE   148
S. pu    HQWGSE....EPDSYAATSILDLLQHLDDIFPKYKQSIRNNGGGFVNHINIYWSTMASNKENVTTMFSGDILGEINESFGSFE   159
E. pa    ADWRESG...QRFDLASKSILEILKNIHEVFENWKEAIRNNGGGFVNHAFYWAVMSPNKAKEPRLFTGKIAQLIDXTYGNFS   153
C. te    SEWRESD...PLDSFSTEPILEILKNLNKVFENFRNAIKNSGGGFVNHALYWACMSPNPTREIRQFTGALLCDIENEFGSFV   158
M. ye    EQWTTDSEVKESVSLAQKPILDILIGLDDVFEKYKSSIQNNGGGYVNHNFYWSVMSPNPNSDDREFINQLGDDIVQEFRNFS   159
S. pi    KSWRESG...NTRQLSTKPLVTILNNLDLIFDQWQLAIRNNGGGYVNHALYWAIMSPNPNNEERRFSGKVAQMIDESFGSYT   156

P. sp    QFKDQFTAEALKLFGSGYVWLNQ.ELSS......GGRFLISITTTANQESFLSDG.LQPILTLDVWEHAYYLKHQLRRPKHV   227
A. ja    EFKEEFTAEALKLFGSGYVWLNQ.EPSP......DGSFHIHITTTANQESFLTDA.LYPILTIDVWEHAYYLKHQLRRPKHV   222
S. pu    EFKNQFTGRALTLFGSGYVWLSRRKPSSPFEEVEEEPASLIISTTANQDTFISNG.LQPILVIDVWEHAYYLKHQNKRANHI   240
E. pa    SFKRWFDNEAISLFGSGYTWICQ.DVT........SGFLSMLNMINQESFVTYK.INPVIVIDIWEHAYYLKHQNKRXGYL   224
C. te    KFSLKFTDRAVSLFGSGYVWISR.NPS.........SGALIITTTVNQDSFISDG.LHPILVIDVWEFSYYLKHQFRRHVYV   229
M. ye    NFKAMFTKEALSLFGSGYVWLSR.DPN.......ENGKIVISKTTNQDSFLTDG.FQRILVIDVWEHAYYLKHQNLRPKYV   231
S. pi    NFQAAFEKAARNLFGSGYAWICL.DVE.........EARLKISATKNQDSFLNKKGLMPILVIDVWEHAYYLKHQNRFSAYI   228

P. sp    EDWKVDWTQVKKISDWWQQ..PSKHEDL   255
A. ja    ADWKAVDWNQVKKISDWWQQ..GSGHDDL   250
S. pu    EDWRVDWSAVANIRDWWSG.DYEVHDDL   269
E. pa    ENWKVVDWDKVNQILEWWEK.QKP.HDDL   252
C. te    ADWKVVDWENVEEIDKWWREVGSYARDDL   259
M. ye    SDWNLVDWSAVEEIDNWWKGPNYGFHDEF   261
S. pi    QTFWNVVNWEAVSNILEWWPR.QEAKHEDL   257
```

Figure 2. Multiple alignment of Ps-Mn-SOD with other invertebrates. Mn-SOD signature sequence is boxed. Triangles point to the active sites for manganese coordination. Asterisk points to the highly conserved Tyr-35 residue.

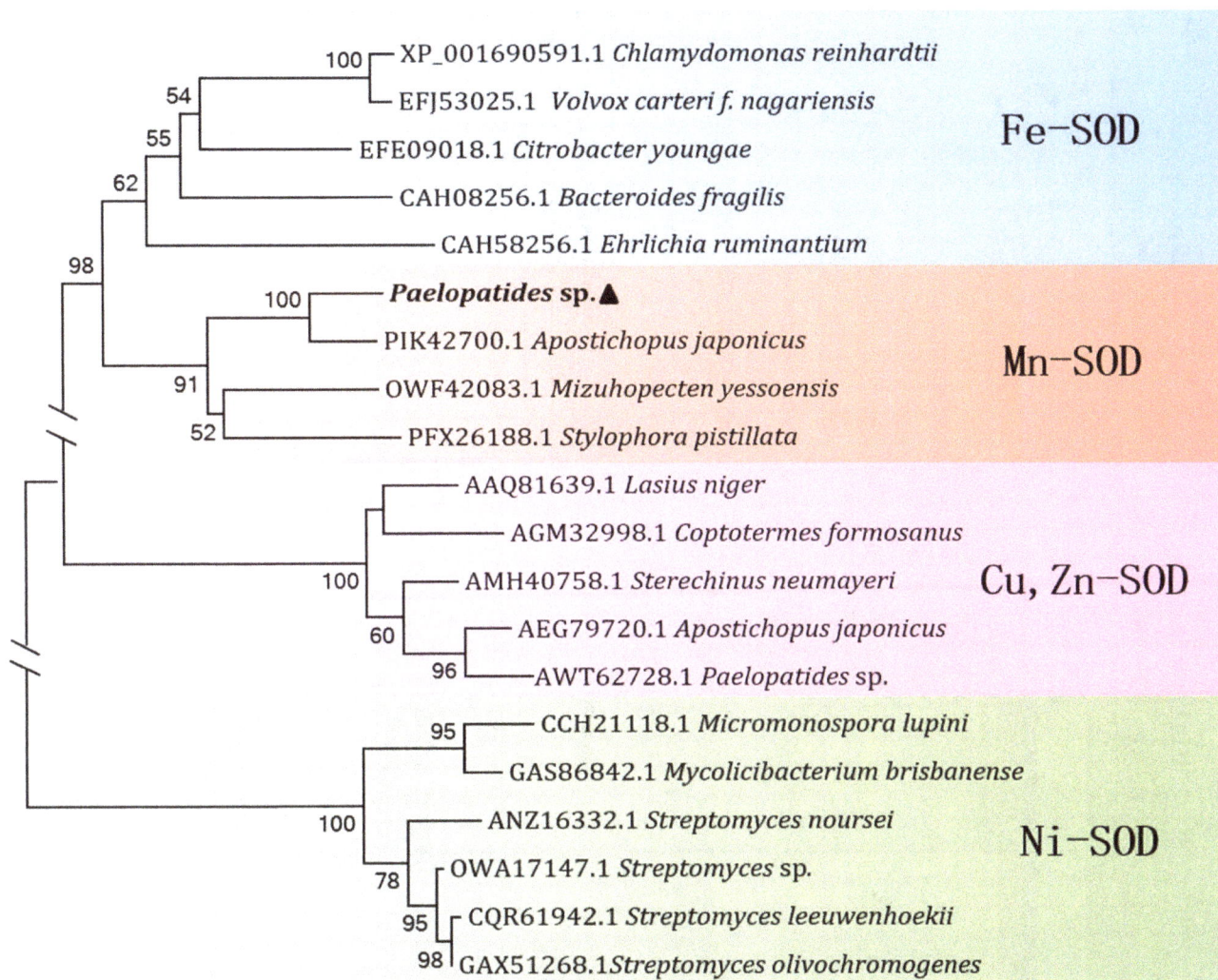

Figure 3. Neighbor-joining phylogenetic tree of SODs based on amino acid sequence homology. Bootstrap values below 50 are cut off. Ps-Mn-SOD is displayed in bold.

2.3. Expression, Purification, and Validation of Ps-Mn-SOD

The Ps-Mn-SOD gene was expressed with a His-tag in *E. coli*. Supplementary Figure S2 shows the SDS-PAGE analysis results. Recombinant Ps-Mn-SOD was expressed under 0.1 mM IPTG at 15 °C for 24 h and produced a distinct band at approximately 30 kDa, consistent with the previously estimated molecular weight (Supplementary Figure S2, lanes 1 and 2). The protein was purified under native conditions due to its highly soluble expression in the supernatant (Supplementary Figure S2, lanes 3 and 4). The maximum protein yield approximated 4.39 mg/L culture. Western blot analysis was performed to verify its successful expression (Supplementary Figure S2, lanes 5 and 6).

2.4. Characterizations of Ps-Mn-SOD

2.4.1. Effects of Temperature on Ps-Mn-SOD

The activity of Ps-Mn-SOD was determined from 0 °C to 80 °C, with the optimum temperature observed at 0 °C. A stable activity was observed at low temperatures, with > 70% activity highlighted from 0 °C to 60 °C. The activity was maintained at 2.53% at 70 °C and lost at 80 °C (Figure 4A).

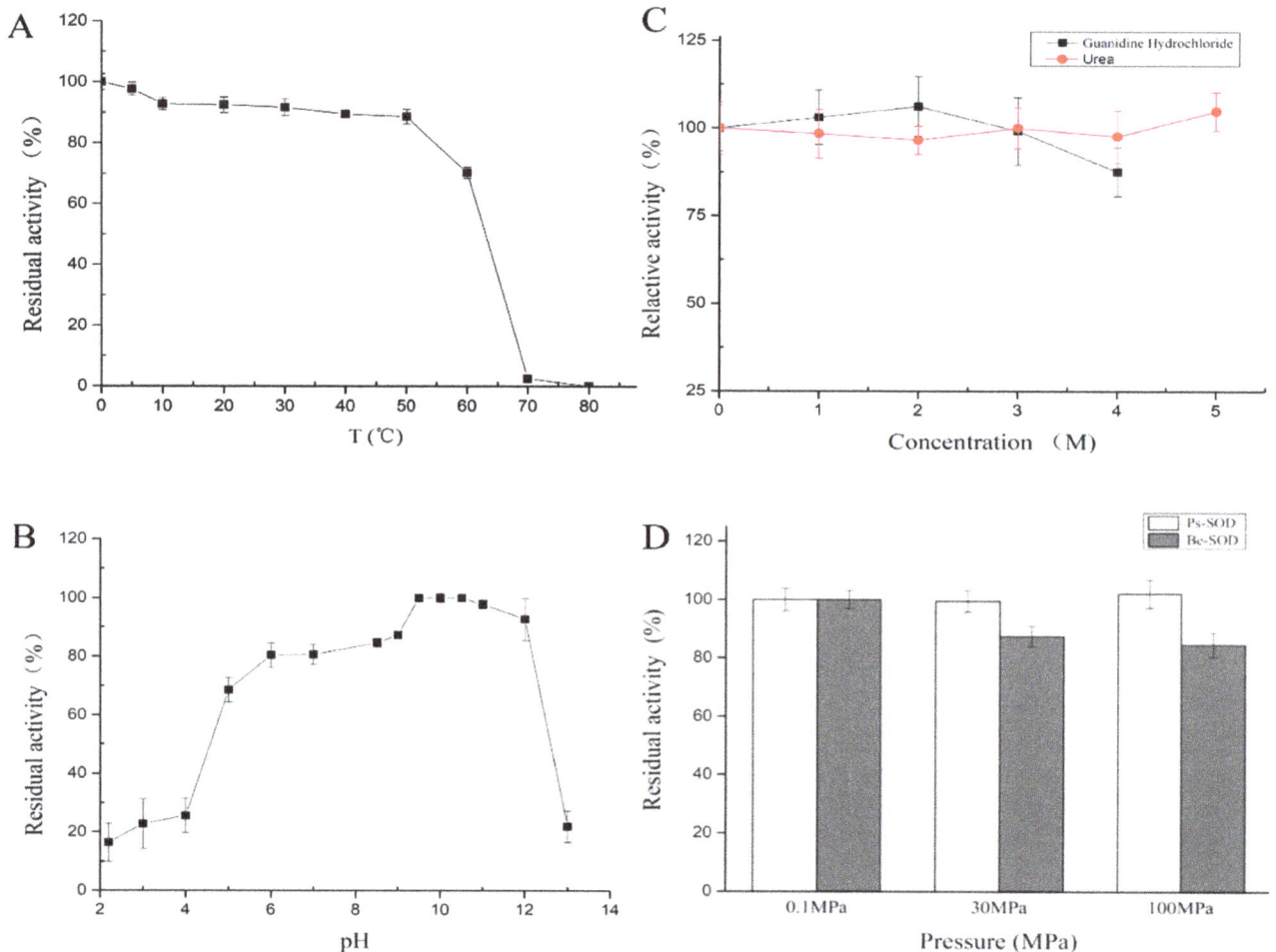

Figure 4. Effects of temperature (**A**), pH (**B**), urea and guanidine hydrochloride (**C**), and high hydrostatic pressure (**D**) on Ps-Mn-SOD. Ps-SOD and Be-SOD represent SOD from *Paelopatides* sp. and bovine erythrocytes, respectively.

2.4.2. Effects of pH on Ps-Mn-SOD

The activity of recombinant Ps-Mn-SOD was measured under pH 2.2–13.0, with an optimum pH observed at 10.5 (Figure 4B). Ps-Mn-SOD could resist extreme pH values (> 20% at pH 3.0–13.0) and showed optimal activity (> 70%) at pH 5.0–12.0.

2.4.3. Effects of Chemicals on Ps-Mn-SOD

The effects of metal ions on Ps-Mn-SOD activity were determined at 0.1 or 1 mM final concentration (Table 1). Ps-Mn-SOD activity was inhibited by Mn^{2+}, Co^{2+}, Ni^{2+}, Zn^{2+}, and 1 mM Cu^{2+} and Ba^{2+}. In particular, Co^{2+} showed more significant inhibition effect on Ps-Mn-SOD activity. Mg^{2+} and Ca^{2+} showed minimal effects.

Table 2 provides the effects of inhibitors, detergents, and denaturants on Ps-Mn-SOD activity. Ps-Mn-SOD activity was strongly inhibited by ethylene diamine tetraacetic acid (EDTA) and SDS and especially sensitive to SDS. Reductant dithiothreitol (DTT) and β-mercaptoethanol (β-ME) minimally affected enzyme activity. Detergents of Tween 20, Triton X-100, and Chaps slightly enhanced enzyme activity at 0.1% concentration.

Table 1. Effects of metal ions on Ps-Mn-SOD. ** $p < 0.01$.

Divalent Metal Ions	Concentration/mmol·L^{-1}	Relative Activity/%
Control	—	100 ± 2.39
Mn^{2+}	0.1	92.89 ± 1.53 **
	1	84.99 ± 2.77 **
Co^{2+}	0.1	80.13 ± 1.23 **
	1	61.49 ± 1.54 **
Ni^{2+}	0.1	95.70 ± 2.38 **
	1	94.42 ± 2.92 **
Zn^{2+}	0.1	90.25 ± 1.76 **
	1	90.99 ± 4.63 **
Cu^{2+}	0.1	98.39 ± 3.97
	1	88.99 ± 5.44 **
Ba^{2+}	0.1	99.16 ± 2.18
	1	95.94 ± 2.40 **
Mg^{2+}	0.1	100.68 ± 3.27
	1	100.61 ± 2.16
Ca^{2+}	0.1	99.71 ± 1.13
	1	100.39 ± 4.48

Table 2. Effects of inhibitors, reductant, and detergents. * $p < 0.05$; ** $p < 0.01$.

Divalent Metal Ions	Concentration	Relative Activity/%
Control	—	100 ± 2.84
EDTA	1 mmol·L^{-1}	64.33 ± 3.08 **
	10 mmol·L^{-1}	58.03 ± 2.59 **
DTT	1 mmol·L^{-1}	96.36 ± 4.65
	10 mmol·L^{-1}	97.00 ± 5.46
β-ME	1 mmol·L^{-1}	96.53 ± 4.47
	10 mmol·L^{-1}	101.85 ± 3.72
Tween 20	0.1%	109.96 ± 6.62 **
	1%	105.01 ± 3.28 **
Chaps	0.1%	103.13 ± 2.32 *
	1%	99.32 ± 3.66
Triton X-100	0.1%	105.02 ± 3.29 **
	1%	99.22 ± 3.79
SDS	0.1%	5.21 ± 3.45 **
	1%	6.11 ± 4.15 **

The enzyme could resist the strong denaturation of urea and guanidine hydrochloride (Figure 4C) and maintain an almost full activity after 1 h treatment in 5 M urea or 4 M guanidine hydrochloride.

Hydrogen peroxide and sodium azide were used to determine the SOD type (Figure 5 and Supplementary Figure S4). After treatment of the recombinant Ps-Mn-SOD using 10 mM hydrogen peroxide and sodium azide at 25 °C for 1 h, the relative activities were 7.73% and 90.39%, respectively. This showed that the SOD from *Paelopatides* sp. belongs to Fe/Mn-SOD family, in accordance with previous phylogenetic analysis and 3D structure prediction.

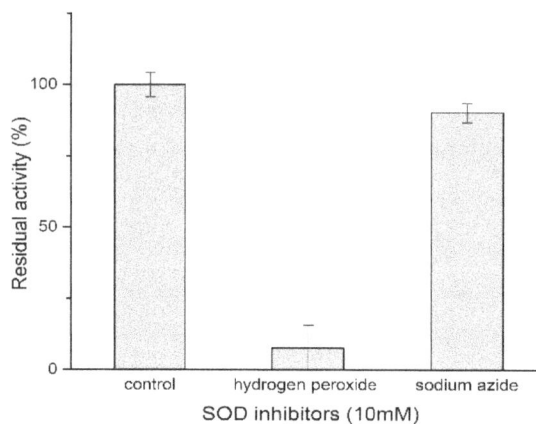

Figure 5. SOD type assay.

2.4.4. Effects of Digestive Enzymes on Ps-Mn-SOD

Digestion experiment was performed to test the stability of recombinant Ps-Mn-SOD in digestive fluid. Residual enzyme activity was measured after different incubation times for 0–4 h at 37 °C and pH 7.4. As shown in Table 3 and Supplementary Table S2, although the Ps-Mn-SOD sequence putatively contains 30 chymotrypsin and 23 trypsin cleavage sites, the enzyme could still maintain intact activity after 4 h treatment at an enzyme/substrate (w/w) ratio of 1/100.

Table 3. Cleavage effect of digestive enzyme on Ps-Mn-SOD at different time periods. Results are shown as mean ($n = 3$) $\pm$ SD. ** $p < 0.01$.

Time (h)	Relative Activity (%)
0	100 ± 2.21
1	108.66 ± 5.70
2	104.46 ± 4.54
3	103.73 ± 3.24
4	106.72 ± 4.80

2.4.5. Effects of High Hydrostatic Pressure on Ps-Mn-SOD

As shown in Figure 4D, the recombinant Ps-Mn-SOD could maintain full activity with increasing hydrostatic pressure until 100 MPa. By contrast, the SOD from bovine erythrocytes exhibited reduced activity of 84.57% when the pressure reached 100 MPa.

2.4.6. Kinetic Parameters

The kinetic parameters of recombinant Ps-Mn-SOD were determined using a series of xanthine (0.006–0.6 mM) concentrations at 37 °C and pH 8.2 (Supplementary Figure S3) based on the Michaelis–Menten equation. The Km and Vmax values of Ps-Mn-SOD were 0.0329 $\pm$ 0.0040 mM and 9112 $\pm$ 248 U/mg, respectively. The R^2 value of the curve fitting was 0.9815.

3. Discussion

Mn-SODs are predominantly found in mitochondria, as the first line of antioxidant defense, which are involved in cellular physiology, such as cell impairment and immune-responsive [8]. The important biological functions of Mn-SODs have attracted increasing attention among researchers. Novel Mn-SODs with remarkable characteristics will have great applications in food, cosmetic, and pharmaceutical industries. In the present study, a novel and kinetically stable Mn-SOD derived from hadal sea cucumber was cloned, expressed, and characterized.

Based on preliminary data, the Ps-Mn-SOD is frigostabile, consistent with the fact that the protein was derived from hadal area, which maintained > 90% activity below 20 °C with the optimum temperature observed at 0 °C. In contrast, Mn-SOD from ark shell, *Scapharca broughtonii*, showed <40% activity below 20 °C [4]. Mn-SOD from seahorse, *Hippocampus abdominalis*, showed <80% and continuously reduced activity below 20 °C [8].

Mn-SODs in several sources have been found to function at wide pH values. For example, a hyperthermostable Mn-SOD from *Thermus thermophilus* HB27 maintained >70% activity at pH 4.0–8.0 [29]; Mn-SOD from deep-sea thermophile *Geobacillus* sp. EPT3 maintained >70% activity at pH 7.0–9.0 [30]; and Mn-SOD from *Thermoascus aurantiacus* var. *levisporus* only maintained >40% activity at pH 6.0–9.0 [31]. In contrast, the present Ps-Mn-SOD could maintain >70% activity at pH 5.0–12.0, showing remarkably wide pH values adaptation. Furthermore, after 1 h treatment in extremely acidic (pH 2.2) or alkaline (pH 13.0) conditions, Ps-Mn-SOD still maintained ~20% activity, showing remarkable stability to extreme pH values. The pH assays also showed that Ps-Mn-SOD is more stable under alkaline (pH 8.5–12.0) than acidic (pH 2.2–5.0) conditions. Metal ligands may undergo protonation at low pH but exhibit stability in alkaline conditions [32]. Similar studies on seahorse and bay scallop SODs were also reported [8,33].

Ps-Mn-SOD is relatively stable in chemicals, such as urea, guanidine hydrochloride, β-ME, DTT, etc. It maintained almost 100% activity after 1 h treatment of 5 M urea or 4 M guanidine hydrochloride at 25 °C, showing excellent resistance to strong protein denaturants. By comparison, the Mn-SOD from deep-sea thermophile *Geobacillus* sp. EPT3 maintained > 70% residual activity in 2.5 M urea or guanidine hydrochloride after 30 min treatment [30]. Fe-SOD from Antarctic yeast *Rhodotorula mucilaginosa* showed relatively low tolerance to urea [34]. However, based on our obtained data (unpublished and [35]), SODs from hadal sea cucumbers constantly exhibited excellent resistance to perturbation of denaturants. In addition, Ps-Mn-SOD maintained 97.00% and 99.22% residual activity after 1 h treatment of 10 mM DTT and 1% Triton X-100, respectively. While Mn-SOD from deep-sea thermophile *Geobacillus* sp. EPT3 only maintained 84.10% and 70.30% activity after 30 min treatment of corresponding chemicals [30].

As expected, Ps-Mn-SOD could also resist the perturbation by high hydrostatic pressure compared to the homolog from atmospheric pressure organism, because it was derived from a hadal field. Given the limitations of our equipment, the experiment was not performed at pressure more than 100 MPa. In fact, Ps-Mn-SOD might resist >100 MPa hydrostatic pressure. Similar results have been reported in other deep-sea enzymes, such as RNA polymerase from *Shewanella violacea* [36], N-acetylneuraminate lyase from *Mycoplasma* sp. [37], and lactate dehydrogenase b from *Corphaenoides armatus* [38]. Nonetheless, the sensitivity of enzymes to high hydrostatic pressure is not always related to the depth where the organisms lived. For example, two polygalacturonases from the hadal yeast *Cryptococcus liquefaciens* strain N6 exhibited an almost constant activity from 0.1 to 100 MPa. While, at the same pressure, polygalacturonase from *Aspergillus japonicus*, which lives under atmospheric pressure, increased by approximately 50% [39]. However, limited studies reported in detail the pressure assays of SODs, proving the difficulty in the interpretation of their pressure tolerance mechanism.

Altogether, these features render Ps-Mn-SOD a potential candidate in the biopharmaceutical and nutraceutical fields.

4. Materials and Methods

4.1. Material and Reagents

Hadal sea cucumber was collected at the depth of 6500 m in the Mariana Trench (10° 57.1693′ N 141° 56.1719′ E). Total RNA was extracted using RNeasy Plus Universal Kits from Qiagen, Hilden, Germany, and reverse-transcribed to cDNA. The transcriptome was obtained by sequencing assembly and annotation by Novogene Company (Tianjin, China). The following reagents were purchased from Takara, Tokyo, Japan: PrimeScript™ II 1st strand cDNA Synthesis Kit, PrimeSTAR® GXL DNA Polymerase, *E. coli* DH5α, and pG-KJE8/BL21 competent cells, pCold II vector, restriction enzymes *BamH* I, and *Pst* I, T4-DNA ligase, and DNA and protein markers. The 1 mL Ni-NTA affinity column, BCA protein assay kit, primers, and trypsin/chymotrypsin complex (2400:400) were obtained from Sangon Biotech Company, Shanghai, China. Polyvinylidene difluoride (PVDF) membrane was obtained from Millipore Company, USA. The primary (ab18184) and secondary antibodies (ab6789) were obtained from Abcam, Cambridge, UK. Pierce™ ECL Plus Western blot analysis substrate was obtained from ThermoFisher, Waltham, MA, USA.

4.2. Cloning and Recombinant

For the manganese SOD (Ps-Mn-SOD) gene, the Mn-SOD sequences of Holothuroidea in GenBank were submitted to the transcriptome database of Paelopatides sp. to run a local blast using Bioedit 7.0 software. The open reading frame (ORF) of Ps-Mn-SOD (deleted signal peptide) was amplified by primers Ps-Mn-SOD-S: CGGGATCCAAGGCTCCGTATGAAGGCCTGGAGA and Ps-Mn-SOD-A: AACTGCAGTCACAATTCTTCATGTTTAGATGGC using the cDNA as template (the underlined restriction enzyme sites). The sequence was submitted to GenBank database with accession numbers MK182093. The purified and digested PCR product was ligated with pCold II vector. The recombinant plasmids, that is, pCold II-Ps-Mn-SOD, were transformed into E. coli DH5α, and positive clones were verified by sequencing.

4.3. Protein Overproduction, Purification, and Confirmation

The recombinant plasmids were transformed into E. coli chaperone competent cells pG-KJE8/BL21, which were inoculated in liquid Luria-Bertani medium (containing 100 μg/mL ampicillin, 20 μg/mL chloramphenicol, 0.5 mg/mL L-arabinose, and 2 ng/mL tetracycline), proliferated at 37 °C until the OD_{600} reached 0.4–0.6, cooled on an ice–water mixture for 40 min, added isopropyl β-D-1-thiogalactopyranoside (IPTG) with a final concentration of 0.1 mM, and then incubated for 24 h at 15 °C to produce the recombinant protein. Cells were harvested, washed with **1** × phosphate-buffered saline, resuspended in binding buffer (50 mM Na_3PO_4, 300 mM NaCl, and 20 mM imidazole, pH 7.4), and then sonicated on ice. The supernatant harboring the recombinant protein was separated from cell debris by centrifugation at 12000 g and 4 °C for 20 min and then applied to 1 mL Ni-NTA column for purification of the target protein based on its 6× His-tag, according to the manufacturer's instructions. The harvested target protein was dialyzed with 1 × tris buffered saline (TBS) at 4 °C for 24 h against three changes of 1 × TBS and finally stored at −80 °C for further experiments. The expression condition was analyzed on 12% sodium dodecyl sulfate polyacrylamide gel electrophoresis (SDS-PAGE) and confirmed using Western blot analysis. The recombinant protein on 12% SDS-PAGE gel was transferred to a PVDF membrane, which was successively incubated with primary (diluted 1:5000) and secondary antibodies (diluted 1:10000), dyed with Pierce™ ECL Plus Western blot analysis substrate, and detected under chemiluminescent imaging system. Additional details were as described by Li et al. [35].

4.4. Bioinformatics Analyses

The amino acid sequence of Ps-Mn-SOD was translated using ExPASy translation tool (http://web.expasy.org/translate/). The signal peptide, secondary structure, motif sequences, and 3D homology model were predicted by SignalP 4.1 Server (http://www.cbs.dtu.dk/services/SignalP/), Scratch Protein Predictor (http://scratch.proteomics.ics.uci.edu/), InterPro Scan (http://www.ebi.ac.uk/InterProScan/), and Swiss model server (http://swissmodel.expasy.org/) [40], respectively. The physicochemical properties of Ps-Mn-SOD were predicted using ExPASy ProtParam tool (http://web.expasy.org/protparam/). The possible cleavage sites of trypsin and chymotrypsin on Ps-Mn-SOD were predicted using the peptide cutter software (http://web.expasy.org/peptide_cutter/). Multiple alignments of Ps-Mn-SOD were processed using DNAMAN 7.0.2 software. Homology analysis was constructed by pairwise alignment tool (https://www.ebi.ac.uk/Tools/psa/emboss_needle/). The neighbor-joining phylogenic tree was generated in MEGA 7.0 with bootstrap values 1000.

4.5. Enzyme Assays

SOD activity was determined via spectrophotometric method using the SOD assay kit from Nanjing Jiancheng Institute of Biology and Engineering (Code No. A001-1-1, Nanjing, China). Each measurement point contained three replicates, and the results are shown as mean (n = 3) ± standard deviation (SD). The 1 × TBS was used as the blank control. One unit of SOD activity was defined as the amount of enzyme that inhibited 50% of chromogen production at 550 nm.

The purified Ps-Mn-SOD was quantified, and residual activities were determined after incubation under different variables, including temperature, pH, chemicals, digestive enzymes, and high hydrostatic pressure. Considering temperature, proteins were treated from 0 °C to 80 °C for 15 min with an interval of 10 °C [33,34]. Proteins were treated at pH 2.2–13 for 1 h at 25 °C [4,34]. The enzymatic activity at optimum temperature and pH was set as 100%. With regard to chemicals, the proteins were mixed with an equal treatment solution at different final concentrations for 40 min at 25 °C [34,41]. The incubation time of urea, guanidine hydrochloride, hydrogen peroxide, and sodium azide was expanded to 1 h. The enzyme activity without chemicals was set as 100%. For proteolytic susceptibility assay, the mass ratio of recombinant Ps-Mn-SOD and trypsin/chymotrypsin complex was 1:100, and the group incubated for 0 h was considered with 100% enzyme activity [34,42]. For high hydrostatic pressure, proteins were treated at 0.1, 30, and 100 MPa for 2 h at 5 °C. The enzyme activity at 0.1 MPa was set as 100%, and bovine erythrocyte SOD was selected for comparison from atmospheric organism. Kinetics of Ps-Mn-SOD were measured as previously described by Li et al. [35].

4.6. Statistical Analysis

Independent sample T-test was used for statistical analysis for each of the two groups using SPSS 21.0 (IBM Company, Armonk, NY, USA); $p < 0.05$ was considered statistically significant.

Supplementary Materials: Figure S1: The predicted 3D model of Ps-Mn-SOD. Red spheres represent manganese ions. (**A**) homodimer.(**B**) close-up of the manganese ion binding site. Figure S2: Analysis of SDS-PAGE. M: protein marker, Lane 1: total proteins before induction, Lane 2: total proteins after induction, Lane 3: inclusion body after ultrasonication, Lane 4: supernatant after ultrasonication, Lane 5: Western blot of recombinant protein, Lane 6: purified protein. Figure S3: The curve of kinetic parameters of Ps-Mn-SOD. Figure S4: SOD type assay. The result is expressed using specific activity. Table S1. Pairwise alignment analysis between Ps-Mn-SOD and other species. Table S2. The prediction of cleavage site of Ps-Mn-SOD.

Author Contributions: H.Z. collected the sample, Y.L. designed and performed the experiments, Y.L. and H.Z. prepared the manuscript, and X.K. gave advice during the experiments.

References

1.   Torres, M.A. ROS in biotic interactions. *Physiol. Plant.* **2010**, *138*, 414–429. [CrossRef] [PubMed]

2.   Kawanishi, S.; Inoue, S. Damage to DNA by reactive oxygen and nitrogen species. *Seikagaku* **1997**, *69*, 1014–1017. [PubMed]

3.   Kim, F.J.; Kim, H.P.; Hah, Y.C.; Roe, J.H. Differential expression of superoxide dismutases containing Ni and Fe/Zn in *Streptomyces coelicolor. Eur. J. Biochem.* **1996**, *241*, 178–185. [CrossRef] [PubMed]

4.   Zheng, L.; Wu, B.; Liu, Z.; Tian, J.; Yu, T.; Zhou, L.; Sun, X.; Yang, A.G. A manganese superoxide dismutase (MnSOD) from ark shell, *Scapharca broughtonii*: Molecular characterization, expression and immune activity analysis. *Fish Shellfish Immunol.* **2015**, *45*, 656–665. [CrossRef] [PubMed]

5.   Zheng, Z.; Jiang, Y.H.; Miao, J.L.; Wang, Q.F.; Zhang, B.T.; Li, G.Y. Purification and characterization of a cold-active iron superoxide dismutase from a psychrophilic bacterium, *Marinomonas* sp. NJ522. *Biotechnol. Lett.* **2006**, *28*, 85–88. [CrossRef] [PubMed]

6. Krauss, I.R.; Merlino, A.; Pica, A.; Rullo, R.; Bertoni, A.; Capasso, A.; Amato, M.; Riccitiello, F.; De Vendittis, E.; Sica, F. Fine tuning of metal-specific activity in the Mn-like group of cambialistic superoxide dismutases. *RSC Adv.* **2015**, *5*, 87876–87887. [CrossRef]

7. Sheng, Y.; Abreu, I.A.; Cabelli, D.E.; Maroney, M.J.; Miller, A.-F.; Teixeira, M.; Valentine, J.S. Superoxide Dismutases and Superoxide Reductases. *Chem. Rev.* **2014**, *114*, 3854–3918. [CrossRef] [PubMed]

8. Ncn, P.; Godahewa, G.I.; Lee, S.; Kim, M.J.; Hwang, J.Y.; Kwon, M.G.; Hwang, S.D.; Lee, J. Manganese-superoxide dismutase (MnSOD), a role player in seahorse (*Hippocampus abdominalis*) antioxidant defense system and adaptive immune system. *Fish Shellfish Immunol.* **2017**, *68*, 435–442.

9. Kim, B.M.; Rhee, J.S.; Park, G.S.; Lee, J.; Lee, Y.M.; Lee, J.S. Cu/Zn- and Mn-superoxide dismutase (SOD) from the copepod *Tigriopus japonicus*: Molecular cloning and expression in response to environmental pollutants. *Chemosphere* **2011**, *84*, 1467–1475. [CrossRef]

10. Li, C.; He, J.; Su, X.; Li, T. A manganese superoxide dismutase in blood clam *Tegillarca granosa*: Molecular cloning, tissue distribution and expression analysis. *Comp. Biochem. Physiol. B Biochem. Mol. Biol.* **2011**, *159*, 64–70. [CrossRef]

11. Wang, H.; Yang, H.; Liu, J.; Yanhong, L.I.; Liu, Z. Combined effects of temperature and copper ion concentration on the superoxide dismutase activity in *Crassostrea ariakensis*. *Acta Oceanol. Sin.* **2016**, *35*, 51–57. [CrossRef]

12. Xie, Z.; Jian, H.; Jin, Z.; Xiao, X. Enhancing the adaptability of the deep-sea bacterium Shewanella piezotolerans WP3 to high pressure and low temperature by experimental evolution under H_2O_2 stress. *Appl. Environ. Microbiol.* **2017**, *84*, e02342-17. [CrossRef] [PubMed]

13. Lebovitz, R.M.; Zhang, H.; Vogel, H.; Cartwright, J.; Dionne, L.; Lu, N.; Huang, S.; Matzuk, M.M. Neurodegeneration, myocardial injury, and perinatal death in mitochondrial superoxide dismutase-deficient mice. *Proc. Natl. Acad. Sci. USA* **1996**, *93*, 9782–9787. [CrossRef] [PubMed]

14. Li, Y.; Huang, T.T.; Carlson, E.J.; Melov, S.; Ursell, P.C.; Olson, J.L.; Noble, L.J.; Yoshimura, M.P.; Berger, C.; Chan, P.H. Dilated cardiomyopathy and neonatal lethality in mutant mice lacking manganese superoxide dismutase. *Nat. Genet.* **1995**, *11*, 376–381. [CrossRef] [PubMed]

15. Delacourte, A.; Defossez, A.; Ceballos, I.; Nicole, A.; Sinet, P.M. Preferential localization of copper zinc superoxide dismutase in the vulnerable cortical neurons in Alzheimer's disease. *Neurosci. Lett.* **1988**, *92*, 247–253. [CrossRef]

16. Kruman, I.I.; Pedersen, W.A.; Springer, J.E.; Mattson, M.P. ALS-linked Cu/Zn-SOD mutation increases vulnerability of motor neurons to excitotoxicity by a mechanism involving increased oxidative stress and perturbed calcium homeostasis. *Exp. Neurol.* **1999**, *160*, 28–39. [CrossRef] [PubMed]

17. Cullen, J.J.; Weydert, C.; Hinkhouse, M.M.; Ritchie, J.; Domann, F.E.; Spitz, D.; Oberley, L.W. The role of manganese superoxide dismutase in the growth of pancreatic adenocarcinoma. *Cancer Res.* **2003**, *63*, 1297–1303.

18. Zhang, Y.; Wang, J.Z.; Wu, Y.J.; Li, W.G. Anti-inflammatory effect of recombinant human superoxide dismutase in rats and mice and its mechanism. *Acta Pharmacol. Sin.* **2002**, *23*, 439–444.

19. Luisa, C.M.; Jorge, J.C.; Van'T, H.R.; Cruz, M.E.; Crommelin, D.J.; Storm, G. Superoxide dismutase entrapped in long-circulating liposomes: Formulation design and therapeutic activity in rat adjuvant arthritis. *BBA Biomembr.* **2002**, *1564*, 227–236. [CrossRef]

20. Cloarec, M.; Caillard, P.; Provost, J.C.; Dever, J.M.; Elbeze, Y.; Zamaria, N. GliSODin, a vegetal sod with gliadin, as preventative agent vs. atherosclerosis, as confirmed with carotid ultrasound-B imaging. *Eur. Ann. Allergy Clin. Immunol.* **2007**, *39*, 45–50.

21. Muth, C.M.; Glenz, Y.; Klaus, M.; Radermacher, P.; Speit, G.; Leverve, X. Influence of an orally effective SOD on hyperbaric oxygen-related cell damage. *Free Radic. Res.* **2004**, *38*, 927–932. [CrossRef] [PubMed]

22. Liu, Y.-X.; Zhou, D.-Y.; Liu, Z.-Q.; Lu, T.; Song, L.; Li, D.-M.; Dong, X.-P.; Qi, H.; Zhu, B.-W.; Shahidi, F. Structural and biochemical changes in dermis of sea cucumber (Stichopus japonicus) during autolysis in response to cutting the body wall. *Food Chem.* **2018**, *240*, 1254–1261. [CrossRef] [PubMed]

23. Jamieson, A.J.; Gebruk, A.; Fujii, T.; Solan, M. Functional effects of the hadal sea cucumber Elpidia atakama (Echinodermata: Holothuroidea, Elasipodida) reflect small-scale patterns of resource availability. *Mar. Biol.* **2011**, *158*, 2695–2703. [CrossRef]

24. Zhang, J.; Lin, S.; Zeng, R. Cloning, expression, and characterization of a cold-adapted lipase gene from an antarctic deep-sea psychrotrophic bacterium, *Psychrobacter* sp. 7195. *J. Microbiol. Biotechnol.* **2007**, *17*, 604. [PubMed]

25. Michels, P.C.; Clark, D.S. Pressure-enhanced activity and stability of a hyperthermophilic protease from a deep-sea methanogen. *Appl. Environ. Microbiol.* **1997**, *63*, 3985–3991. [PubMed]

26. Borgstahl, G.E.O.; Parge, H.E.; Hickey, M.J.; Beyer, W.F., Jr.; Hallewell, R.A.; Tainer, J.A. The structure of human mitochondrial manganese superoxide dismutase reveals a novel tetrameric interface of two 4-helix bundles. *Cell* **1992**, *71*, 107–118. [CrossRef]

27. Liu, P.; Ewis, H.E.; Huang, Y.J.; Lu, C.D.; Tai, P.C.; Weber, I.T. Crystal Structure of the Bacillus subtilis Superoxide Dismutase. *Acta Crystallogr.* **2008**, *63 Pt 12*, 1003–1007.

28. Benkert, P.; Biasini, M.; Schwede, T. Toward the estimation of the absolute quality of individual protein structure models. *Bioinformatics* **2011**, *27*, 343–350. [CrossRef]

29. Liu, J.; Yin, M.; Hu, Z.; Lu, J.; Cui, Z. Purification and characterization of a hyperthermostable Mn-superoxide dismutase from *Thermus thermophilus* HB27. *Extremophiles* **2011**, *15*, 221–226. [CrossRef]

30. Zhu, Y.B.; Wang, G.H.; Ni, H.; Xiao, A.F.; Cai, H.N. Cloning and characterization of a new manganese superoxide dismutase from deep-sea thermophile *Geobacillus* sp. EPT3. *World J. Microbiol. Biotechnol.* **2014**, *30*, 1347–1357. [CrossRef]

31. Song, N.N.; Zheng, Y.; Shi-Jin, E.; Li, D.C. Cloning, expression, and characterization of thermostable Manganese superoxide dismutase from *Thermoascus aurantiacus* var. *levisporus*. *J. Microbiol.* **2009**, *47*, 123–130. [CrossRef] [PubMed]

32. Dolashki, A.; Abrashev, R.; Stevanovic, S.; Stefanova, L.; Ali, S.A.; Velkova, L.; Hristova, R.; Angelova, M.; Voelter, W.; Devreese, B. Biochemical properties of Cu/Zn-superoxide dismutase from fungal strain *Aspergillus niger* 26. *Spectrochim. Acta Part A Mol. Biomol. Spectrosc.* **2009**, *71*, 975–983. [CrossRef] [PubMed]

33. Bao, Y.; Li, L.; Xu, F.; Zhang, G. Intracellular copper/zinc superoxide dismutase from bay scallop *Argopecten irradians*: Its gene structure, mRNA expression and recombinant protein. *Fish Shellfish Immunol.* **2009**, *27*, 210–220. [CrossRef] [PubMed]

34. Kan, G.; Wen, H.; Wang, X.; Zhou, T.; Shi, C. Cloning and characterization of iron-superoxide dismutase in Antarctic yeast strain Rhodotorula mucilaginosa AN5. *J. Basic Microbiol.* **2017**, *57*, 680–690. [CrossRef] [PubMed]

35. Li, Y.; Kong, X.; Chen, J.; Liu, H.; Zhang, H. Characteristics of the Copper, Zinc Superoxide Dismutase of a Hadal Sea Cucumber (*Paelopatides* sp.) from the Mariana Trench. *Mar. Drugs* **2018**, *16*, 169. [CrossRef] [PubMed]

36. Kawano, H.; Nakasone, K.; Matsumoto, M.; Yoshida, Y.; Usami, R.; Kato, C.; Abe, F. Differential pressure resistance in the activity of RNA polymerase isolated from *Shewanella violacea* and *Escherichia coli*. *Extremophiles* **2004**, *8*, 367–375. [CrossRef]

37. Wang, S.L.; Li, Y.L.; Han, Z.; Chen, X.; Chen, Q.J.; Wang, Y.; He, L.S. Molecular Characterization of a NovelN-Acetylneuraminate Lyase from a Deep-Sea Symbiotic Mycoplasma. *Mar. Drugs* **2018**, *16*, 80. [CrossRef]

38. Brindley, A.A.; Pickersgill, R.W.; Partridge, J.C.; Dunstan, D.J.; Hunt, D.M.; Warren, M.J. Enzyme sequence and its relationship to hyperbaric stability of artificial and natural fish lactate dehydrogenases. *PLoS ONE* **2008**, *3*, e2042. [CrossRef]

39. Abe, F.; Minegishi, H.; Miura, T.; Nagahama, T.; Usami, R.; Horikoshi, K. Characterization of cold- and high-pressure-active polygalacturonases from a deep-sea yeast, *Cryptococcus liquefaciens* strain N6. *J. Agric. Chem. Soc.* **2006**, *70*, 296–299.

40. Sujiwattanarat, P.; Pongsanarakul, P.; Temsiripong, Y.; Temsiripong, T.; Thawornkuno, C.; Uno, Y.; Unajak, S.; Matsuda, Y.; Choowongkomon, K.; Srikulnath, K. Molecular cloning and characterization of Siamese crocodile (*Crocodylus siamensis*) copper, zinc superoxide dismutase (CSI-Cu,Zn-SOD) gene. *Comp. Biochem. Physiol. A Mol. Integr. Physiol.* **2016**, *191*, 187–195. [CrossRef]

41. Zhu, Y.; Li, H.; Ni, H.; Liu, J.; Xiao, A.; Cai, H. Purification and biochemical characterization of manganesecontaining superoxide dismutase from deep-sea thermophile *Geobacillus* sp. EPT3. *Acta Oceanol. Sin.* **2014**, *33*, 163–169. [CrossRef]

42. Ken, C.F.; Hsiung, T.M.; Huang, Z.X.; Juang, R.H.; Lin, C.T. Characterization of Fe/Mn−Superoxide Dismutase from Diatom *Thallassiosira weissflogii*: Cloning, Expression, and Property. *J. Agric. Food Chem.* **2005**, *53*, 1470–1474. [CrossRef] [PubMed]

14

Purification and Characterization of a Novel Alginate Lyase from the Marine Bacterium *Bacillus* sp. Alg07

Peng Chen [1,2], **Yueming Zhu** [1], **Yan Men** [1], **Yan Zeng** [1] **and Yuanxia Sun** [1,*]

[1] National Engineering Laboratory for Industrial Enzymes, Tianjin Institute of Industrial Biotechnology, Chinese Academy of Sciences, Tianjin 300308, China; chen_p@tib.cas.cn (P.C.); zhu_ym@tib.cas.cn (Y.Z.); men_y@tib.cas.cn (Y.M.); zeng_y@tib.cas.cn (Y.Z.)

[2] University of Chinese Academy of Sciences, Beijing 100049, China

* Correspondence: Sun_yx@tib.cas.cn.

Abstract: Alginate oligosaccharides with different bioactivities can be prepared through the specific degradation of alginate by alginate lyases. Therefore, alginate lyases that can be used to degrade alginate under mild conditions have recently attracted public attention. Although various types of alginate lyases have been discovered and characterized, few can be used in industrial production. In this study, AlgA, a novel alginate lyase with high specific activity, was purified from the marine bacterium *Bacillus* sp. Alg07. AlgA had a molecular weight of approximately 60 kDa, an optimal temperature of 40 °C, and an optimal pH of 7.5. The activity of AlgA was dependent on sodium chloride and could be considerably enhanced by Mg^{2+} or Ca^{2+}. Under optimal conditions, the activity of AlgA reached up to 8306.7 U/mg, which is the highest activity recorded for alginate lyases. Moreover, the enzyme was stable over a broad pH range (5.0–10.0), and its activity negligibly changed after 24 h of incubation at 40 °C. AlgA exhibited high activity and affinity toward poly-β-D-mannuronate (polyM). These characteristics suggested that AlgA is an endolytic polyM-specific alginate lyase (EC 4.2.2.3). The products of alginate and polyM degradation by AlgA were purified and identified through fast protein liquid chromatography and electrospray ionization mass spectrometry, which revealed that AlgA mainly produced disaccharides, trisaccharides, and tetrasaccharide from alginate and disaccharides and trisaccharides from polyM. Therefore, the novel lysate AlgA has potential applications in the production of mannuronic oligosaccharides and poly-α-L-guluronate blocks from alginate.

Keywords: alginate lyase; marine bacterium; *Bacillus* sp. Alg07; purification; alginate oligosaccharides

1. Introduction

Alginate is a linear copolymer that is composed of homopolymeric blocks of (1–4)-linked α-L-guluronic acid (G) and its C5 epimer β-D-mannuronic acid (M), which forms three types of blocks: poly-α-L-guluronate (polyG), poly-β-D-mannuronate (polyM), and random heteropolymeric sequences (polyMG) [1]. Alginate is the most abundant carbohydrate in brown algae, and it accounts for up to 10–45% of the dry weight of brown algae [2]. Some bacteria that belong to the genera *Azotobacter* [3] and *Pseudomonas* [4] produce alginate as an extracellular polysaccharide. In contrast to algal alginate, bacterial alginate is acetylated. Commercial alginate manufactured from brown algae has been used as a thickening agent or gelling agent in the food and pharmaceutical industries [5]. Alginate can be degraded into alginate oligosaccharides (AOS) through a chemical process or by alginate lyase. Given that AOS can stimulate the growth of endothelial cells [6] and the production of multiple cytokines [7], they may be applied as growth-promoting agents in some plants [8] and bifidobacteria [9]. Furthermore, AOS demonstrate excellent antioxidant activity [10] and have potential uses in protection against pathogens [11].

In alginate degradation, alginate lyases cleave the (1–4)-linked glucosidic bond of alginate via a β-elimination mechanism and generate unsaturated oligosaccharides with 4-deoxy-alpha-L-erythrohex-4-enopyranuronosyl uronate as the nonreducing terminal residue [12]. Numerous alginate lyases have been isolated from various organisms, such as marine algae [13], marine mollusks [14], marine and terrestrial bacteria [15,16], marine fungi [17], and viruses [18]. Alginate lyases can be categorized into polyM-specific, polyG-specific, and polyMG-specific lyases on the basis of their substrate preferences [19] or into endo- or exo-alginate lyases on the basis of their cleavage mode [19]. In the carbohydrate-active enzyme database, alginate lyases belong to the polysaccharide lyase family [20]. The structures of some alginate lyases have been elucidated.

Alginate lyases are widely used in many fields. For example, alginate lyases have been employed to explain the fine structures of alginate [21] and to prepare red and brown algal protoplasts [22]. These enzymes may be utilized in the treatment of cystic fibrosis [23] and have been used as catalysts for AOS production [24]. The application of alginate lyases in alginate degradation under mild conditions has recently attracted public attention given the high efficiency and specifi of these enzymes. Nevertheless, present studies on alginate lyases remain in development, and the low catalytic efficiency and poor thermostability of alginate lyases limit their utility in AOS production. Therefore, high-efficiency and thermostable alginate lyases should be identified for use in AOS production.

In our work, we isolated and identified *Bacillus* sp. Alg07, a novel marine bacterium. AlgA, the alginate lyase secreted by this strain, showed extremely high activity. Hence, we purified and characterized AlgA to confirm that it has potential applications in AOS production.

2. Results and Discussion

2.1. Screening and Identification of Strain Alg07

Twenty-one strains with alginate lyase activity were isolated using alginate as the sole carbon source. Alg07 secreted the alginate lyase with the highest activity in the fermentation culture.

The 16S rRNA gene of Alg07 was cloned, sequenced, and submitted to GeneBank (accession number KM040772) for strain identification. The alignment of 16S rRNA gene sequences from different *Bacillus* species showed that strain Alg07 is closely related to *Bacillus litoralis* S20409 (97%) and *Bacillus simplex* J2S3 (97%). However, the low similarity shared by the 16s rRNA gene sequence of Alg07 with that of other known *Bacillus* species indicated that Alg07 may be a novel *Bacillus* species. In accordance with the neighbor-joining phylogenetic tree, the strain was assigned to the genus *Bacillus* and designated as *Bacillus* sp. Alg07 (Figure 1).

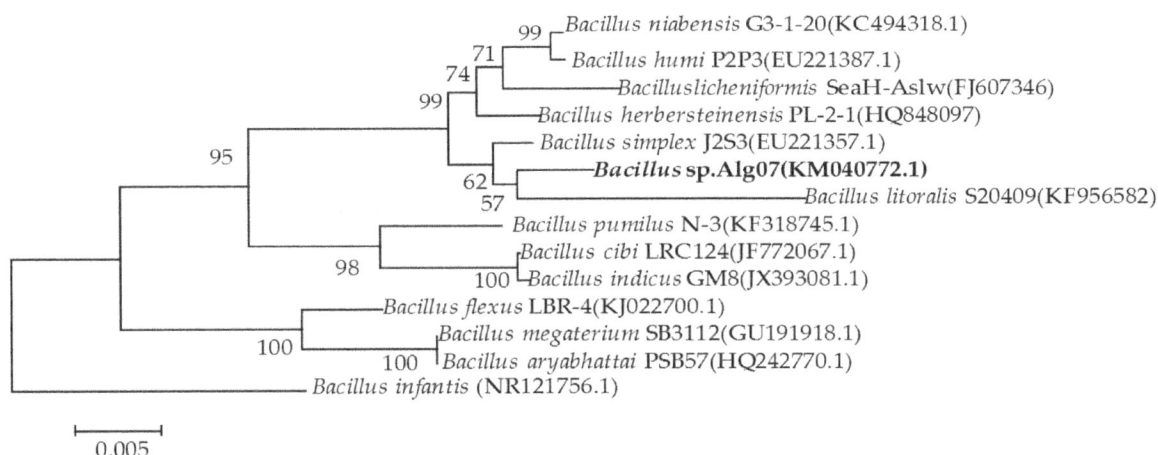

Figure 1. Neighbor-joining phylogenetic tree generated on the basis of the 16S rRNA gene sequences of strain Alg07 and other known *Bacillus* species.

2.2. Purification of Alginate Lyase from Bacillus sp. Alg07

Bacillus sp. Alg07 was cultured in optimized liquid medium for 24 h until its alginate lyase reached the highest activity. The supernatant containing 510 U/mL of alginate lyase was subjected to further purification through the two simple steps of tangential flow filtration concentration and anion exchange chromatography with Source 15Q (Figure S1). After purification, the alginate lyase was purified 8.34-fold with a yield of 62.4% (Table 1). The final specific activity of the purified alginate lyase was 8306.7 U/mg. The purified alginate lyase from *Bacillus* sp. Alg07 was designated as AlgA.

Table 1. Summary of the purification of AlgA.

Step	Total Protein (mg)	Total Activity (U)	Specific Activity (U/mg)	Folds	Yield (%)
Culture broth	23.82	23,724.72	996.01	1.00	100
Vivaflow50	10.30	22,672.14	2201.18	2.21	95.4
Source 15Q	1.78	14,785.9	8306.71	8.34	62.4

The activities of AlgA and those of other well-studied strains are shown in Table 2, which shows that AlgA has the highest activity among all reported alginate lyases. The simple purification, high recovery, and high specific enzyme activity of AlgA indicated that it may be produced industrially on a large scale.

Table 2. Comparison of the properties of AlgA with those of alginate lyases from different microorganisms.

Enzyme	Source	Specific Activity (U/mg)	Molecular Mass (kDa)	Optimal Temperature (°C)	Optimal pH	Cation Activators	Substrate Specificity
AlgA	This study	8306.71	60	40	7.5	Na^+, Mg^{2+}, Ca^{2+}, Mn^{2+}	PM
Oal17	*Shewanella* sp.Kz7 [25]	32	82	50	6.2	Na^+	PM
Cel32	*Cellulophaga* sp. NJ-1 [15]	2417.8	32	50	8.0	Ca^{2+}, Mg^{2+}, K^+	PM, PG
Alg7D	*Saccharophagus* degradans [26]	4.6	63.2	50	7.0	Na^+	PM, PG
AlySY08	*Vibrio* sp. Aly08 [27]	1070.2	33	40	7.6	Na^+, K^+, Ca^{2+}, Mg^{2+}	PM, PG
AlyV5	*Vibrio* sp. QY105 [28]	2152	37	38	7.0	Na^+, Mg^{2+}, Ca^{2+}, Mn^{2+}	PM, PG
Alm	*Agarivorans* sp. JAM-Alm [29]	108.5	31	30	10.0	Na^+, K^+	PG, PMG
FlAlyA	*Flavabacterium* sp. UMI-01 [16]	2347.8	30	55	7.7	Na^+, K^+, Ca^{2+}, Mg^{2+}	PM

Figure 2 shows that the purified AlgA exhibited a clear and unique band on sodium dodecyl sulfate polyacrylamide gel electrophoresis (SDS-PAGE). This result suggested that the two-step purification process is successful. The molecular weight of AlgA was approximately 60 kDa, which is similar to molecular weights of alginate lyases from *Vibrio* sp.YKW-34 [30] and *Saccharophagus degradans* [26]. AlgA belongs to the class of large alginate lyases on the basis of its molecular weight [31].

The N-terminal amino acid sequence of the purified AlgA was analyzed. The sequence was identified as Glutamic acid–Glutamic acid–Glutamic acid–Glutamic acid–Aspartic acid–Valine–Threonine–Tyrosine (Figure S2). The results from homology search using BLASTp and CLUSTAL X indicated that the N-terminal amino acid sequence of AlgA is absent from the sequences of the previously reported alginate lyases. Furthermore, a protein with four Glu residues at its N-terminal is unusual. Therefore, AlgA may be a novel alginate lyase.

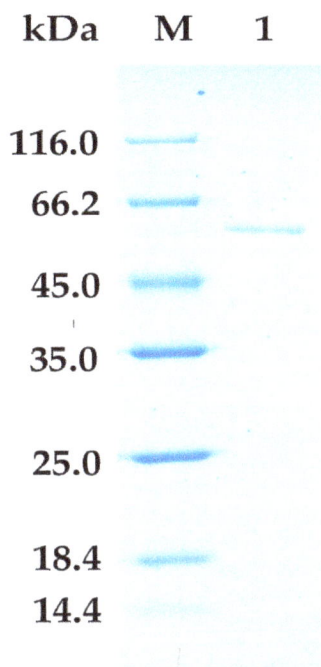

Figure 2. Sodium dodecyl sulfate polyacrylamide gel electrophoresis (SDS-PAGE) result for AlgA. Lane M, protein ladder; Lane 1, purified AlgA.

2.3. Biochemical Characterization of AlgA

The optimal temperature of AlgA is 40 °C, which is similar to alginate lyases derived from *Vibrio* sp.YKW-34 [30], *Flavobacterium* sp. LXA [11], *Pseudomonas aeruginosa* PA1167 [32], and *Stenotrophomnas maltophilia* KJ-2 [33]. The activity of AlgA significantly decreased under temperatures exceeding 50 °C (Figure 3A). Thermostability analysis indicated that the activity of AlgA remained relatively unchanged after 5 h of incubation at 40 °C and did not decrease even after 24 h of incubation at 40 °C (data not shown). However, the enzyme was less stable under high temperatures, and the half-life of AlgA is approximately 3 h under 45 °C and 0.75 h under 50 °C (Figure 3B). This result indicated that at its optimal temperature, AlgA might be the most stable enzyme among all reported alginate lyases. Alginate lyases from *Favobacterium* sp. LXA [11], *Vibrio* sp. W13 [34], and *Zobellia galactanivorans* [35] are stable only under temperatures less than 40 °C. The activities of many alginate lyases considerably decrease after incubation at 40 °C. AlyL2 from *Agarivorans* sp. L11 has a half-life of 125 min at 40 °C [36], and OalS17 from *Shewanella* sp. Kz7 retains 88% of its activity after 1 h of incubation at 40 °C [25]. Furthermore, alginate lyases from *Flammeovirga* sp. MY04 [37] and *Vibrio* sp. SY08 [27] retain approximately 80 and 75%, respectively, of their activities after 2 h of incubation at 40 °C. Thermostable enzymes are more advantageous than thermolabile counterparts due totheir long half-lives and low production costs. Thus, AlgA has potential industrial applications given its excellent thermostability.

The optimal pH of AlgA was 7.5 in 20 mM Tris-HCl buffer. AlgA exhibited more than 90% of its maximal activity in pH 8.0 buffer. These results suggested that AlgA is basophilic (Figure 3C). Similarly, the optimal activities of most alginate lyases from marine bacteria are observed at pH 7.5–8.0 (Table 2). The results of the pH stability assay showed that AlgA presented the highest stability for an extended period over a pH range of 5.0–10.0 (Figure 3D). The good stability of AlgA over a wide pH range indicated its suitability for industrial application. Furthermore, the use of AlgA may decrease production costs given that pH adjustment will be unnecessary even when various alginates from different sources are employed as substrates.

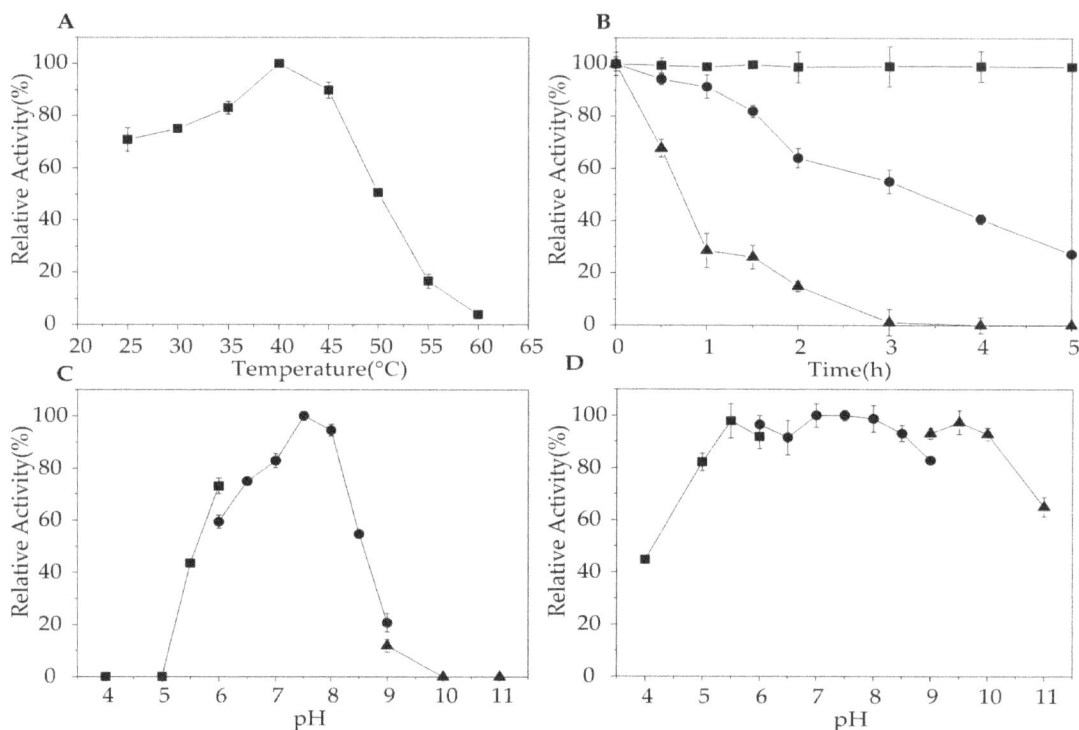

Figure 3. Effects of temperature and pH on the relative activity of AlgA. (**A**) Optimal temperature of AlgA. (**B**) Thermostability of AlgA at 40 °C (filled square), 45 °C (filled circle), and 50 °C (filled triangle). (**C**) Optimal pH for the relative activity of AlgA was determined in 20 mM CH$_3$COOH-CH$_3$COONa buffer (filled square), 20 mM Tris-HCl buffer (filled circle), or 20 mM Glycine-NaOH buffer (filled triangle). (**D**) pH stability of AlgA in 20 mM CH$_3$COOH-CH$_3$COONa buffer (filled square), 20 mM Tris-HCl (filled circle), and 20 mM Glycine-NaOH (filled triangle).

To determine the effect of NaCl on AlgA, the activity of AlgA in the presence of various NaCl concentrations was measured. As shown in Figure 4A, the optimal NaCl concentration for AlgA activity was 200 mM, and no activity was detected in the absence of NaCl. These results indicated that the activity of AlgA is dependent on NaCl. Thus, NaCl concentration is crucial for the activity of AlgA, which is a salt-activated alginate lyase. However, high NaCl concentrations decreased the activity of AlgA. Similarly, the activities of marine bacterial alginate lyases, such as AlyV5 from *Vibrio* sp. QY105 [28] and AlyYKW-34 from *Vibrio* sp. YKW-34 [30], are dependent on NaCl.

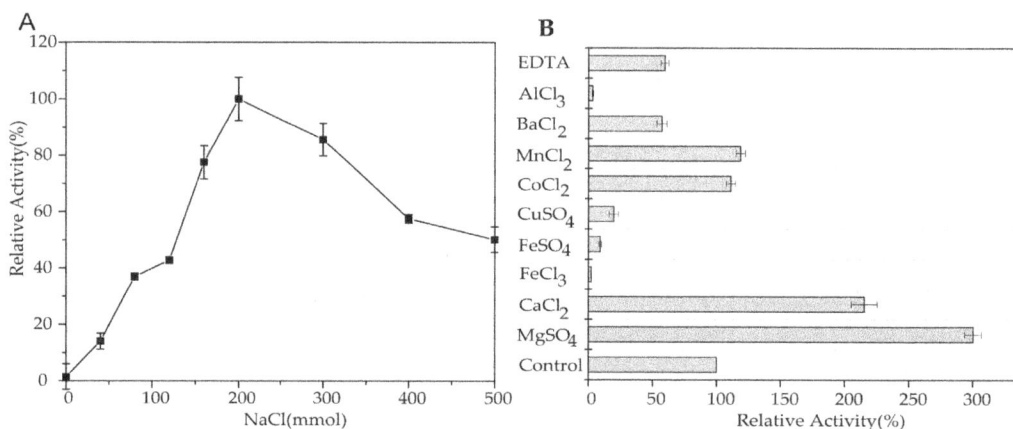

Figure 4. Effect of NaCl (**A**) and metal ions (**B**) on the activity of AlgA.

The effects of metal ions and EDTA on the activity of AlgA were determined in the presence of 200 mM NaCl. Mg^{2+} or Ca^{2+} significantly enhanced enzymatic activity by 300% or 215%, and Co^{2+} and Mn^{2+} slightly increased enzymatic activity (Figure 4B). By contrast, Hg^{2+}, Fe^{3+}, Fe^{2+}, and Cu^{2+} completely inhibited lyase activity. Ba^{2+} and EDTA partially inhibited lyase activity. Ca^{2+} and Mg^{2+} increase the activity of many alginate lyases, such as aly-SJ02 from *Pseudoalteromonas* sp. SM0524 [38] and AlyV5 from *Vibrio* sp. QY105 [28]. However, Mg^{2+} and Ca^{2+} decrease the activity of Alg7D from *S. degradans* [26].

2.4. Substrate Specificity and Kinetic Parameters of AlgA

AlgA exhibited activity toward alginate but not toward pectin, hyaluronan, chitin, or agar. This behavior suggested that AlgA is indeed an alginate lyase. As shown in Figure 5, the relative activities of AlgA toward alginate, polyM, and polyG blocks are 100 ± 4.3, 87.2 ± 6.9, and $10.5 \pm 1.7\%$, respectively. The slight activity toward polyG block might result from the presence of a few M residues in polyG substrates. These results indicated that the polyM block of substrates is the preferred substrate of AlgA. Thus, AlgA is a mannuronate lyase. Alginate lyases from *Flavobacterium* sp. UMI-01 [16], *Vibrio* sp. JAM-A9m [39], and *Pseudomonas* sp. QD03 [40] also belong to the mannuronate lyase class of enzymes.

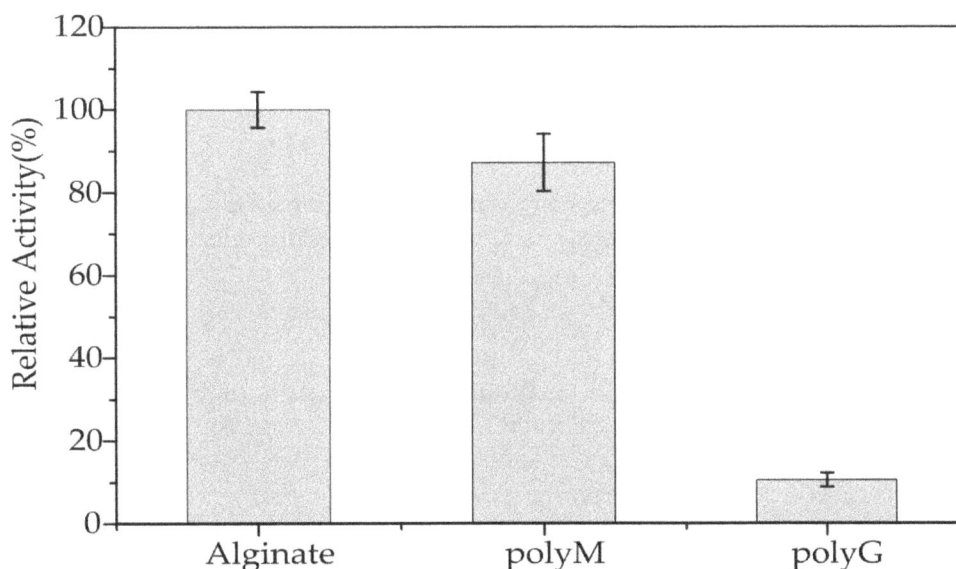

Figure 5. Relative activities of AlgA toward alginate, polyM, and polyG.

The kinetic parameters of AlgA were determined through nonlinear regression analysis (Figure S3) and are shown in Table 3. AlgA has a lower Km value for polyM than for sodium alginate. This result suggested that AlgA has high affinity for polyM blocks and further confirmed that AlgA is a mannuronate lyase. However, the $kcat$ values of AlgA for sodium alginate were higher than those for polyM. Therefore, AlgA has equivalent catalytic efficiency for sodium alginate and polyM.

Table 3. Kinetic parameters of the activity of AlgA toward sodium alginate and polyM blocks.

Parameter	Sodium Alginate	PolyM
$Vmax$ (U mg of protein^{-1})	1052.0 ± 214.6	547.6 ± 22.4
Km (mg mL^{-1})	9.0 ± 3.3	3.6 ± 0.4
$kcat$ (s^{-1})	911.7 ± 185.9	474.6 ± 19.4
$kcat/Km$ (mg^{-1} mL s^{-1})	101.3 ± 20.7	132.0 ± 5.4

2.5. Fast Protein Liquid Chromatography and Electrospray Ionization Mass Spectrometry Analysis of the Degradation Products of AlgA

To investigate the action patterns of AlgA, degradation of alginate by this enzyme was performed. The degradation products at different time intervals were analyzed through gel chromatography. Fast protein liquid chromatography (FPLC) analysis indicated that AOS with different degrees of polymerization (DP) gradually accumulated (Figure 6). Therefore, AlgA is an endo-type alginate lyase.

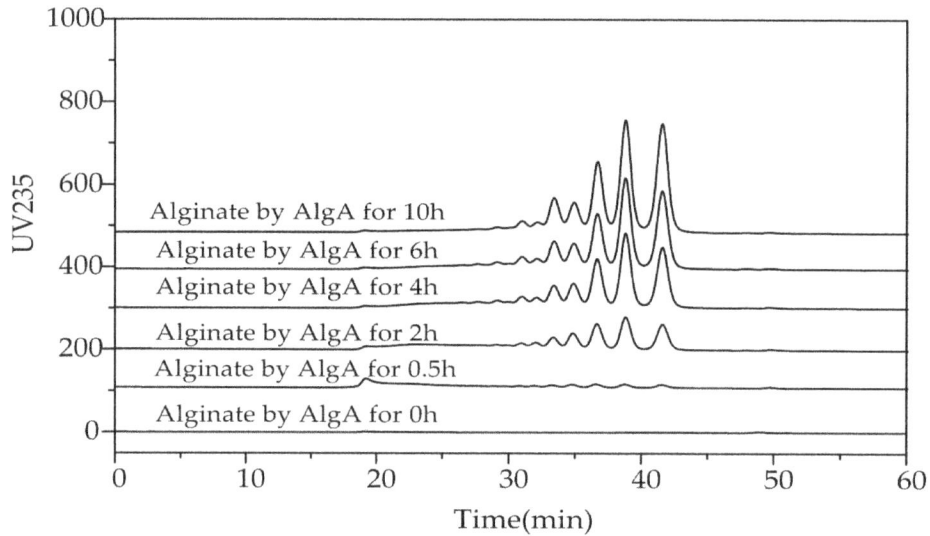

Figure 6. Patterns of the polysaccharide degradation products of AlgA. Enzymatic degradation products collected at 0.5, 2, 4, 6, and 10 h were subjected to gel filtration with a Superdex peptide 10/300 GL column. The absorbances of the products were monitored at 235 nm.

In addition, the alginate was completely digested with an excess of AlgA at 40 °C for 24 h. The products were separated through gel chromatography. The elution profiles (Figure 7) of the degradation products presented three major fractions (peaks 1, 2, and 3).

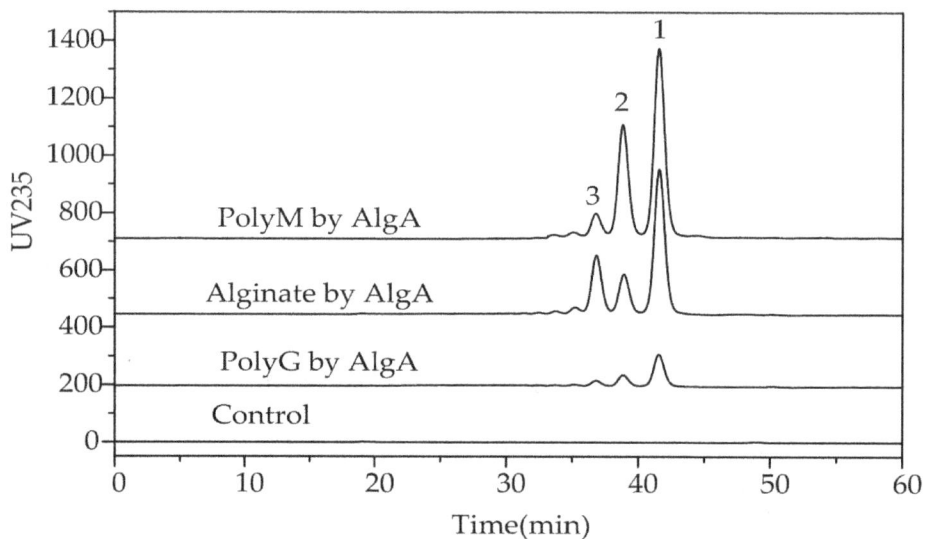

Figure 7. Final products of alginate, polyM, and polyG after degradation by AlgA. Oligosaccharide products were gel-filtered through a Superdex peptide 10/300 GL column and monitored at a wavelength of 235 nm.

To identify the final oligosaccharide products of AlgA degradation and to determine their DP, three major fractions were subjected to electrospray ionization mass spectrometry (ESI-MS). The molecular masses of oligosaccharides in peaks 1, 2, and 3 were determined to be 351.06, 527.09, and 703.12, respectively (Figure 8). These results indicated that the main degradation products are di-, tri- and tetra-saccharides. The relative contents of di-, tri- and tetra-saccharides in alginate were 61.19%, 15.59%, and 23.22%, respectively, whereas those of di-, tri- and tetra-saccharides in the polyM substrate were 58.30%, 34.26%, and 7.43%, respectively. Meanwhile, AlgA has limited activity toward polyG. Therefore, AlgA may be used in the production of mannuronic oligosaccharides from polyM blocks and the preparation of polyG blocks from sodium alginate via the degradation of polyM blocks. The products of mannuronic oligosaccharides and polyG blocks possess special biological activity and have potential applications in many fields. For example, mannuronate oligosaccharides can promote the secretion of multiple cytokines [7], and polyG demonstrates higher macrophage-stimulation activity than polyM [41].

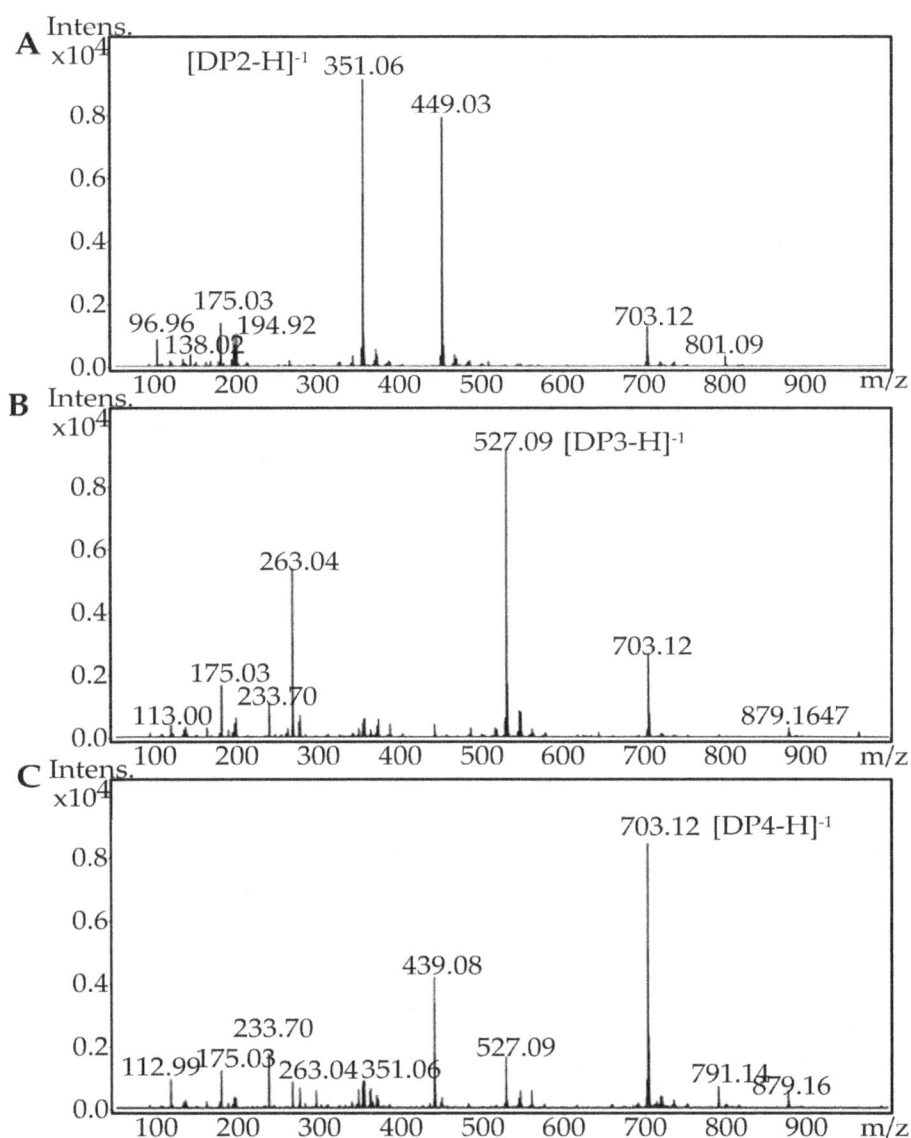

Figure 8. Electrospray ionization mass spectrometry (ESI-MS) analysis of the final oligosaccharide products. (**A**) Fraction peak 1 separated through fast protein liquid chromatography (FPLC), (**B**) fraction peak 2 separated through FPLC, and (**C**) fraction peak 3 separated through FPLC.

3. Materials and Methods

3.1. Materials

Sodium alginate derived from brown seaweed was purchased from Sigma (St. Louis, MO, USA). SOURCE™ 15Q 4.6/100 PE and Superdex peptide 10/300 gel filtration columns were purchased from GE HealthCare Bio-Sciences (Uppsala, Sweden). DNA polymerase, protein molecular weight markers, and polyacrylamide were purchased from New England Biolabs (Ipswich, MA, USA). Other chemicals and reagents used in this study were of analytical grade.

3.2. Screening and Identification of Strain Alg07

Sea mud and rotten kelp samples were collected from a seaweed farm in Weihai, China. Five grams of the samples were added to 45 mL of modified marine broth 2216 medium containing 5 g/L $(NH_4)_2SO_4$, 19.45 g/L NaCl, 12.6 g/L $MgCl_2 \cdot 6H_2O$, 6.64 g/L $MgSO_4 \cdot 7H_2O$, 0.55 g/L KCl, 0.16 g/L $NaHCO_3$, 1 g/L ferric citrate, and 10 g/L sodium alginate. After 48 h of enrichment at 30 °C, the culture was serially diluted with deionized water and spread on modified marine broth 2216 agar containing 10 g/L sodium alginate. The plates were incubated at 30 °C for two days until colonies appeared. Single colonies were inoculated into marine broth 2216 medium containing 10 g/L sodium alginate and incubated for 48 h at 30 °C. Then, the activities of alginate lyase in the supernatants were determined. The alginate lyase from one of the isolates, strain Alg07, showed the highest activity among all lyases.

To identify the Alg07 strain, the 16S rRNA gene of the strain was amplified through PCR by using universal primers. The purified PCR fragment was sequenced and compared with reported 16s rRNA sequences in Genbank by using BLAST. A phylogenetic tree was constructed using CLUSTAL X and MEGA 6.0 [42] through neighbor-joining method [43].

3.3. Production and Purification of AlgA

Strain Alg07 was cultured for 24 h at 30 °C and 180 rpm in the optimized liquid medium, which contained 1g/L peptone, 3 g/L yeast extract, 9 g/L sodium alginate, 5 g/L NaCl, 1 g/L $MgSO_4 \cdot 7H_2O$, 5 g/L KCl, and 4 g/L $CaCl_2$ (pH 6.5). The supernatant was collected after 30 min of centrifugation at 10,000 rpm and 4 °C and then concentrated and desalted using a tangential flow filtration system (Vivaflow 50, Sartorius, Goettingen, Germany). The concentrated solution was subjected to AKTA FPLC (GE Healthcare Life Science, Marlborough, MA, USA) equipped with a SOURCE™ (Barrie, ON, Canada) 15Q 4.6/100 PE column that had been equilibrated with 20 mM Tris–HCl buffer (pH 7.0). Adsorbed proteins were eluted with a linear gradient of 0–0.5 M NaCl in equilibrating buffer under a flow rate of 1 mL/min. Fractions possessing the highest specific activity among all fractions were pooled and dialyzed against 20 mM Tris-HCl buffer (pH 7.0) for further enzyme characterization. All purification procedures were performed at 4 °C. Protein concentration was determined with a protein quantitative analysis kit (Solarbio, Beijing, China) using bovine serum albumin as the calibration standard. The purity of the isolated alginate lyase was analyzed through 12.5% SDS-PAGE in accordance with the method of Laemmli (1970) [44].

3.4. Enzyme Activity Assay

To determine the activity of alginate lyase, 100 μL of appropriately diluted enzyme was added to 1900 μL of substrate solution containing 10 g/L sodium alginate, 20 mM Tris-HCl, and 200 mM NaCl (pH 7.5). The reaction was allowed to proceed for 20 min at 40 °C and terminated by the addition of 20 μL of 10 M NaOH. Absorbance at 235 nm was recorded. One unit (U) was defined as the amount of enzyme required to increase the absorbance at 235 nm by 0.1 per min. For kinetic parameter analysis, the 3,5-dinitrosalicylic acid method was performed to determine alginate lyase activity based on the release of reducing sugars from substrates [45]. One unit (U) was defined as the amount of enzyme required to release 1 μmol of reducing sugar per min. All enzyme reactions were performed in triplicate, and reaction parameters were expressed as mean ± standard deviation.

3.5. Characterization of AlgA

Enzyme reactions were carried out at different temperatures (25 °C–60 °C) to determine the effect of temperature on AlgA. To evaluate the thermal stability of AlgA, the enzyme was incubated for different intervals at 40 °C, 45 °C, and 50 °C. Then, the residual activities of the enzyme were tested. The activity of the enzyme stored at 4 °C was used to represent 100% enzyme activity. The effect of different pH values on AlgA was determined by calculating the residual activities of AlgA after 20 h of incubation at 4 °C in 20 mM sodium acetate (pH 3.0–6.0), Tris-HCl (pH 6.0–9.0), or glycine-NaOH (pH 9.0–11.0) buffers. The initial activities in different pH buffers represented 100% enzyme activity.

Enzyme reactions were performed in the presence of different concentrations of NaCl (0–500 mM) to evaluate the effect of NaCl on AlgA. To determine the effects of metal ions and EDTA on the activity of AlgA, the highest enzyme activity was considered as 100% enzyme activity. AlgA was subjected to an activity assay after 12 h of incubation at 4 °C in the presence of 2 mM of different metal ions and EDTA.

3.6. Analysis of the N-Terminal Amino Acid Sequence of AlgA

After SDS-PAGE, the protein band of AlgA was electro-transferred onto a polyvinylidene difluoride membrane (Imobulon; Millipore, Darmstadt, Germany). Amino acid sequences were determined with a PPSQ-31A protein sequencer (Shimadzu Corporation; Kyoto, Japan).

3.7. Substrate Specificity and Kinetic Parameters of AlgA

A standard enzymatic assay was performed to test the substrate preference of AlgA by using pectin, hyaluronan, chitin, agar, sodium alginate, polyM, and polyG as substrates. PolyM and polyG were prepared in accordance with the method of Haug et al. [46].

The kinetic parameters of AlgA toward alginate and polyM were determined by measuring the initial velocities of enzyme activity under various substrate concentrations and were calculated on the basis of the nonlinear regression fitting of the Michaelis–Menten equation using Prism 6 (GraphPad Software, Inc., La Jolla, CA, USA).

3.8. FPLC and ESI-MS Analysis of the Degradation Products of AlgA

To elucidate the mode of action of AlgA toward alginate, alginate degradation was performed at 40 °C with 10 g/L sodium alginate as a substrate. The reaction was initiated by the addition of 2 μg of purified AlgA in a 10 mL reaction volume. Reaction solutions were withdrawn at appropriate time intervals, and AlgA was inactivated by 5 min of boiling. The samples were analyzed through FPLC with a Superdex peptide 10/300 gel filtration column (GE Health, Marlborough, MA, USA) with 0.2 M ammonium bicarbonate as the mobile phase at a flow rate of 0.4 mL/min [34]. The reaction was monitored at 235 nm.

To determine the oligosaccharide compositions of the final digests, an excess of AlgA was used to completely degrade 10 g/L sodium alginate, polyM, or polyG. The reaction was carried out in a 10 mL reaction volume at 40 °C for 24 h and then terminated by boiling for 5 min. Oligosaccharides were separated through gel filtration as described above. Peak fractions containing unsaturated oligosaccharide products were collected and repeatedly freeze-dried to remove NH_4HCO_3 for ESI-MS analysis. The molecular weight of each oligosaccharide fraction was determined using the ESI-MS method on microTOF-Q II equipment (Bruker, Billerica, MA, USA) with the following conditions: capillary voltage of 4 kV, dry temperature of 180 °C, gas flow rate of 4.0 L/min, and scan range of 50–1000 m/z.

4. Conclusions

An alginate lyase–producing marine bacterium was isolated and identified as *Bacillus* sp. Alg07. AlgA, the alginate lyase derived from *Bacillus* sp. Alg07, was purified through the two simple steps of tangential flow filtration concentration and anion-exchange chromatography. The results of SDS-PAGE indicated that the molecular mass of AlgA is approximately 60 kDa. The optimal temperature and pH for the activity of AlgA is 40 °C and 7.5, respectively. The activity of AlgA is dependent on NaCl and is promoted by the addition of Mg^{2+} and Ca^{2+}. Under optimal conditions, the specific activity of AlgA reaches up to 8306.7 U/mg, which is the highest activity recorded for all reported alginate lyases. Moreover, AlgA is stable over a broad pH range (5.0–10.0) and under its optimal temperature (40 °C). AlgA is an endolytic polyM-specific alginate lyase and mainly produces disaccharides, trisaccharides, and tetrasaccharides from alginate and disaccharides and trisaccharides from polyM. The highly efficient and thermostable AlgA can have potential applications in the production of mannuronic oligosaccharides and polyG blocks from alginate.

Acknowledgments: This work was supported by the National key R&D Program of China (2017YFD200900), the Science Technology Planning Project of Tianjin (16YFXTNC00160), and Youth Innovation Promotion Association, Chinese Academy of Sciences (to Yueming Zhu).

Author Contributions: Peng Chen and Yueming Zhu conceived and designed the experiments; Peng Chen performed the experiments; Peng Chen and Yan Zeng analyzed the data; Yan Men contributed reagents/materials/analysis tools; Peng Chen wrote the paper. Yuanxia Sun edited the paper.

References

1. Gacesa, P. Enzymic degradation of alginates. *Int. J. Biochem.* **1992**, *24*, 545–552. [CrossRef]

2. Mabeau, S.; Kloareg, B. Isolation and Analysis of the Cell Walls of Brown Algae: *Fucus spiralis, F. ceranoides, F. vesiculosus, F. serratus, Bifurcaria bifurcata* and *Laminaria digitata. J. Exp. Bot.* **1987**, *38*, 1573–1580. [CrossRef]

3. Gorin, P.A.J.; Spencer, J.F.T. Exocellular alginic acid from *Azotobacter vinelandii. Can. J. Chem.* **1966**, *44*, 993–998. [CrossRef]

4. Evans, L.R.; Linker, A. Production and Characterization of the Slime Polysaccharide of *Pseudomonas aeruginosa. J. Bacteriol.* **1973**, *116*, 915–924. [PubMed]

5. Scott, C.D. Immobilized cells: A review of recent literature. *Enzym. Microb. Technol.* **1987**, *9*, 66–72. [CrossRef]

6. Kawada, A.; Hiura, N.; Tajima, S.; Takahara, H. Alginate oligosaccharides stimulate VEGF-mediated growth and migration of human endothelial cells. *Arch. Dermatol. Res.* **1999**, *291*, 542–547. [CrossRef] [PubMed]

7. Yamamoto, Y.; Kurachi, M.; Yamaguchi, K.; Oda, T. Stimulation of multiple cytokine production in mice by alginate oligosaccharides following intraperitoneal administration. *Carbohydr. Res.* **2007**, *342*, 1133–1137. [CrossRef] [PubMed]

8. Iwasaki, K.I.; Matsubara, Y. Purification of Alginate Oligosaccharides with Root Growth-promoting Activity toward Lettuce. *Biosci. Biotechnol. Biochem.* **2000**, *64*, 1067–1070. [CrossRef] [PubMed]

9. Wang, Y.; Han, F.; Hu, B.; Li, J.; Yu, W. In vivo prebiotic properties of alginate oligosaccharides prepared through enzymatic hydrolysis of alginate. *Nutr. Res.* **2006**, *26*, 597–603. [CrossRef]

10. Falkeborg, M.; Cheong, L.Z.; Gianfico, C.; Sztukiel, K.M.; Kristensen, K.; Glasius, M.; Xu, X.; Guo, Z. Alginate oligosaccharides: enzymatic preparation and antioxidant property evaluation. *Food Chem.* **2014**, *164*, 185–194. [CrossRef] [PubMed]

11. An, Q.D.; Zhang, G.L.; Wu, H.T.; Zhang, Z.C.; Zheng, G.S.; Luan, L.; Murata, Y.; Li, X. Alginate-deriving oligosaccharide production by alginase from newly isolated *Flavobacterium* sp. LXA and its potential application in protection against pathogens. *J. Appl. Microbiol.* **2009**, *106*, 161–170. [CrossRef] [PubMed]

12. Kim, H.S.; Lee, C.G.; Lee, E.Y. Alginate lyase: Structure, property, and application. *Biotechnol. Bioprocess Eng.* **2011**, *16*, 843–851. [CrossRef]

13. Inoue, A.; Mashino, C.; Uji, T.; Saga, N.; Mikami, K.; Ojima, T. Characterization of an Eukaryotic PL-7 Alginate Lyase in the Marine Red *AlgaPyropia yezoensis. Curr. Biotechnol.* **2015**, *4*, 240–258. [CrossRef] [PubMed]

14. Suzuki, H.; Suzuki, K.; Inoue, A.; Ojima, T. A novel oligoalginate lyase from abalone, *Haliotis discus hannai*, that releases disaccharide from alginate polymer in an exolytic manner. *Carbohydr. Res.* **2006**, *341*, 1809–1819. [CrossRef] [PubMed]

15. Zhu, B.; Chen, M.; Yin, H.; Du, Y.; Ning, L. Enzymatic Hydrolysis of Alginate to Produce Oligosaccharides by a New Purified Endo-Type Alginate Lyase. *Mar. Drugs.* **2016**, *14*, 108. [CrossRef] [PubMed]

16. Inoue, A.; Takadono, K.; Nishiyama, R.; Tajima, K.; Kobayashi, T.; Ojima, T. Characterization of an Alginate Lyase, FlAlyA, from *Flavobacterium* sp. Strain UMI-01 and Its Expression in *Escherichia coli. Mar. Drugs* **2014**, *12*, 4693–4712. [CrossRef] [PubMed]

17. Singh, R.P.; Gupta, V.; Kumari, P.; Kumar, M.; Reddy, C.R.K.; Prasad, K.; Jha, B. Purification and partial characterization of an extracellular alginate lyase from *Aspergillus oryzae* isolated from brown seaweed. *J. Appl. Phycol.* **2011**, *23*, 755–762. [CrossRef]

18. Suda, K.; Tanji, Y.; Hori, K.; Unno, H. Evidence for a novel *Chlorella* virus-encoded alginate lyase. *FEMS Microbiol. Lett.* **1999**, *180*, 45–53. [CrossRef] [PubMed]

19. Wong, T.Y.; Preston, L.A.; Schiller, N.L. ALGINATE LYASE: Review of major sources and enzyme characteristics, structure-function analysis, biological roles, and applications. *Annu. Rev. Microbiol.* **2000**, *54*, 289–340. [CrossRef] [PubMed]

20. Cantarel, B.L.; Coutinho, P.M.; Rancurel, C.; Bernard, T.; Lombard, V.; Henrissat, B. The Carbohydrate-Active EnZymes database (CAZy): an expert resource for Glycogenomics. *Nucleic Acids Res.* **2009**, *37*, 233–238. [CrossRef] [PubMed]

21. Aarstad, O.A.; Tøndervik, A.; Sletta, H.; Skjåkbræk, G. Alginate Sequencing: An Analysis of Block Distribution in Alginates Using Specific Alginate Degrading Enzymes. *Biomacromolecules* **2012**, *13*, 106–116. [CrossRef] [PubMed]

22. Inoue, A.; Mashino, C.; Kodama, T.; Ojima, T. Protoplast preparation from *Laminaria japonica* with recombinant alginate lyase and cellulase. *Mar. Biotechnol.* **2011**, *13*, 256–263. [CrossRef] [PubMed]

23. Islan, G.A.; Martinez, Y.N.; Illanes, A.; Castro, G.R. Development of novel alginate lyase cross-linked aggregates for the oral treatment of cystic fibrosis. *RSC Adv.* **2014**, *4*, 11758–11765. [CrossRef]

24. Li, L.; Jiang, X.; Guan, H.; Wang, P. Preparation, purification and characterization of alginate oligosaccharides degraded by alginate lyase from *Pseudomonas* sp. HZJ 216. *Carbohydr. Res.* **2011**, *346*, 794–800. [CrossRef] [PubMed]

25. Wang, L.; Li, S.; Yu, W.; Gong, Q. Cloning, overexpression and characterization of a new oligoalginate lyase from a marine bacterium, *Shewanella* sp. *Biotechnol. Lett.* **2015**, *37*, 665–671. [CrossRef] [PubMed]

26. Kim, H.T.; Ko, H.J.; Kim, N.; Kim, D.; Lee, D.; Choi, I.G.; Woo, H.C.; Kim, M.D.; Kim, K.H. Characterization of a recombinant endo-type alginate lyase (Alg7D) from *Saccharophagus degradans. Biotechnol. Lett.* **2012**, *34*, 1087–1092. [CrossRef] [PubMed]

27. Li, S.; Wang, L.; Hao, J.; Xing, M.; Sun, J.; Sun, M. Purification and Characterization of a New Alginate Lyase from Marine Bacterium *Vibrio* sp. SY08. *Mar. Drugs* **2017**, *15*, 1. [CrossRef] [PubMed]

28. Wang, Y.; Guo, E.W.; Yu, W.G.; Han, F. Purification and characterization of a new alginate lyase from a marine bacterium *Vibrio* sp. *Biotechnol. Lett.* **2013**, *35*, 703–708. [CrossRef] [PubMed]

29. Kobayashi, T.; Uchimura, K.; Miyazaki, M.; Nogi, Y.; Horikoshi, K. A new high-alkaline alginate lyase from a deep-sea bacterium *Agarivorans* sp. *Extremophiles* **2009**, *13*, 121–129. [CrossRef] [PubMed]

30. Xiao, T.F.; Hong, L.; Sang, M.K. Purification and characterization of a Na^+/K^+ dependent alginate lyase from turban shell gut *Vibrio* sp. YKW-34. *Enzym. Microb. Technol.* **2007**, *41*, 828–834.

31. Zhu, B.; Yin, H. Alginate lyase: Review of major sources and classification, properties, structure-function analysis and applications. *Bioengineered* **2015**, *6*, 125–131. [CrossRef] [PubMed]

32. Yamasaki, M.; Moriwaki, S.; Miyake, O.; Hashimoto, W.; Murata, K.; Mikami, B. Structure and function of a hypothetical *Pseudomonas aeruginosa* protein PA1167 classified into family PL-7. *J. Biol. Chem.* **2004**, *279*, 31863–31872. [CrossRef] [PubMed]

33. Lee, S.I.; Choi, S.H.; Lee, E.Y.; Kim, H.S. Molecular cloning, purification, and characterization of a novel polyMG-specific alginate lyase responsible for alginate MG block degradation in *Stenotrophomas maltophilia* KJ-2. *Appl. Microbiol. Biotechnol.* **2012**, *95*, 1643–1653. [CrossRef] [PubMed]

34. Zhu, B.; Tan, H.; Qin, Y.; Xu, Q.; Du, Y.; Yin, H. Characterization of a new endo-type alginate lyase from *Vibrio* sp. W13. *Int. J. Biol. Macromol.* **2015**, *75*, 330–337. [CrossRef] [PubMed]

35. Thomas, F.; Lundqvist, L.C.E.; Jam, M.; Jeudy, A.; Barbeyron, T.; Sandström, C.; Michel, G.; Czjzek, M. Comparative Characterization of Two Marine Alginate Lyases from *Zobellia galactanivorans* Reveals Distinct Modes of Action and Exquisite Adaptation to Their Natural Substrate. *J. Biol. Chem.* **2013**, *288*, 23021–23037. [CrossRef] [PubMed]

36. Li, S.; Yang, X.; Bao, M.; Wu, Y.; Yu, W.; Han, F. Family 13 carbohydrate-binding module of alginate lyase from *Agarivorans* sp. L11 enhances its catalytic efficiency and thermostability, and alters its substrate preference and product distribution. *FEMS Microbiol. Lett.* **2015**, *362*. [CrossRef] [PubMed]

37. Han, W.; Gu, J.; Cheng, Y.; Liu, H.; Li, Y.; Li, F. Novel Alginate Lyase (Aly5) from a Polysaccharide-Degrading Marine Bacterium, *Flammeovirga* sp. Strain MY04: Effects of Module Truncation on Biochemical Characteristics, Alginate Degradation Patterns, and Oligosaccharide-Yielding Properties. *Appl. Environ. Microbiol.* **2015**, *82*, 364–374. [CrossRef] [PubMed]

38. Li, J.W.; Dong, S.; Song, J.; Li, C.B.; Chen, X.L.; Xie, B.B.; Zhang, Y.Z. Purification and Characterization of a Bifunctional Alginate Lyase from *Pseudoalteromonas* sp. SM0524. *Mar. Drugs* **2011**, *9*, 109–123. [CrossRef] [PubMed]

39. Uchimura, K.; Miyazaki, M.; Nogi, Y.; Kobayashi, T.; Horikoshi, K. Cloning and sequencing of alginate lyase genes from deep-sea strains of *Vibrio* and *Agarivorans* and characterization of a new *Vibrio* enzyme. *Mar. Biotechnol.* **2010**, *12*, 526–533. [CrossRef] [PubMed]

40. Xiao, L.; Han, F.; Yang, Z.; Lu, X.Z.; Yu, W.G. A Novel Alginate Lyase with High Activity on Acetylated Alginate of *Pseudomonas aeruginosa* FRD1 from *Pseudomonas* sp. QD03. *World J. Microbiol. Biotechnol.* **2006**, *22*, 81–88. [CrossRef]

41. Ueno, M.; Hiroki, T.; Takeshita, S.; Jiang, Z.; Kim, D.; Yamaguchi, K.; Oda, T. Comparative study on antioxidative and macrophage-stimulating activities of polyguluronic acid (PG) and polymannuronic acid (PM) prepared from alginate. *Carbohydr. Res.* **2012**, *352*, 88–93. [CrossRef] [PubMed]

42. Tamura, K.; Stecher, G.; Peterson, D.; Filipski, A.; Kumar, S. MEGA6: Molecular Evolutionary Genetics Analysis Version 6.0. *Mol. Biol. Evol.* **2013**, *30*, 2725–2729. [CrossRef] [PubMed]

43. Saitou, N.; Nei, M. The neighbor-joining method: A new method for reconstructing phylogenetic trees. *Mol. Biol. Evol.* **1987**, *4*, 406–425. [PubMed]

44. Laemmli, U.K. Cleavage of Structural Proteins during the Assembly of the Head of Bacteriophage T4. *Nature* **1970**, *227*, 680–685. [CrossRef] [PubMed]

45. Miller, G.L. Use of Dinitrosalicylic Acid Reagent for Determination of Reducing Sugar. *Anal. Biochem.* **1959**, *31*, 426–428. [CrossRef]

46. Haug, A.; Larsen, B. A study on the constitution of alginic acid by partial acid hydrolysis. *Acta Chem. Scand.* **1966**, *20*, 183–190. [CrossRef]

15

Characterization of the Specific Mode of Action of a Chitin Deacetylase and Separation of the Partially Acetylated Chitosan Oligosaccharides

Xian-Yu Zhu [1,2], Yong Zhao [1], Huai-Dong Zhang [1,3], Wen-Xia Wang [1], Hai-Hua Cong [2] and Heng Yin [1,*]

[1] Liaoning Provincial Key Laboratory of Carbohydrates, Dalian Institute of Chemical Physics, Chinese Academy of Sciences, Dalian 116023, China; zhuxy0721@126.com (X.-Y.Z.); zhaoyong_2019@163.com (Y.Z.); huaidongzhang@yahoo.com (H.-D.Z.); wangwx@dicp.ac.cn (W.-X.W.)
[2] College of Food Science and Engineering, Dalian Ocean University, Dalian 116023, China; haihuacong780@gmail.com
[3] Engineering Research Center of Industrial Microbiology, Ministry of Education; College of Life Sciences, Fujian Normal University, Fuzhou 350117, China
*   Correspondence: yinheng@dicp.ac.cn.

Abstract: Partially acetylated chitosan oligosaccharides (COS), which consists of N-acetylglucosamine (GlcNAc) and glucosamine (GlcN) residues, is a structurally complex biopolymer with a variety of biological activities. Therefore, it is challenging to elucidate acetylation patterns and the molecular structure-function relationship of COS. Herein, the detailed deacetylation pattern of chitin deacetylase from *Saccharomyces cerevisiae*, *Sc*CDA$_2$, was studied. Which solves the randomization of acetylation patterns during COS produced by chemical. *Sc*CDA$_2$ also exhibits about 8% and 20% deacetylation activity on crystalline chitin and colloid chitin, respectively. Besides, a method for separating and detecting partially acetylated chitosan oligosaccharides by high performance liquid chromatography and electrospray ionization mass spectrometry (HPLC-ESI-MS) system has been developed, which is fast and convenient, and can be monitored online. Mass spectrometry sequencing revealed that *Sc*CDA$_2$ produced COS with specific acetylation patterns of DAAA, ADAA, AADA, DDAA, DADA, ADDA and DDDA, respectively. *Sc*CDA$_2$ does not deacetylate the GlcNAc unit that is closest to the reducing end of the oligomer furthermore *Sc*CDA$_2$ has a multiple-attack deacetylation mechanism on chitin oligosaccharides. This specific mode of action significantly enriches the existing limited library of chitin deacetylase deacetylation patterns. This fully defined COS may be used in the study of COS structure and function.

Keywords: chitin deacetylase; deacetylation patterns; chitooligosaccharides; separating; detecting

1. Introduction

Chitin, which consists of β-1,4-linked N-acetyl-D-glucosamine residues, is the main component of crustacean shells, such as shrimp, crab and shellfish [1,2]. Chitin, a renewable raw material whose annual production is about 10^{11} tons, is the second most abundant natural biopolymer after cellulose [3,4]. As a new type of functional material, chitin has attracted wide attention in various fields [5]. However, it is insoluble in water and most organic solvents, this property severely restricts its development and application [3]. On the other hand, chitosan, the deacetylation product of chitin, is soluble in dilute acid solution and has been widely used in agriculture, biomedicine, environmental science and other fields, as a plant inducer, biodegradable hydrogel and sewage treatment agent in antitumor drugs and in other green products [2,6–10]. Chitooligosaccharide (COS), the hydrolytic product of chitosan, has broader biological activities, such as immunological, antitumor, antioxidant and antibacterial activity [11–13]. Due to its water-soluble ability and broad biological activity, COS has attracted more attention than chitosan. The biological activity of COS are believed to be strongly dependent on the degree of polymerization (DP), the degree of acetylation (DA) and the pattern of acetylation (PA) [14]. Vander et al. reported that COS with different degrees of deacetylation is involved in the induction of phenylalanine ammonia lyase and peroxidase activities, both of which must be activated for lignin biosynthesis [15]. It has previously been observed that the specific recognition of the N-acetyl moiety allows AtCERKl to distinguish chitin and chitosan, which then activate plant immune receptors and elicit a plant immune response [16].

In order to investigate the specific biological activity of COS in a particular acetylation pattern, COS with a completely known structure is required. However, chitosan and COS produced by chemical methods usually exhibit a randomized pattern of acetylation, making them difficult to control and predict their biological activity [17]. Moreover, chitosan produced by chemical methods requires high energy consumption and causes environmental pollution [1]. In contrast, chitin deacetylase (CDA, E.C. 3.5.1.41) is able to hydrolyse the N-acetamido groups of N-acetyl-D-glucosamine residues in chitin, chitin oligosaccharides, chitosan and chitosan oligosaccharides under mild conditions by a specific mode of action. Previous studies have identified CDAs from bacteria, fungi and insects, such as *Bacillus thuringiensis* [18], *Bacillus amyloliquefaciens* [19], *Colletotrichum gloeosporioides* [20] *Mucor rouxii* [21] *Aspergillus nidulans* [22] *Saccharomyces cerevisiae* [23] *Bombyx mori* [24], *Drosophila melanogaster* [25], *Encephalitozoon cuniculi* [26], *Mamestra configurata* [27]. Although some CDAs have been reported, the deacetylation patterns of deacetylases are poorly understood. CDA from different sources can modify their substrates in different ways: Some being specific for a single position [28], others show showing multiple-attack [29,30]. In addition, COS with specific deacetylation patterns can be produced by enzymatic deacetylation of chitin oligomers, but the diversity is limited by the available CDA.

Two genes encoding chitin deacetylases (CDA_1 and CDA_2) have been identified in *Saccharomyces cerevisiae* in previous reports. And these genes have been proved to be involved in the formation of the ascospores wall of *Saccharomyces cerevisiae* [31]. However, it is very interesting that the deletion of each gene will result in activity decrease of CDA, and the functions of the two genes cannot be replaced by each other [31]. Therefore, the deacetylation mechanisms of these two different chitin deacetylases may be different. However, detailed deacetylation mechanisms of chitin deacetylase from *Saccharomyces cerevisiae* have not been reported so far.

In this study, the chitin deacetylase (CDA_2) from *Saccharomyces cerevisiae* ($ScCDA_2$) with a specific mode of action has been characterized and a fast, convenient and online monitoring method has been developed that can be used to separate and detect partially acetylated chitosan oligosaccharides. Mass spectrometry sequencing showed that $ScCDA_2$ can hydrolyze N-acetamido groups rather than the reducing ends of chitin oligosaccharides, producing fully defined chitosan oligosaccharides by a multiple attack mode of action. Furthermore, $ScCDA_2$ is able to remove about 8% and 20% of the acetyl groups from crystalline chitin and colloidal chitin.

2. Results and Discussion

2.1. Bioinformatic Analysis and Expression of $ScCDA_2$

CDA belongs to the carbohydrate esterase family 4 (CE4) according to the classification of the CAZY database (www.cazy.org) [32]. The presence of divalent metal ions, such as Zn^{2+}, Ca^{2+} and Co^{2+}, have been proved to increase the catalytic activity and stability of the CDAs [30]. The *Colletotrichum lindemuthianum*'s CDA crystal structure indicates that there is a zinc-binding triad (His-His-Asp) around Zn^{2+} [33].

The sequence of $ScCDA_2$ aligned with deacetylase sequences from marine *Arthrobacter* (*Ar*CE4A, 34%) [34], *Streptomyces lividans* (*Sl*CE4, 33%) [35] and *Streptococcus pneumoniae* (*Sp*PgdA, 29%) [36]

(Figure 1) [37]. The structure-based sequence alignments of *Ar*CE4A, *Sl*CE4 and *Sp*PgdA showed different levels of sequence identities according to their source from different genera and enabled identification the key residues that may contribute to catalysis function, including active site residues (Asp102, His250) and zinc-binding residues (Asp103, His149, His153) (Figure 1). Asp103, His149 and His153 form a zinc-binding triplet (His-His-Asp) around Zn^{2+}, which is similar to chitin deacetylase from *Colletotrichum lindemuthianum* [33], although the CDA sequence from *C. lindemuthianum* only has a 30% similarity to *Sc*CDA$_2$. The full-length open reading frame encoding the N-acetylglucosamine deacetylase sequence from *Saccharomyces cerevisiae* was successfully cloned and transformed into *Pichia pastoris* X-33 for highly efficient secretion expression (Figure 2). The molecular weight of *Sc*CDA$_2$, which was digested by N-glycosidase F (PNGase F), decreased by about 10 kDa. PNGase F is an amidase working by cleaving between the innermost GlcNAc and asparagine residues of high mannose, hybrid, and complex oligosaccharides from N-linked glycoproteins and glycopeptides. This results in a deaminated protein or peptide and a free glycan [38,39]. Therefore, there are N-glycosylation post-translational modifications in *Sc*CDA$_2$. Glycosylation is one of the most common post-translational modifications of proteins in fungi. It plays an important role in protein activity, thermal stability, proteolytic resistance, folding and secretion [40]. Mass spectrometry showed that *Sc*CDA$_2$ have N-glycosylation post-translational modification at positions Asn 181, Asn 199 and Asn 203 (Figure S5).

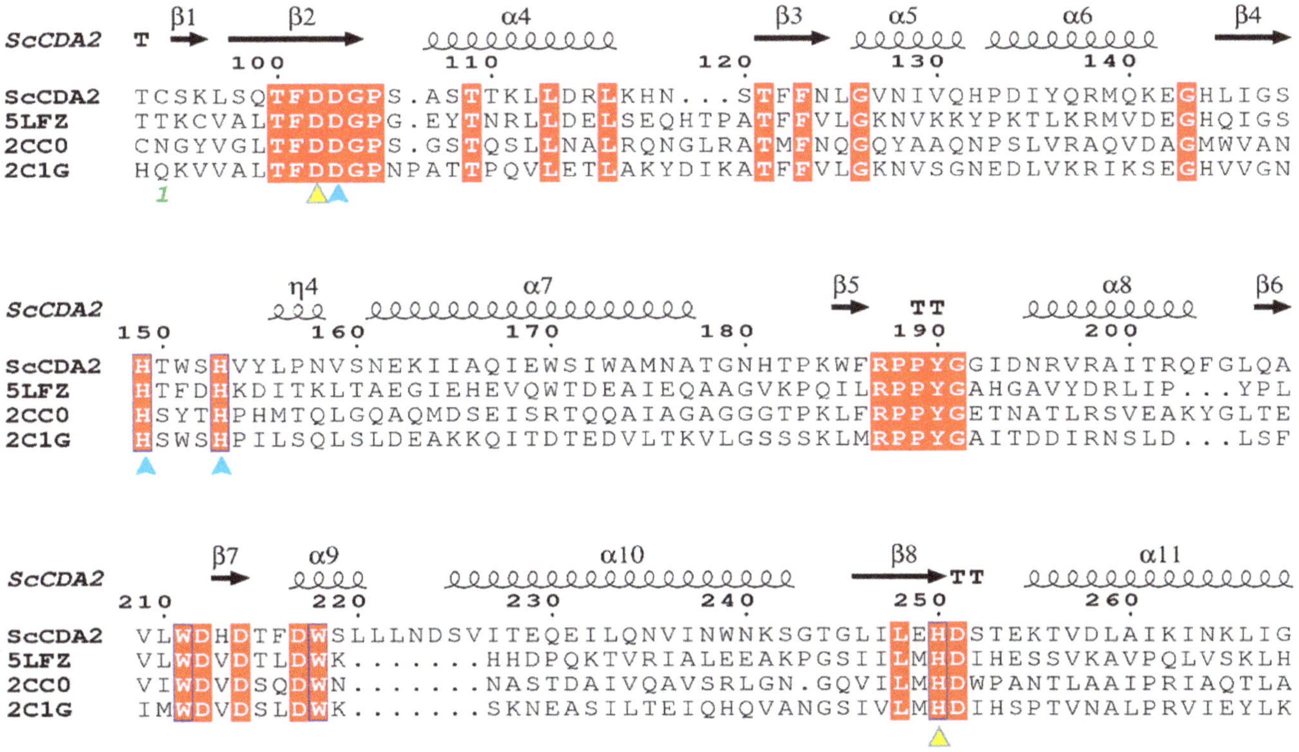

Figure 1. Structure-based on sequence alignments between four chitin deacetylases (CDAs). The sequence of chitin deacetylase from *Saccharomyces cerevisiae* (*Sc*CDA$_2$) was aligned with *Ar*CE4A sequences from a marine *Arthrobacter* species (PDB ID: 5LFZ), the *Sl*CE4 sequence from *Streptomyces lividans* (PDB ID: 2CC0) and the *Sp*PgdA sequence from *Streptococcus pneumoniae* (PDB ID: 2C1G). The conserved motifs are highlighted by a red background and the catalytic amino acids are marked with a yellow triangle. Amino acids capable of forming coordinate bonds with Zn^{2+} are marked with blue triangles. The symbol above the sequence represents the secondary structure, helices represent α-helices, and the arrow represents the beta fold.

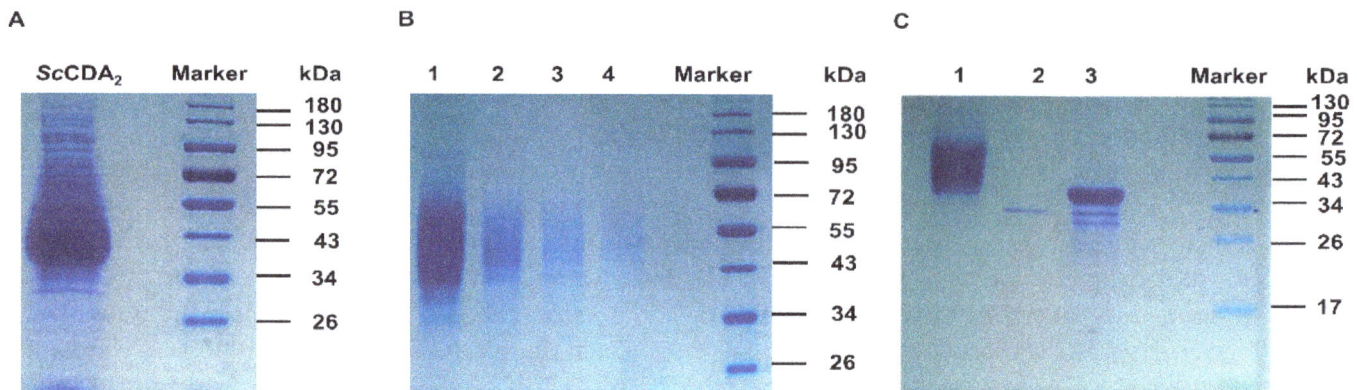

Figure 2. Analysis of molecular weight of $ScCDA_2$ protein by 12% SDS-PAGE. (**A**) $ScCDA_2$ crude enzyme; (**B**) purified $ScCDA_2$10 μM, 5 μM, 2.5 μM and 1.0 μM, marked as lanes 1, 2, 3 and 4, respectively; (**C**) PNGase F digestion confirmed that the enzyme is glycoprotein. Lane 1, $ScCDA_2$ before being digested by PNGase F; lane 2, PNGase F; lane 3, $ScCDA_2$ has been digested by PNGase F.

2.2. Homology Modeling and Substrate Binding Specificity of $ScCDA_2$

The crystal structures of several CDAs have already been determined, while CDA/substrate complex structure determination is less well defined and the interaction between the enzyme and the substrate is poorly understood.

To study the characteristics of $ScCDA_2$ and chitin molecule interactions, we performed molecular docking to understand the binding mechanism of $ScCDA_2$. Homology modelling of $ScCDA_2$ (Figure 3A) revealed that the secondary structure consists of a conserved $(\alpha/\beta)_8$ folded barrel structure and six loops.

The model was further evaluated for protein geometry by SAVES. Evaluation report shows that 97.3% residues in additional allowed regions and 85.57% of the residues have averaged 3D-1D score ≥0.2, and the quality factor is 91.2214, indicating that the structure quality was acceptable (Figure 4).

The docking results (Figure 3B) show that chitin lies in the substrate-binding pocket which is surrounded by six loops, His250, Asp102, Asp103, His149 and His153. Asp103, His149 and His153 form a coordinate bond with Zn^{2+}, and the metal ion serves as a Lewis acid to assist the water affinity attack on the carbon atom on the amide bond.

The adjacent His250 and Asp102 play a catalytic role through protonation, and the common action of these amino acids leads to the cleaving of the acetyl group. In addition, the structural superposition of ArCE4A (PDB ID: 5LFZ), SLCE4 (PDB ID: 2CC0), SpPgdA (PDB ID: 2C1G) and model of $ScCDA_2$ reveal that there are six conserved loop domains in $ScCDA_2$ (Figure 3C).

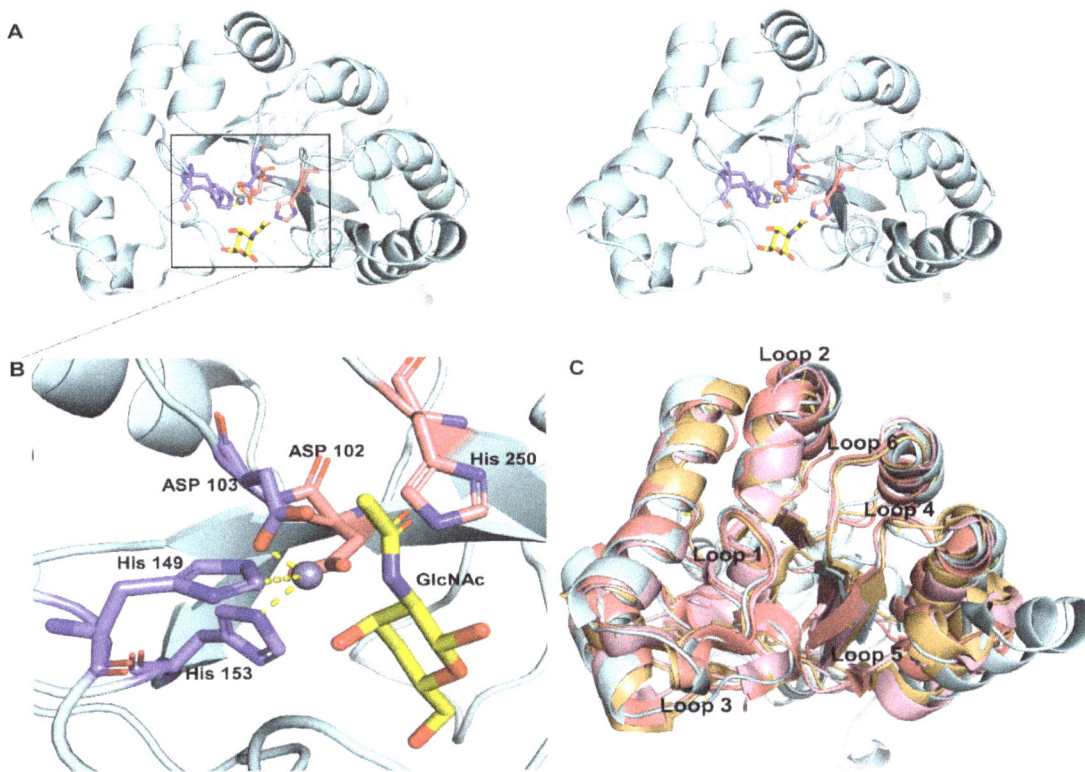

Figure 3. Catalytic binding mode resulting from homologous modelling and molecular docking. (**A**) The stereo view of the overall structure of *Sc*CDA$_2$ (**B**) Highlights the binding pocket of *Sc*CDA$_2$ docked with GlcNAc. The pink sticks represent a catalytic amino acid, and the blue sticks represent the amino acid that forms a coordinate bond with Zn^{2+}. The substrate GlcNAc is represented by a yellow stick. (**C**) Conservative loops were found through multiple structure superposition. The model of *Sc*CDA2 was superposed with an ArCE4A structure from a marine *Arthrobacter* species (PDB ID: 5LFZ), a CE4 carbohydrate esterase structure from *Streptomyces lividans* (PDB ID: 2CC0) and an *Sp*PgdA structure from *Streptococcus pneumoniae* (PDB ID: 2C1G). The conservative loops also have been marked.

2.3. Biochemical Characterization of ScCDA$_2$

The investigation of substrate specificity could provide important information for the potential applications of deacetylase. Using a coupled enzyme assay measure the amount of acetate released has been reported to be successfully applied to quantitatively determine the deacetylation activity of a recombinant chitin deacetylase [14]. When determining activity and substrate specific, interestedly, *Sc*CDA$_2$ was observed that it is able to remove about 8% and 20% of the acetyl groups from crystalline chitin, alpha-chitin and beta-chitin (Figure 4). In addition, A$_n$ (A = GlcNAc; n = 1, 2, 3, 4, 5 or 6) as substrates also have been measured (Figure S3). To promote the application of *Sc*CDA$_2$ in industry, more detailed physical and chemical properties characterization of CDA is essential. The optimal PH and metal ions of *Sc*CDA$_2$ are pH = 8.0 and 50 °C when A4 was used as a substrate (Figure S4). When Co^{2+} is present, *Sc*CDA$_2$ exhibits the maximum activity on A4. Despite the existence of a conserved zinc-binding triad in the *Sc*CDA$_2$, biochemical data (Figure S4C) and structure-based on sequence alignments (Figure 1) indicate that *Sc*CDA$_2$ as a metal-dependent metalloenzyme with a Co^{2+} dependence greater than Zn^{2+}. The peptidoglycan deacetylase from *Streptococcus pneumoniae* also shows that the peptidoglycan deacetylase is more metal-dependent on Co^{2+} than Zn^{2+}. Besides, the reported structures of two distinct acetylxylan esterases of CE4 from *Streptomyces lividans* and *Clostridium thermocellum*, in native and complex forms, show that the enzymes are sugar-specific and metal ion-dependent and possess a single metal (Zn^{2+}) center however with a chemical preference for Co^{2+} [35].

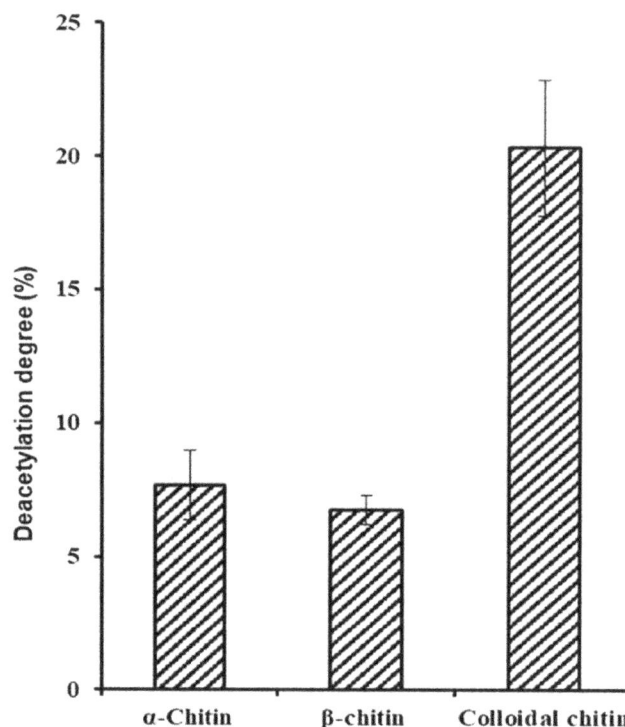

Figure 4. $ScCDA_2$ substrate specificity on chitin. $ScCDA_2$ activity on colloidal chitin, alpha-chitin and beta-chitin. 0.5 mg/mL substrates were incubated with 0. 75 μM $ScCDA_2$ at 37 °C for 30 min. The data represent the mean SD values of the results from three independent experiments.

Most of the reported CDAs show only minimal activity or no activity on chitin *in vitro*. For example, CDA from *Cyclobacterium marinum* has been reported to be able to convert acetylglucosamine to glucosamine only with the cooperation of chitinase [17]. However, $ScCDA_2$ can release up to 20.33% of acetyl groups from colloid chitin, as well as 9.16% and 7.29% of acetyl groups from insoluble alpha-chitin and Betabeta-chitin (Figure 4). Previous reported CDAs have no activity or low activity on insoluble chitin, which may be due to poor accessibility of chitin substrates [41]. However, the charge distribution on the surface of $ScCDA_2$ indicates that $ScCDA_2$ has an excessive negative charge in the region that interacts with the longer substrate, which may lead to enhanced substrate accessibility of $ScCDA_2$ to chitin (Figure S2).

2.4. Isolation and Identification of Partially Acetylated Chitooligosaccharides

Due to its special biological activity, partially acetylated chitosan oligosaccharides have attracted wide interest, and these potential activities are significantly correlated with the degree of polymerization and degree of acetylation of chitooligosaccharides [14,42]. However, the method of preparing and isolating high-purity chitooligosaccharides is time consuming and labor intensive, which severely limits the large-scale production of partially acetylated chitooligosaccharides [43]. Much research into the separation of chitosan oligosaccharides has so far limited to the separation and identification of chitosan oligosaccharides of different degrees of polymerization [44–46]. As far as we know, the method for isolation and identification of partially acetylated chitosan oligosaccharides with a degree of polymerization of four has not been reported.

We have separated and identified the partially acetylated chitosan oligosaccharides with a degree of polymerization of 4. Chitin oligomers were deacetylated with recombinant $ScCDA_2$ to form partially acetylated chitosan oligosaccharides. Three different partially acetylated chitosan oligosaccharides (A1D3, A2D2, A1D3) were obtained. These partially acetylated chitosan oligosaccharides were separated and detected by HPLC-ESI-MS (Figure 5).

A

B

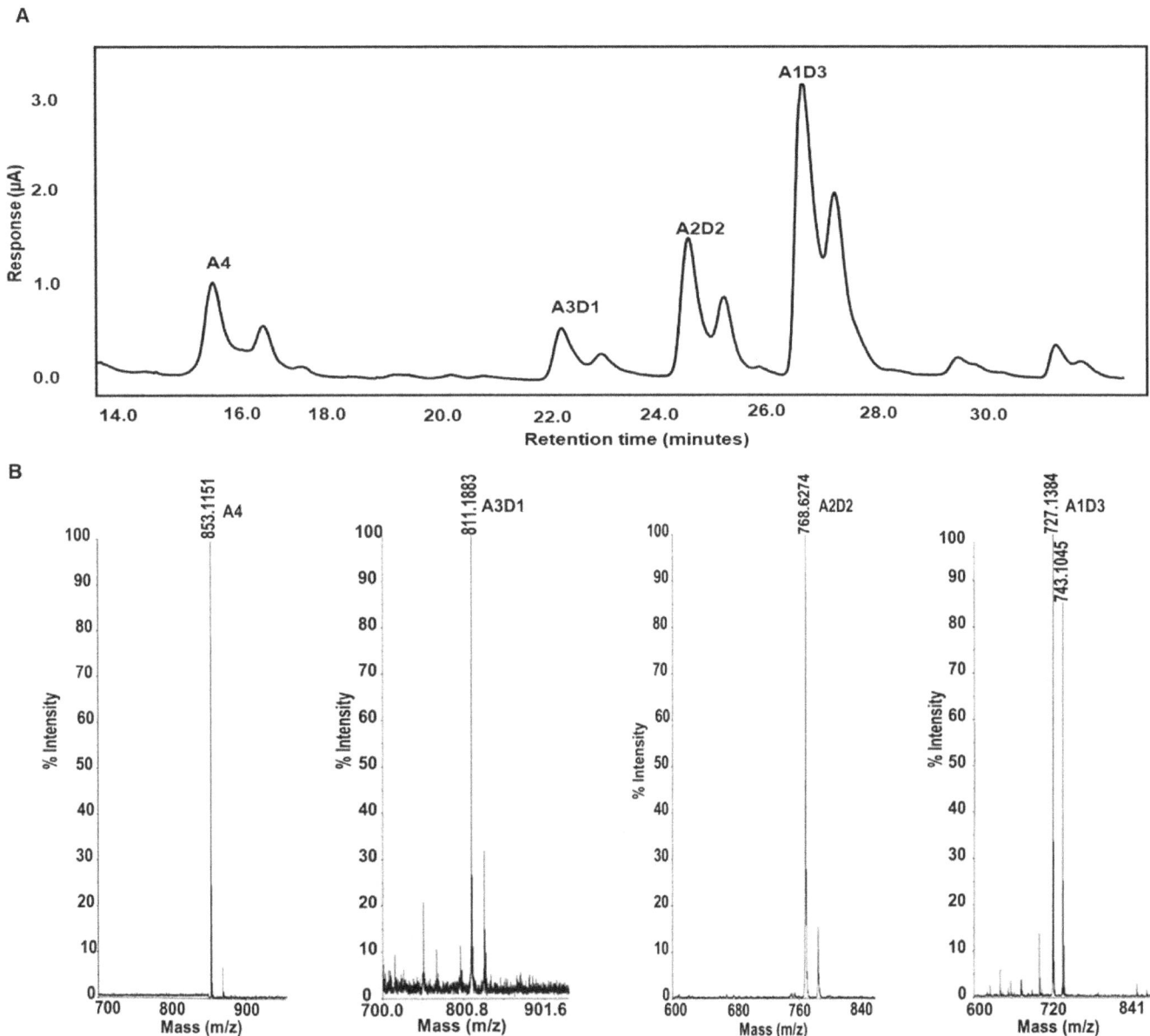

Figure 5. HPLC-ESI-MS analysis of chitin tetramer (A4) treated with $ScCDA_2$. (**A**) The target peak of the UHPLC-ESI chromatogram began to appear after 14 min, and the deacetylation peak was mainly concentrated between 20 and 26 min. (**B**) The m/z ratio in the MS spectrum corresponds to the mass of the substrate (A4; m/z 853.24), its mono-deacetylated products A3D1 (m/z 811.25), A2D2 (m/z 768.62) and A1D3 (m/z 727.13).

2.5. Partially Acetylated Chitooligosaccharides Production Processes

Exploring the partially acetylated chitooligosaccharides production process (simultaneously or in some order) is important to aid in understanding the action mode of CDA deacetylation. Therefore, the effects of enzyme concentration on the production process of partially acetylated COS have also been determined. As is shown in Figure 6, partially acetylated chitosan oligosaccharides (A1D3, A2D2, A1D3) are gradually produced according to the degree of deacetylation. With the amount of enzymes in the system increases, the types of enzyme reaction products gradually increase. From almost no product generation, to the production of the A3D1 and A2D2, the final substrate is completely consumed at the same time producing A1D3.

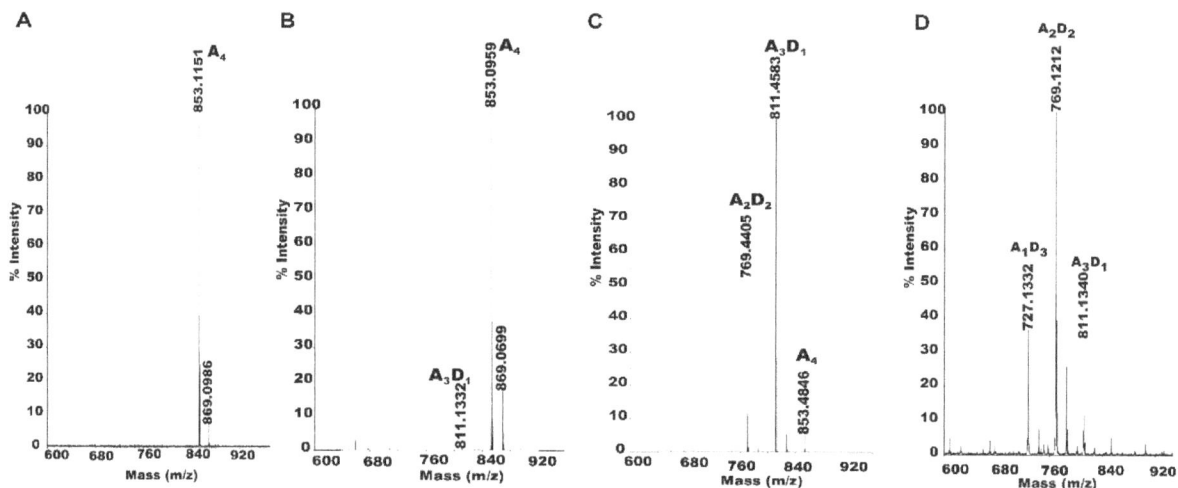

Figure 6. Partially acetylated chitooligosaccharides production processes. To explore the production processes of partially acetylated chitooligosaccharides 0.25 μM, 0.5 μM, 0.75 μM and 1 μm enzymes were incubated with A4 in 20 mM Tris-Cl buffer (pH 8.0) for 30 min. Then determined by MALDI-TOF-MS.

2.6. Specific Mode of Action of ScCDA₂ on A4

The partially acetylated chitooligosaccharide derivatized with a reducing amine showed molecular weights of 1005.96 Da (A3D1), 963.82 Da (A2D2) and 921.80 Da (A1D3) in MALDI-TOF mass spectrometry (Figure 7).

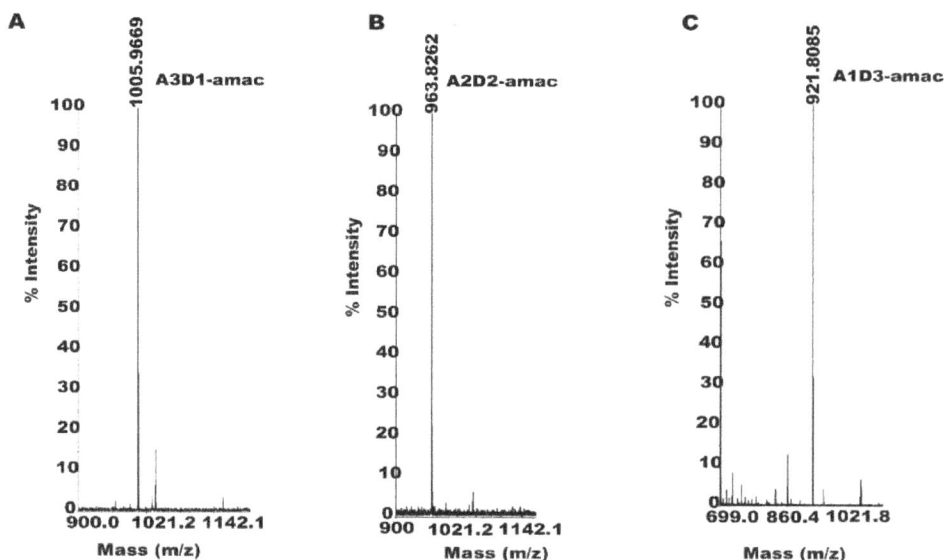

Figure 7. Analysis of partially acetylated COS by reductive amine derivatization with mass spectrometry. The reaction product generated after $ScCDA_2$ treatment (GlcNAc)$_4$ was labelled with AMAC and analyzed by MALDI-TOF-MS. (**A**). MS1 spectrum of A3D1 labelled with AMAC (m/z 1005.96). (**B**). MS1 spectrum of A2D2 labelled with AMAC m/z 963.82). (**C**). MS1 spectrum of A1D3 labelled with AMAC (m/z 921.80).

Then, by applying MALDI-TOF-MS analysis in MS2 mode, we were able to identify and analyze $ScCDA_2$'s partially acetylated products and determine specific acetylation pattern of partially acetylated chitooligosaccharides. As is shown in Figure 8, the A4 is first deacetylated to DAAA, ADAA and AADA (Figure 8A), followed by further deacetylation products to the intermediate product DDAA, DADA and ADDA (Figure 8B). Finally, the end product DDDA was obtained, due to the third deacetylation (Figure 8C). Therefore, deacetylation occurred mainly at the non-reducing end, and the acetyl at the reducing end was always present. No matter how we prolonged the reaction time or increased the concentration of the enzyme, the acetyl group at the reducing end could not be removed. After comparing the intermediate and end products generated by the deacetylation of the chitin tetramer, we found that the deacetylation occurred at any position except for the reducing end, indicating that $ScCDA_2$ has a multiple attack mechanism like $ClCDA$ and $SpPgdA$ [33,36]. However, $ScCDA_2$ cannot deacetylate at the reducing end to form completely deacetylated COS (DDDD). Therefore, the deacetylation pattern of $ScCDA_2$ is significantly different from the reported CDA derived from *C. lindemuthianum*, *Mucor rouxii*, *Aspergillus nidulans*, *Vibrio cholera*, *Puccinia graminis*, etc [14,29,30,41,47]. A "subsite-capping model" has been proposed to explain the differentiation of the deacetylation process and product patterns of CDA [30]. This subsite-capping model states that the position and dynamics of loops play an important role in substrate preference and regioselectivity of deacetylation. So, the difference in the deacetylation mode of $ScCDA_2$ may be due to the loop length, position and dynamic effects [47].

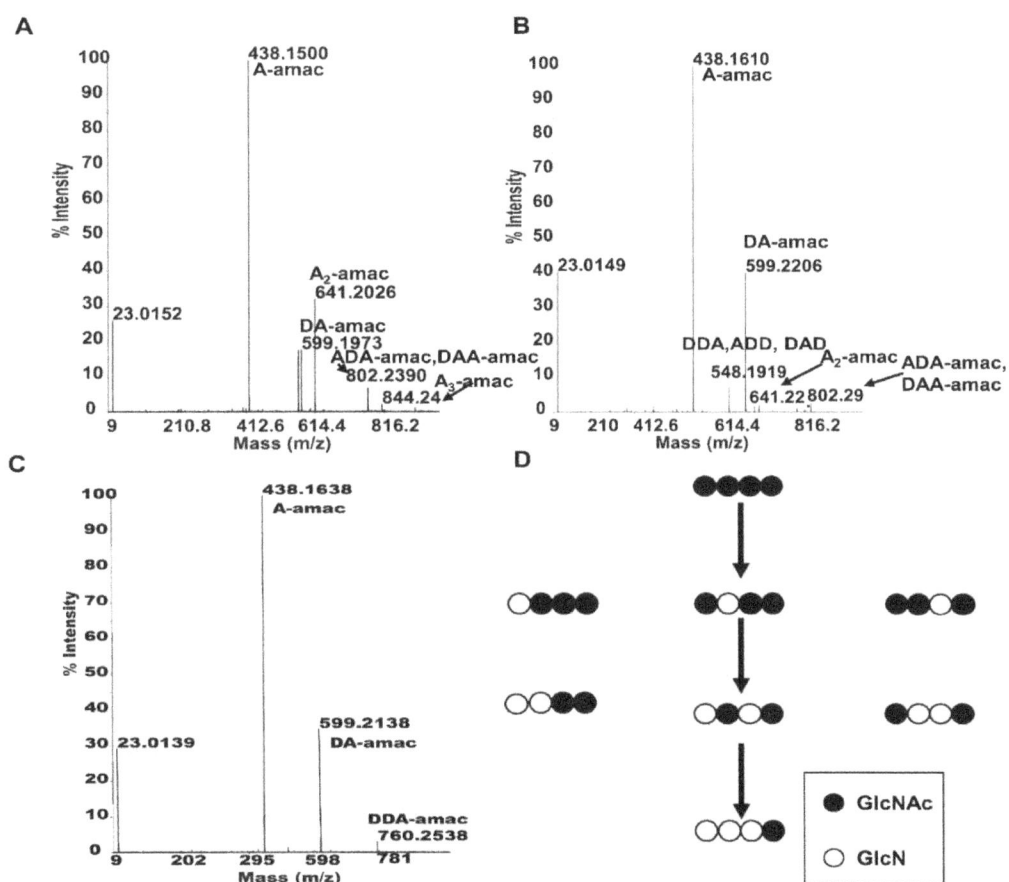

Figure 8. MALDI-TOF-MS2 determines the acetylation pattern of partially acetylated COS. (**A**) The MS_2 spectrum of A3D1 labelled with AMAC and the resulting ion fragments: A-amac, DA-amac, AA-amac, ADA-amac, DAA-amac, AAA-amac (m/z 438.15, 599.19, 641.20, 802.23, 844.24); so, the acetylation pattern of A3D1 is DAAA, ADAA and AADA. (**B**) MS_2 spectrum of A2D2 labelled with AMAC and the resulting ion fragments: A-amac, DDA, DAD, ADD, DA-amac, AA-amac and ADA-amac, (m/z 438.16, 548.19, 599.22, 641.22, 802.29); so, the acetylation pattern of A2D2 is DDAA, ADDA and DADA. (**C**) MS_2 spectrum of A1D3 labelled with AMAC, resulting in ion fragmentation of A-amac, DA-amac and DDA-amac (m/z 438.16, 599.21, 760.25); so, the acetylation pattern of A1D3 is DDDA. (**D**) The deacetylation process of $ScCDA_2$ when A4 is used as a substrate.

3. Materials and Methods

3.1. Materials

Escherichia coli TOP10, plasmid pPICZαA, T4 DNA ligase and DNA polymerase were purchased from Takara Biotechnology (Dalian). Pichia pastoris X-33 was stored in our laboratory. Beta-chitin was purchased from Sigma (St. Louis, MO, USA). Alpha-chitin was purchased from Seikagaku (Tokyo, Japan) [33]. Colloidal chitin was prepared according to the previously reported method [48]. AMAC, chitin oligosaccharides (GlcNAc) n (n = 2–6, dimers to hexamers (A_2 to A_6; A, GlcNAc)) were purchased from Sigma-Aldrich (Munich, Germany). Unless otherwise noted, all reagents were analytical grade. Acetate release was measured using an acetate kit from R-Biopharm (Darmstadt, Germany) [14].

3.2. Cloning, Expression and Purification of ScCDA$_2$

The cda_2 gene from Saccharomyces cerevisiae S288c (GenBank: NP_013411.1) was amplified with upstream primer ScCDA$_2$-F:(5'-CATGCCATGGGAAGCTAATAGGGAAGATTTA-3') and downstream primer: ScCDA$_2$-R (5'-CCGCTCGAGGGACAAGAATTCTTTTATGTAATC-3'). The target gene was digested and then ligated into a pPICZαA expression vector containing the N-terminal hexa histidine fusion tag coding region.

The cda_2 gene was recombined into a Pichia pastoris X-33 chromosome. Then the recombinant Pichia pastoris X-33 was induced by 0.5% methanol for 4 days, and methanol was added in batches every 24 h. The culture supernatant was collected by centrifugation at $8000 \times g$, 4 °C for 20 min. The crude enzyme from the supernatant was concentrated using a 10 kDa ultrafiltration membrane. Then, the concentrated supernatant was purified by Ni-NTA Sepharose excel column (GE Healthcare). The pre-equilibrated buffer was subjected to Ni-NTA with buffer containing 20 mM PBS, pH 7.4, 300 mM NaCl and then washed with 50 mM PBS, pH 7.4, 300 mM NaCl, 20 mM imidazole. Finally, the target protein, eluted with 20 mM PBS, pH 7.4, 300 mM NaCl, 250 mM imidazole was obtained. Protein concentration was determined by using the Pierce™ BCA Protein Assay Kit (Thermo Fisher Scientific).

3.3. ScCDA$_2$ Activity Assay and Biochemistry Properties

The purified ScCDA$_2$ was studied to determine its enzymatic properties and deacetylation patterns. The colloidal chitin (water-soluble chitin), colloidal chitin and chitin oligomers dimers to hexamers (A_2 to A_6) were used as substrates [14]. The reaction mixture for ScCDA$_2$ enzyme activity assay containing 20 mM Tris-HCl buffer (pH8.0), including 1 mM CoCl$_2$, 0.5 mg/mL substrate and 0.75 µM purified soluble protein (ScCDA$_2$) or distilled water as a control was incubated at 37 °C for 30 min. The reaction was terminated by the addition of 10 µL 5% formic acid [14]. Determination of CDA activity by measuring the amount of acetate released by a coupled enzyme assay using an acetate assay kit [14]. The total reaction volume of the coupled enzyme reaction was 266 µL, which was measured spectrophotometrically at 340 nm [14].

In order to determine the optimum pH of ScCDA$_2$, protein in different buffers (final concentration 20 mM) was incubated at 37 °C for 30 min at the pH range of 4.0 to 10.0, in either sodium citrate disodium hydrogen phosphate buffer (pH 3.0–5.0), sodium phosphate dibasic sodium dihydrogen phosphate buffer (pH 6.0–7.0), Tris-HCl buffer (pH 8.0) or sodium carbonate sodium bicarbonate buffer (pH 9.0–10.0). The optimum temperature was determined in the 20 mM Tris-HCl buffer, at the optimum pH of 8.0, and each protein solution was incubated at 37 °C, 50 °C and 65 °C for 60 min. Subsequently, the remaining enzyme activity was measured using standard activity assays. The effects of different metal ions on enzyme activity were verified by adding 1 mM of different metal ion solutions (NaCl, NH$_4$Cl, BaCl$_2$, CoCl$_2$, MnCl$_2$, ZnCl$_2$, CuCl$_2$, MgCl$_2$ and FeCl$_3$) to the reaction mixtures [17].

3.4. Identification of ScCDA₂ Products by MALDI-TOF-MS

To determine the effect of different enzyme concentrations on the enzyme reaction products, four concentration gradients (0.25 µM, 0.5 µM, 0.75 µM, 1 µM) were prepared under 20 mM pH = 8.0 Tris-HCl. MS spectra were obtained using an Ultraflex™ ToF/ToF mass spectrometer (Bruker Daltonik GmbH, Bremen, Germany) to analyze the degree of acetylation, as previously described [49].

3.5. Preparation of Partially Acetylated COS

To analysis the deacetylation pattern of ScCDA₂, 20 mM Tris-HCl buffer (pH8.0), including 1 mM CoCl₂, 0.5 mg/mL substrate and 0.75 µM purified soluble protein (ScCDA₂) was incubated at 37 °C for 30 min. Then, 50 µL of the sample was injected into an X-Amide (4.6 mm × 250 mm) column for separation. The column was eluted with 0.3% formic acid and 50 mM ammonium formate buffer at a flow rate of 2 mL/min. The separated sample was analyzed by electrospray ionization mass spectrometry (ESI-MS).

3.6. Acetylation Pattern Analysis of COS

Reductive amine derivatisation of partially acetylated COS was performed as previously described [50]. In brief, 0.5 mg of the partially acetylated product was dissolved in 10 µL of 0.1 mol/L solution of 2-aminoacridone (AMAC) in acetic acid/DMSO (v/v, 3/17) and stirred manually for 30 s; then 10 µL of 1 M sodium cyanoborohydride solution was added and stirred for a further 30 s. The mixture was heated at 90 °C for 30 min, cooled to −20 °C and then completely freeze-dried. The dried sample was dissolved in 200 µL of methanol/water (v/v, 50/50) solution and sufficiently centrifuged at 12,000× g, 4 °C for 10 min. Then the supernatant was immediately analyzed by mass spectrometry or stored at −20 °C for one month. The method of mass spectrometry to detect the results of reductive amine derivatization was the same as the method of mass spectrometry detection of the enzyme reaction product mentioned previously [51]. MS² spectra were used to analyze the acetylation pattern of COS [52].

3.7. Homology Modelling and Molecular Docking

YASARA software (version 14.12.2) was used to build the homology model of ScCDA₂ with three crystal structures (PDB ID: 5LFZ, 2CC0 and 2C1G) as templates, the similarity between ScCDA₂ and templates is 34%, 33%, 29%, respectively [53], which are highly homologous to ScCDA₂, based on BLAST results using. The 3D structural model was visualized using VMD software (version 1.9.3, University of Illinois; Urbana–Champaign, IL, USA) [54]. The model was further evaluated for protein geometry by SAVES (http://services.mbi.ucla.edu/SAVES/), PROCHECK, ERRAT and VERIFY3D [55]. The chitin molecule structure was acquired from the zinc database (http://zinc.docking.org/). Molecular docking was performed using LeDock software (http://www.lephar.com/) with default parameters [56]. The dimensions of the binding box were set as 10 Å around the active site. The docking center was set at the Zn²⁺. The number of binding poses of the ligand was 100. Finally, the docking pose that fulfilled the catalytic criteria was chosen as the initial conformation for analysis.

4. Conclusions

In this study, we firstly report the detailed deacetylation patterns of chitin deacetylase from *Saccharomyces cerevisiae* (ScCDA₂). Fully defined chitooligosaccharides (DAAA, ADAA, AADA, DDAA, DADA, ADDA and DDDA) have been produced by ScCDA₂ through multiple attack catalytic mechanisms. In addition, a fast, convenient and online monitoring method has been developed that can be used to separate and detect partially acetylated chitosan oligosaccharides. Enzymatic production of fully defined chitooligosaccharides and on-line monitoring and separation chitooligosaccharides, which solves the time-consuming and labor-intensive problem of isolating high purity chitooligosaccharides. This bio-enzymatic application could avoid the use of irritating chemicals and allows the production of functional chitosan and COS from crustacean waste chitin.

Supplementary Materials

Figure S1: The model was further evaluated for protein geometry by SAVES (A comprehensive measurement website for the quality of a protein structure). 97.3% Residues in additional allowed regions and 85.57% of the residues have averaged 3D-1D score $\geq$ 0.2, and the quality factor is 91.2214. **Figure S2:** Compare deacetylase charge distribution. These pictures show the surface charge distributions of chitin deacetylase from *Saccharomyces cerevisiae* (*Sc*CDA$_2$) and chitin deacetylase from *Aspergillus Nidulans* (*An*CDA, PDB ID: 2Y8U) calculated using ABPS (The Adaptive Poisson-Boltzmann Solver to generate electrostatic surface displayed) in VMD. Red represents a negative charge and blue represents a positive charge. **Figure S3:** *Sc*CDA$_2$ substrate specificity on chitin oligomers. 0.5 mg/mL chitin oligomers as substrates were incubated with 0. 75 μM *Sc*CDA$_2$ at 37 °C for 30 min. The data represents the mean SD values of the results from three independent experiments. **Figure S4:** Effects of pH, temperature and metal ion on enzyme activity. (**A**) The optimum pH was determined by incubating the 0.75 μM *Sc*CDA$_2$ with A4 chitin oligomer (0.5 mg/mL) for 60 min at pH 3–11 in universal buffer. (**B**) To obtain the optimal temperature, the enzyme (075 μmol) was incubated for 60 min at different temperatures in 50 mM Tris-HCl buffer (pH 8.0) containing chitin oligomer mixture (0.5 mg/mL) as a substrate. (**C**) Relative activity with different metal cations. Proteins were incubated with 1 mm metallized cations, and activity was determined in 50 mM Tris-HCl buffer (pH 8.0) using 0.5 mg/mL A4 as a substrate. **Figure S5:** Spectra of N-glycosylation of *Sc*CDA$_2$. Mass spectrometry showed that *Sc*CDA$_2$ have N-glycosylation post-translational modification at Asn 181, Asn 199 and Asn 203.

Author Contributions: General concept and design of studies: H.Y. Experimental concept and data analysis: X.-Y.Z., Y.Z., Experimental conduct and manuscript writing: X.-Y.Z. Manuscript review: H.Y., H.-D.Z., W.-X.W. and H.-H.C. Page: 12. Manuscript finalization: H.Y.

Acknowledgments: The authors would like to thank L.H.W. for the MALDI-TOF-MS analyses.

References

1.   Kaur, S.; Dhillon, G.S. Recent trends in biological extraction of chitin from marine shell wastes: A review. *Crit. Rev. Biotechnol.* **2015**, *35*, 44–61. [CrossRef]

2.   Yan, N.; Chen, X. Sustainability: Don't waste seafood waste. *Nature* **2015**, *524*, 155–157. [CrossRef] [PubMed]

3.   Hamed, I.; Ozogul, F.; Regenstein, J.M. Industrial applications of crustacean by-products (chitin, chitosan, and chitooligosaccharides): A review. *Trends Food Sci. Technol.* **2016**, *48*, 40–50. [CrossRef]

4.   Revathi, M.; Saravanan, R.; Shanmugam, A. Production and characterization of chitinase from Vibrio species, a head waste of shrimp Metapenaeus dobsonii (Miers, 1878) and chitin of Sepiella inermis Orbigny, 1848. *Adv. Biosci. Biotechnol.* **2012**, *3*, 392–397. [CrossRef]

5.   Kumar, M.N.V.R. A review of chitin and chitosan applications. *React. Funct. Polym.* **2000**, *46*, 1–27. [CrossRef]

6.   Chambon, R.; Despras, G.; Brossay, A.; Vauzeilles, B.; Urban, D.; Beau, J.M.; Armand, S.; Cottaz, S.; Fort, S. Efficient chemoenzymatic synthesis of lipo-chitin oligosaccharides as plant growth promoters. *Green Chem.* **2015**, *17*, 3923–3930. [CrossRef]

7.   Zhang, X.D.; Xiao, G.; Wang, Y.Q.; Zhao, Y.; Su, H.J.; Tan, T.W. Preparation of chitosan-TiO2 composite film with efficient antimicrobial activities under visible light for food packaging applications. *Carbohydr. Polym.* **2017**, *169*, 101–107. [CrossRef] [PubMed]

8.   Zhang, M.; Tan, T.W. Insecticidal and fungicidal activities of chitosan and oligo-chitosan. *J. Bioact. Compat. Polym.* **2003**, *18*, 391–400. [CrossRef]

9.   Zhang, H.; Li, P.; Wang, Z.; Cui, W.W.; Zhang, Y.; Zhang, Y.; Zheng, S.; Zhang, Y. Sustainable Disposal of Cr(VI): Adsorption-Reduction Strategy for Treating Textile Wastewaters with Amino-Functionalized Boehmite Hazardous Solid Wastes. *ACS Sustain. Chem. Eng.* **2018**, *6*, 6811–6819. [CrossRef]

10.   Yu, P.; Wang, H.-Q.; Bao, R.-Y.; Liu, Z.; Yang, W.; Xie, B.-H.; Yang, M.-B. Self-Assembled Sponge-like Chitosan/Reduced Graphene Oxide/Montmorillonite Composite Hydrogels without Cross-Linking of Chitosan for Effective Cr(VI) Sorption. *ACS Sustain. Chem. Eng.* **2017**, *5*, 1557–1566. [CrossRef]

11. Xu, G.; Liu, P.; Pranantyo, D.; Neoh, K.-G.; Kang, E.T. Dextran- and Chitosan-Based Antifouling, Antimicrobial Adhesion, and Self-Polishing Multilayer Coatings from pH-Responsive Linkages-Enabled Layer-by-Layer Assembly. *ACS Sustain. Chem. Eng.* **2018**, *6*, 3916–3926. [CrossRef]

12. Duri, S.; Harkins, A.L.; Frazier, A.J.; Tran, C.D. Composites Containing Fullerenes and Polysaccharides: Green and Facile Synthesis, Biocompatibility, and Antimicrobial Activity. *ACS Sustain. Chem. Eng.* **2017**, *5*, 5408–5417. [CrossRef]

13. Fang, Y.; Zhang, R.; Duan, B.; Liu, M.; Lu, A.; Zhang, L. Recyclable Universal Solvents for Chitin to Chitosan with Various Degrees of Acetylation and Construction of Robust Hydrogels. *ACS Sustain. Chem. Eng.* **2017**, *5*, 2725–2733. [CrossRef]

14. Naqvi, S.; Cord-Landwehr, S.; Singh, R.; Bernard, F.; Kolkenbrock, S.; El Gueddari, N.E.; Moerschbacher, B.M. A Recombinant Fungal Chitin Deacetylase Produces Fully Defined Chitosan Oligomers with Novel Patterns of Acetylation. *Appl. Environ. Microbiol.* **2016**, *82*, 6645–6655. [CrossRef]

15. Vander, P.; Km, V.R.; Domard, A.; Eddine, E.G.N.; Moerschbacher, B.M. Comparison of the ability of partially N-acetylated chitosans and chitooligosaccharides to elicit resistance reactions in wheat leaves. *Plant Physiol.* **1998**, *118*, 1353. [CrossRef]

16. Liu, T.; Liu, Z.; Song, C.; Hu, Y.; Han, Z.; She, J.; Fan, F.; Wang, J.; Jin, C.; Chang, J.; Zhou, J.-M.; Chai, J. Chitin-Induced Dimerization Activates a Plant Immune Receptor. *Science* **2012**, *336*, 1160–1164. [CrossRef]

17. Lv, Y.M.; Laborda, P.; Huang, K.; Cai, Z.P.; Wang, M.; Lu, A.M.; Doherty, C.; Liu, L.; Flitsch, S.L.; Voglmeir, J. Highly efficient and selective biocatalytic production of glucosamine from chitin. *Green Chem.* **2017**, *19*, 527–535. [CrossRef]

18. Hu, K.; Yang, H.; Liu, G.; Tan, H. Identification and characterization of a polysaccharide deacetylase gene from Bacillus thuringiensis. *Can. J. Microbiol.* **2006**, *52*, 935–941. [CrossRef]

19. Zhou, G.; Zhang, H.; He, Y.; He, L. Identification of a chitin deacetylase producing bacteria isolated from soil and its fermentation optimization. *Afr. J. Microbiol. Res.* **2010**, *4*, 2597–2603.

20. Pacheco, N.; Trombotto, S.; David, L.; Shirai, K. Activity of chitin deacetylase from Colletotrichum gloeosporioides on chitinous substrates. *Carbohydr. Polym.* **2013**, *96*, 227–232. [CrossRef]

21. Araki, Y.; Ito, E. Pathway of chitosan formation in mucor-rouxii—Enzymatic deacetylation of chitin. *Eur. J. Biochem.* **1975**, *55*, 71–78. [CrossRef] [PubMed]

22. Alfonso, C.; Nuero, O.M.; Santamaria, F.; Reyes, F. Purification of a heat-stable chitin deacetylase from aspergillus-nidulans and its role in cell-wall degradation. *Curr. Microbiol.* **1995**, *30*, 49–54. [CrossRef] [PubMed]

23. Martinou, A.; Koutsioulis, D.; Bouriotis, V. Expression, purification, and characterization of a cobalt-activated chitin deacetylase (Cda2p) from Saccharomyces cerevisiae. *Protein Expr. Purif.* **2002**, *24*, 111–116. [CrossRef] [PubMed]

24. Zhong, X.-W.; Wang, X.-H.; Tan, X.; Xia, Q.-Y.; Xiang, Z.-H.; Zhao, P. Identification and Molecular Characterization of a Chitin Deacetylase from Bombyx mori Peritrophic Membrane. *Int. J. Mol. Sci.* **2014**, *15*, 1946–1961. [CrossRef] [PubMed]

25. Wang, S.Q.; Jayaram, S.A.; Hemphala, J.; Senti, K.A.; Tsarouhas, V.; Jin, H.N.; Samakovlis, C. Septate-junction-dependent luminal deposition of chitin deacetylases restricts tube elongation in the Drosophila trachea. *Curr. Biol.* **2006**, *16*, 180–185. [CrossRef]

26. Brosson, D.; Kuhn, L.; Prensier, G.; Vivares, C.P.; Texier, C. The putative chitin deacetylase of Encephalitozoon cuniculi: A surface protein implicated in microsporidian spore-wall formation. *FEMS Microbiol. Lett.* **2005**, *247*, 81–90. [CrossRef]

27. Zhao, Y.; Park, R.-D.; Muzzarelli, R.A.A. Chitin Deacetylases: Properties and Applications. *Mar. Drugs* **2010**, *8*, 24–46. [CrossRef] [PubMed]

28. John, M.; Rohrig, H.; Schmidt, J.; Wieneke, U.; Schell, J. Rhizobium nodb protein involved in nodulation signal synthesis is a chitooligosaccharide deacetylase. *Proc. Natl. Acad. Sci. USA* **1993**, *90*, 625–629. [CrossRef] [PubMed]

29. Hekmat, O.; Tokuyasu, K.; Withers, S.G. Subsite structure of the endo-type chitin deacetylase from a Deuteromycete, Colletotrichum lindemuthianum: An investigation using steady-state kinetic analysis and MS. *Biochem. J.* **2003**, *374*, 369–380. [CrossRef]

30. Andres, E.; Albesa-Jove, D.; Biarnes, X.; Moerschbacher, B.M.; Guerin, M.E.; Planas, A. Structural basis of chitin oligosaccharide deacetylation. *Angew. Chem. Int. Ed. Engl.* **2014**, *53*, 6882–6887. [CrossRef] [PubMed]

31. Christodoulidou, A.; Bouriotis, V.; Thireos, G. Two sporulation-specific chitin deacetylase-encoding genes are required for the ascospore wall rigidity of Saccharomyces cerevisiae. *J. Biol. Chem.* **1996**, *271*, 31420–31425. [CrossRef] [PubMed]

32. Cantarel, B.L.; Coutinho, P.M.; Rancurel, C.; Bernard, T.; Lombard, V.; Henrissat, B. The Carbohydrate-Active EnZymes database (CAZy): An expert resource for Glycogenomics. *Nucleic Acids Res.* **2009**, *37*, D233–D238. [CrossRef] [PubMed]

33. Blair, D.E.; Hekmat, O.; Schuttelkopf, A.W.; Shrestha, B.; Tokuyasu, K.; Withers, S.G.; van Aalten, D.M.F. Structure and mechanism of chitin deacetylase from the fungal pathogen Colletotrichum lindemuthianum. *Biochemistry* **2006**, *45*, 9416–9426. [CrossRef]

34. Tuveng, T.R.; Rothweiler, U.; Udatha, G.; Vaaje-Kolstad, G.; Smalås, A.; Eijsink, V.G.H. Structure and function of a CE4 deacetylase isolated from a marine environment. *PLoS ONE* **2017**, *12*, e0187544. [CrossRef] [PubMed]

35. Taylor, E.J.; Gloster, T.M.; Turkenburg, J.P.; Vincent, F.; Brzozowski, A.M.; Dupont, C.; Shareck, F.; Centeno, M.S.J.; Prates, J.A.M.; Puchart, V.; et al. Structure and activity of two metal ion-dependent acetylxylan esterases involved in plant cell wall degradation reveals a close similarity to peptidoglycan deacetylases. *J. Biol. Chem.* **2006**, *281*, 10968–10975. [CrossRef] [PubMed]

36. Blair, D.E.; Schuttelkopf, A.W.; MacRae, J.I.; van Aalten, D.M.F. Structure and metal-dependent mechanism of peptidoglycan deacetylase, a streptococcal virulence factor. *Proc. Natl. Acad. Sci. USA* **2005**, *102*, 15429–15434. [CrossRef] [PubMed]

37. Schäffer, A.; Aravind, L.; Madden, T.; Shavirin, S.; Spouge, J.; Wolf, Y.; Koonin, E.; Altschul, S. Improving the accuracy of PSI-BLAST protein database searches with composition-based statistics and other refinements. *Nucleic Acids Res.* **2001**, *29*, 2994–3005. [CrossRef]

38. Tarentino, A.L.; Trimble, R.B.; Plummer, T.H., Jr. Enzymatic approaches for studying the structure, synthesis, and processing of glycoproteins. *Methods Cell Biol.* **1989**, *32*, 111–139. [PubMed]

39. Tarentino, A.L.; Plummer, T.H. Enzymatic deglycosylation of asparagine-linked glycans—Purification, properties, and specificity of oligosaccharide-cleaving enzymes from flavobacterium-meningosepticum. In *Guide to Techniques in Glycobiology*; Lennarz, W.J., Hart, G.W., Eds.; Elsevier Academic Press Inc.: San Diego, CA, USA, 1994; Volume 230, pp. 44–57.

40. Amore, A.; Knott, B.; Supekar, N.; Shajahan, A.; Azadi, P.; Zhao, P.; Wells, L.; Linger, J.; Hobdey, S.; Vander Wall, T.; et al. Distinct roles of N- and O-glycans in cellulase activity and stability. *Proc. Natl. Acad. Sci. USA* **2017**, *114*, 13667–13672. [CrossRef] [PubMed]

41. Liu, Z.; Gay, L.M.; Tuveng, T.R.; Agger, J.W.; Westereng, B.; Mathiesen, G.; Horn, S.J.; Vaaje-Kolstad, G.; van Aalten, D.M.F.; Eijsink, V.G.H. Structure and function of a broad-specificity chitin deacetylase from Aspergillus nidulans FGSC A4. *Sci. Rep.* **2017**, *7*, 1746. [CrossRef] [PubMed]

42. Li, K.C.; Xing, R.G.; Liu, S.; Qin, Y.K.; Li, P.C. Access to N-Acetylated Chitohexaose with Well-Defined Degrees of Acetylation. *Biomed. Res. Int.* **2017**, *2017*, 2486515. [CrossRef] [PubMed]

43. Wu, Y.X.; Lu, W.P.; Wang, J.N.; Gao, Y.H.; Guo, Y.C. Rapid and Convenient Separation of Chitooligosaccharides by Ion-Exchange Chromatography. In Proceedings of the 5th Annual International Conference on Material Science and Engineering, Xiamen, China, 20–22 October 2017; Aleksandrova, M., Szewczyk, R., Eds.; Iop Publishing Ltd.: Bristol, UK, 2018; Volume 275.

44. Cao, L.; Wu, J.; Li, X.; Zheng, L.; Wu, M.; Liu, P.; Huang, Q. Validated HPAEC-PAD Method for the Determination of Fully Deacetylated Chitooligosaccharides. *Int. J. Mol. Sci.* **2016**, *17*, 1699. [CrossRef] [PubMed]

45. Li, K.C.; Liu, S.; Xing, R.G.; Qin, Y.K.; Li, P.C. Preparation, characterization and antioxidant activity of two partially N-acetylated chitotrioses. *Carbohydr. Polym.* **2013**, *92*, 1730–1736. [CrossRef] [PubMed]

46. Li, K.C.; Xing, R.G.; Liu, S.; Li, R.F.; Qin, Y.K.; Meng, X.T.; Li, P.C. Separation of chito-oligomers with several degrees of polymerization and study of their antioxidant activity. *Carbohydr. Polym.* **2012**, *88*, 896–903. [CrossRef]

47. Aragunde, H.; Biarnes, X.; Planas, A. Substrate Recognition and Specificity of Chitin Deacetylases and Related Family 4 Carbohydrate Esterases. *Int. J. Mol. Sci.* **2018**, *19*, 30. [CrossRef] [PubMed]

48. Hirano, S.; Nagao, N. An Improved Method for the Preparation of Colloidal Chitin by Using Methanesulfonic-Acid. *Agric. Biol. Chem. Tokyo* **1988**, *52*, 2111–2112.

49. Wang, S.; Liu, C.; Liang, T. Fermented and enzymatic production of chitin/chitosan oligosaccharides by extracellular chitinases from Bacillus cereus TKU027. *Carbohydr. Polym.* **2012**, *90*, 1305–1313. [CrossRef]

50. Bahrke, S.; Einarsson, J.M.; Gislason, J.; Haebel, S.; Letzel, M.C.; Peterkatalinić, J.; Peter, M.G. Sequence analysis of chitooligosaccharides by matrix-assisted laser desorption ionization postsource decay mass spectrometry. *Biomacromolecules* **1900**, *3*, 696–704. [CrossRef]

51. Chen, M.; Zhu, X.; Li, Z.; Guo, X.; Ling, P. Application of matrix-assisted laser desorption/ionization time-of-flight mass spectrometry (MALDI-TOF-MS) in preparation of chitosan oligosaccharides (COS) with degree of polymerization (DP) 5–12 containing well-distributed acetyl groups. *Int. J. Mass Spectrom.* **2010**, *290*, 94–99. [CrossRef]

52. Lee, J.H.; Ha, Y.W.; Jeong, C.S.; Kim, Y.S.; Park, Y. Isolation and tandem mass fragmentations of an anti-inflammatory compound from Aralia elata. *Arch. Pharm. Res.* **2009**, *32*, 831–840. [CrossRef]

53. Krieger, E.; Koraimann, G.; Vriend, G. Increasing the precision of comparative models with YASARA NOVA—A self-parameterizing force field. *Proteins-Struct. Funct. Genet.* **2002**, *47*, 393–402. [CrossRef] [PubMed]

54. Humphrey, W.; Dalke, A.; Schulten, K. VMD: Visual molecular dynamics. *J. Mol. Graph. Model.* **1996**, *14*, 33–38. [CrossRef]

55. Eisenberg, D.; Luthy, R.; Bowie, J.U. VERIFY3D: Assessment of protein models with three-dimensional profiles. In *Macromolecular Crystallography, Pt B*; Carter, C.W., Sweet, R.M., Eds.; Academic Press: Cambridge, MA, USA, 1997; Volume 277, pp. 396–404.

56. Wang, Z.; Sun, H.; Yao, X.; Li, D.; Xu, L.; Li, Y.; Tian, S.; Hou, T. Comprehensive evaluation of ten docking programs on a diverse set of protein-ligand complexes: The prediction accuracy of sampling power and scoring power. *Phys. Chem. Chem. Phys.* **2016**, *18*, 12964–12975. [CrossRef] [PubMed]

16

Characterization of an Alkaline Alginate Lyase with pH-Stable and Thermo-Tolerance Property

Yanan Wang [1,†], **Xuehong Chen** [1,†], **Xiaolin Bi** [2], **Yining Ren** [1], **Qi Han** [1], **Yu Zhou** [1], **Yantao Han** [1,*], **Ruyong Yao** [3] **and Shangyong Li** [1,*]

[1]   Department of Pharmacology, College of Basic Medicine, Qingdao University, Qingdao 266071, China; sunshine4581@163.com (Y.W.); chen-xuehong@163.com (X.C.); Renyn796@163.com (Y.R.); xiaoyu19990727@163.com (Q.H.); zy18339956716@163.com (Y.Z.)

[2]   Department of Rehabilitation Medicine, Qingdao University, Qingdao 266071, China; 18661809159@163.com

[3]   Central Laboratory of Medicine, Qingdao University, Qingdao 266071, China; yry0303@163.com

*   Correspondence: hanyt@qdu.edu.cn (Y.H.); lisy@qdu.edu.cn (S.L.).

†   These authors contributed equally to this paper.

Abstract: Alginate oligosaccharides (AOS) show versatile bioactivities. Although various alginate lyases have been characterized, enzymes with special characteristics are still rare. In this study, a polysaccharide lyase family 7 (PL7) alginate lyase-encoding gene, *aly08*, was cloned from the marine bacterium *Vibrio* sp. SY01 and expressed in *Escherichia coli*. The purified alginate lyase Aly08, with a molecular weight of 35 kDa, showed a specific activity of 841 U/mg at its optimal pH (pH 8.35) and temperature (45 °C). Aly08 showed good pH-stability, as it remained more than 80% of its initial activity in a wide pH range (4.0–10.0). Aly08 was also a thermo-tolerant enzyme that recovered 70.8% of its initial activity following heat shock treatment for 5 min. This study also demonstrated that Aly08 is a polyG-preferred enzyme. Furthermore, Aly08 degraded alginates into disaccharides and trisaccharides in an endo-manner. Its thermo-tolerance and pH-stable properties make Aly08 a good candidate for further applications.

Keywords: Alginate lyase; Thermo-tolerant; pH-stability; Endo-manner; *Vibrio* sp. SY01

---

1. Introduction

Alginate is an acidic hetero-polysaccharide extracted from brown algae, which accounting for 22–44% of its dry weight [1–3]. Alginate mainly contains two different uronic acids, including α-L-guluronic acid (G) and β-D-mannuronic acid (M). They are arranged into three different kinds of blocks by (1→4)-linked monosaccharides: homopolymeric G blocks, polyguluronate (PolyG); homopolymeric M blocks, polymannuronate (PolyM); and random or heteropolymeric blocks of alternating M and G units (PolyMG) [4,5].

Alginate lyase (E.C. 4.2.2.3 and E.C. 4.2.2.11) is a kind of polysaccharide lyase that degrades alginate by β-eliminating the glycoside 1-4 O-bonds between C4 and C5 at the non-reducing end, thus producing unsaturated alginate oligosaccharides (UAOS) as main products [6,7]. Due to its high efficiency, specificity and mild degradation function, alginate lyases have attracted widespread attention in industrial applications, especially in the preparation of alginate oligosaccharides [8,9].

According to the Carbohydrate-Active enZYmes (CAZy) databases, alginate lyases belong to PL families 5, 6, 7, 14, 15, 17, and 18 based on the analysis of their amino acid sequences [10–12]. Based on the substrate specificity, alginate lyases can be further classified into two types, one type is the G block-specific lyases (polyG lyases, EC 4.2.2.11), and the other type is the M block-specific lyases (polyM lyases, EC 4.2.2.3) [13,14]. In PL families 5, 7, 14, 17, and 18, most of the reported alginate lyases are polyM lyases. Only the alginate lyase reported in PL family 6 is mainly comprised of polyG lyases. Thus far, hundreds of alginate lyases have been purified, cloned, and characterized from marine microorganisms, brown seaweeds, and mollusks [15–18]. However, these reported enzymes with characteristics specific for commercial use are rare. Cold-adapted alginate lyases can run biocatalytic processes at low temperature and reduce the danger of contamination. Thermo-tolerant enzymes persist at high temperatures, thereby not only improving degradation efficiency but also reducing production costs. Meanwhile, high proportion product in a mixture of products will be propitious to the purification of oligosaccharide. Therefore, there is an urgency to obtain an alginate lyase with the optimal characteristics (e.g., pH-stability, thermo-tolerance, and single product distribution) needed for industrial applications.

In this study, a new alginate lyase-encoding gene, *aly08*, was cloned from *Vibrio* sp. SY01, and expressed in *Escherichia coli* BL21 (DE3). The recombinant enzyme Aly08 degraded alginate, yielding alginate disaccharides and trisaccharides as main products. This study also revealed that Aly08 was a polyG-preferred enzyme with special characteristics, such as wide pH-stability, thermo-tolerance, and single product distribution. These special features suggest that Aly08 may play essential roles in saccharification processes of alginate and carbon cycling.

2. Results and Discussion

2.1. Sequence Analysis of Aly08

The marine bacterium *Vibrio* sp. SY01 was isolated from Yellow sea sediment, China. It grew rapidly in the alginate sole-carbon medium and efficiently degraded brown seaweed with a high alginate lyase activity (more than 50 U/mL). The genome sequence analysis of *Vibrio* sp. SY01 showed that it contained the putative alginate lyase-encoding gene, *aly08*, consisting of 897 bp of an open reading frame (ORF). The identified alginate lyase, Aly08, contained 299 amino acid residues. Signal peptide analysis showed that Aly08 predicted a putative signal peptide (Met[1] to Phe[22]) in its N-terminal. Furthermore, the theoretically isoelectric point (pI) and theoretical molecular weight (Mw) of mature Aly08 were 4.57 and 32.89 kDa, respectively. According to a search of Conserved Domain Database (CDD) of NCBI, Aly08 is a new alginate lyase with a single-domain belonging to the alginate lyase superfamily 2.

Based on the sequences of Aly08 and other reported PL family 7 alginate lyases, phylogenetic trees were created. Pectate lyase (Genbank number CAD56882) from *Bacillus licheniformis* 14A was included as a control (Figure 1). Among all of the reported alginate lyase, Aly08 had the highest identity sequence of amino acids (78%) with a PL family 7 alginate lyase, AlyL2 (Genbank number MH791447), from *Agarivorans* sp. L11 [19].

A deeply branched cluster was formed in the phylogenetic tree among the enzymes Aly08 (Genbank number MH791447), AlyL2 (Genbank number AJO61885), AlgMsp (Genbank number BAJ62034), and Alg7D (Genbank number ABD81807). According to the multiple sequence alignment (Figure 2), the enzyme contains three conserved regions "RTELREMLR", "QIH", and "MYFKAG" which are related to substrate binding and catalytic activity in PL family 7 [20]. These results identified Aly08 as a new member of the PL family 7. Thus far, several alginate lyases have been identified from various bacteria, such as *Pseudoalteromona*, *Flavobacterium*, *Nitratiruptor*, *Agarivorans*, and *Vibrio* [7,17,21–23]. After determining their various properties, most of the alginate lyases showed a preference towards polyM blocks. In this study, the purified alginate lyase Aly08 is a polyG-preferred alginate lyase containing the "QIH" conserved region, according to their sequence analysis (Figure 2). The conserved region, "QIH" or "QVH", plays a key role in the substrate preferences of alginate lyases. Aly08, along with AlgMsp, A1-II and Alg2A containing the "QIH" conserved region showed preferences for polyG [24–26]. In addition, other alginate lyases derived from PL family 7, such as AlxM from *Photobacterium* sp. ATCC 43367, and A9mT from *Vibrio* sp. A9m, possess "QVH" regions show preference for degrading polyM blocks (Table 1) [27,28]. Through further sequence screening, it was found that "QIH" or "QVH" may be an indicator for substrate-preferred analysis. Furthermore,

Aly08 can be used for the next part of combining a polyM-preferred alginate lyase for synergetic degradation alginate or brown seaweeds.

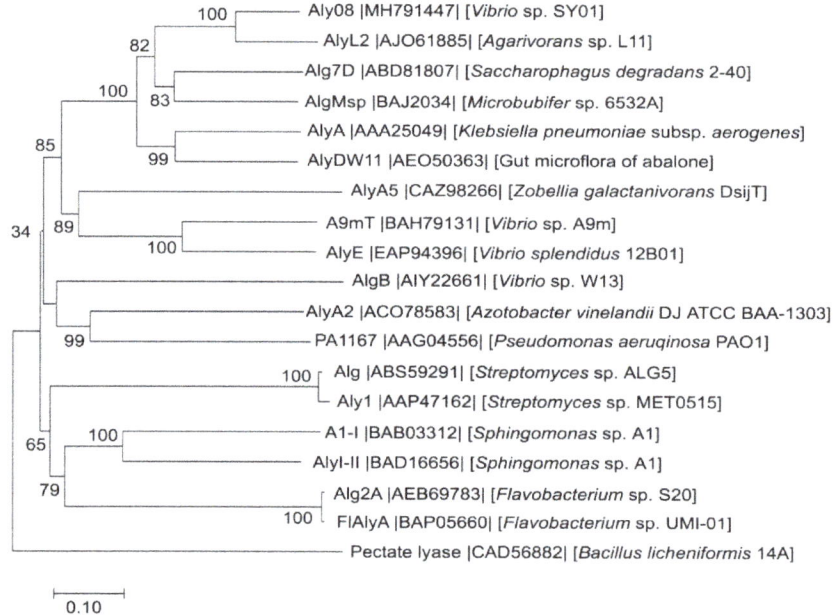

```
            100 ┌── Aly08 |MH791447| [Vibrio sp. SY01]
         82 ┌───┤
        ┌───┤   └── AlyL2 |AJO61885| [Agarivorans sp. L11]
  100   │   │
 ┌──────┤   └──── Alg7D |ABD81807| [Saccharophagus degradans 2-40]
 │      │ 83
 │      └───── AlgMsp |BAJ2034| [Microbubifer sp. 6532A]
85       ┌─── AlyA |AAA25049| [Klebsiella pneumoniae subsp. aerogenes]
     ┌───┤ 99
     │   └── AlyDW11 |AEO50363| [Gut microflora of abalone]
 34  │
─────┤     ──── AlyA5 |CAZ98266| [Zobellia galactanivorans DsijT]
     │ 89 ┌── A9mT |BAH79131| [Vibrio sp. A9m]
     │  ┌─┤ 100
     │  │ └── AlyE |EAP94396| [Vibrio splendidus 12B01]
     │  │     ── AlgB |AIY22661| [Vibrio sp. W13]
     │  │        ── AlyA2 |ACO78583| [Azotobacter vinelandii DJ ATCC BAA-1303]
     │  └ 99 ── PA1167 |AAG04556| [Pseudomonas aeruqinosa PAO1]
     │        100 ┌ Alg |ABS59291| [Streptomyces sp. ALG5]
     │       ┌────┤
  65 │       │    └ Aly1 |AAP47162| [Streptomyces sp. MET0515]
─────┤  100  │ ── A1-I |BAB03312| [Sphingomonas sp. A1]
     │      ┌┤
  79 │      └── AlyI-II |BAD16656| [Sphingomonas sp. A1]
     │         ┌ Alg2A |AEB69783| [Flavobacterium sp. S20]
     └─────100─┤
               └ FlAlyA |BAP05660| [Flavobacterium sp. UMI-01]
        ── Pectate lyase |CAD56882| [Bacillus licheniformis 14A]

 ├──────┤ 0.10
```

Figure 1. Phylogenetic analysis of Aly08 with other reported alginate lyases. The reliability of the phylogenetic reconstructions was determained by boot-strapping values (1000 replicates). Branch-related numbers are bootstrap values (confidence limits) representing the substitution frequency of each amino acid residue. A pectate lyase (CAD56882) from *Bacillus licheniformis* 14A was taken as control.

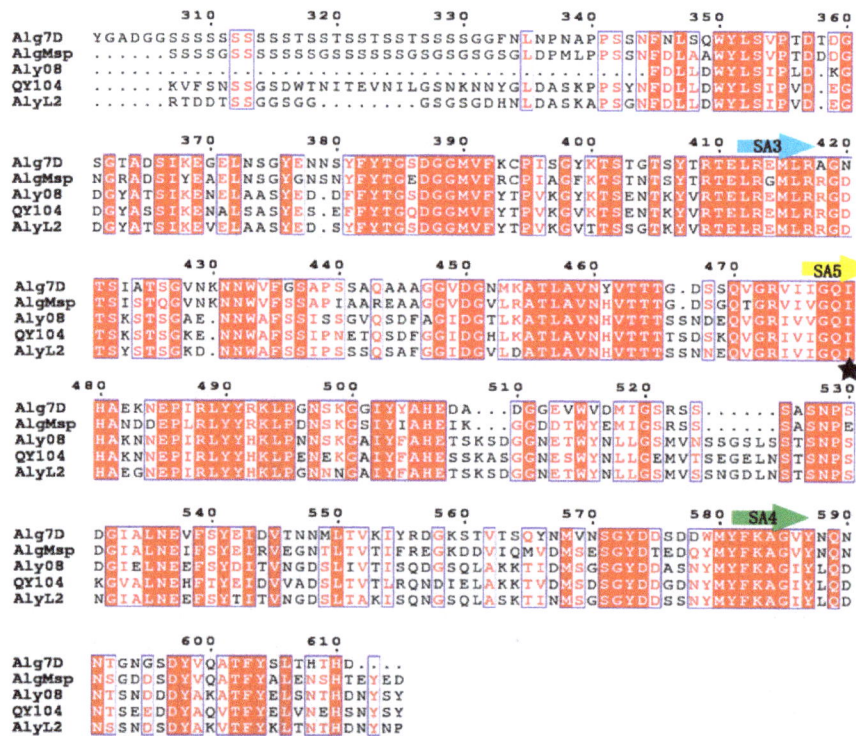

```
              310       320       330       340       350       360
Alg7D  YGADGGSSSSSSSSSTSSTSSTSSTSSSSGGFNLNPNAPPSSNFNLSQWYLSVPTDTDG
AlgMsp .....SSSSSGSSSSSSSGSSSSSSGSGSGSGSGSGLDPMLPPSSNFDLAAWYLSVPTDDDG
Aly08  ...........................................FDLLDWYLSIPLD.KG
QY104  .....KVFSNSSGSDWTNITEVNILGSNKNNYGLDASKPPSYNFDLLDWYLSIPVD.EC
AlyL2  .....RTDDTSSGGSGG........GSGSGDHNLDASKAPSGNFDLLDWYLSIPVD.EC

              370       380       390       400       410  SA3  420
Alg7D  SGTADSIKEGELNSGYENNSYFYTGSDGGMVFKCPISGYKTSIGTSYTRTELREMLRAGN
AlgMsp NGRADSIYEAELNSGYGNSNYFYTGEDGGMVFRCPIAGFKTSTNTSYTRTELRGMLRRGD
Aly08  DGYATSIKENELAASYED.DFFYTGSDGGMVFYTPVKCYKTSIKENTKYVRTELREMLRRGD
QY104  DGYASSIKENALSASYES.EFFYTGQDGGMVFYTPVKCVKTSENTKYVRTELREMLRRGD
AlyL2  DGYATSIKEVELAASYED.SYFYTGSDGGMVFYTPVKCVTTSSGTKYVRTELREMLRRGD

              430       440       450       460       470  SA5  480
Alg7D  TSIATSGVNKNNWVFGSAPSSAQAAAGGVDGNMKATLAVNYVTTTG.DSSQVGRVIIGQI
AlgMsp TSISTQGVNKNNWVFSSAPIAAREAAGGVDGVLRATLAVNHVTTTG.DSGQTGRVIVGQI
Aly08  TSKSTSGAE.NNWAFSSISSGVQSDFAGIDGTLKATLAVNHVTTTSSNDEQVGRIVVGQI
QY104  TSKSTSGKE.NNWAFSSIPNETQSDFGGIDGHLKATLAVNHVTTTSDSKQVGRIVIGQI
AlyL2  TSYSTSGKD.NNWAFSSIPSSQSAFGGIDGVLDATLAVNHVTTTSSNNEQVGRIVIGQI

              490       500       510       520          530
Alg7D  HAEKNEPIRLYYRKLPGNSKGGIYYAHEDA...DGGEVWVDMIGSRSS......SASNPS
AlgMsp HANDDEPLRLYYRKLPDNSKGSIYIAHEIK...GGDDTWYEMIGSRSS......SASNPE
Aly08  HAKNNEPIRLYYHKLPNNSKGAIYFAHETSKSDGGNETWYNLLGSMVNSSGSLSSTSNPS
QY104  HAKNNEPIRLYYHKLPENEKCALYFAHESSKASGGNESWYNLLGSEMVTSEGELNSTSNPS
AlyL2  HAEGNEPIRLYYHKLPGNNNCAIYFAHETSKSDGGNETWYNLLGSMVSSNGDLNSTSNPS

              540       550       560       570  SA4  580     590
Alg7D  DGIALNEVSYEIDVTNNMDTVKIYRDGKSTVTSQYNMVNSGYDSDDWMYFKAGVYNCN
AlgMsp DGIALNEISYEIRVEGNTLTVTIFREGKDDVIQMVDMSESGYDTEDQYMYFKAGVYNCN
Aly08  DGIELNEESYDITVNGDSLIVTISQDGSQLAKKTIDMSGSGYDDASNYMYFKAGIYLQD
QY104  KGVALNEHPIYEIDVVADSLTVTLRQNDIELAKKTVDMSDSGYDDGDNYMYFKAGIYLQD
AlyL2  NGIALNEESYTITVNGDSLTAKISQNGSQLASKTINLMSGSGYDDSSNYMYFKAGIYLQD

              600       610
Alg7D  NTGNGSDYVQATFYSLTHIHD.|...
AlgMsp NSGDDSDYVQATFYALENSHTEYED
Aly08  NTSNDDDYAKATFYEDSNIHDNYSY
QY104  NTSEEDDYAQVTFYEDVNEHSNYSY
AlyL2  NSSNDQSDYAKVTFYKLTNTHDNYNP
```

Figure 2. Sequence comparison of Aly08 with related alginate lyases from PL family 7: Alg7D (ABD81807) from *Saccharophagus degradans* 2–40, AlgMsp (BAJ62034) from *Microbulbifer* sp. 6532A, AlyV4 (AGL7859) from *Vibrio* sp. QY104, and AlyL2 (AJO61885) from *Agarivorans* sp. L11. The conserved regions and identical residues are marked with bands and black star, respectively.

Table 1. Comparison of the properties of Aly08 with other alginate lyases.

Protein Name	Optimal pH/ Temperature (°C)	Conserved Region QIH/QVH	Substrate Specificity	Products (DP)	Source	References
Aly08	8.35/45	QIH	PolyG	2,3	*Vibrio* sp. SY01	This study
AlgNJU-03	7.0/30	QIH	PolyG, polyM,alginate	2,3,4	*Vibrio* sp. NJU-03	[20]
AlgMsp	8.0/40	QIH	PolyG	2–5	*Microbulbifer* sp. 6532A	[24]
A1-II'	7.5/40	QIH	polyG,polyM	3,4	*Sphingomonas* sp.A1	[29]
Aly2	6.0/40	QIH	polyG	2,3,4	*Flammeovirga* sp. strain MY04	[30]
AlyPI	7.0/40	QIH	polyG,polyM	-	*Pseudoalteromonas* sp. CY24.	[31]
Alg2A	8.3/40	QIH	polyG	5,6,7	*Flavobacterium* sp. S20	[26]
AlxM	-	QVH	polyM	-	*Photobacterium* sp. ATCC 43367	[32]
A1m	9.0/30	QIH	polyG	-	*Agarivorans* sp. JAM-A1m	[33]
A9mT	7.5/30	QVH	polyM	-	*Vibrio* sp. A9m	[27]
FlAlyA	7.7/55	QIH	polyM, polyG	2–5	*Flavobacterium* sp. strain UMI-01	[34]
AlyH1	7.5/40	QIH	polyG, alginate	2,3,4	*Vibrio furnissii* H1	[35]

2.2. Expression, Purification, and Characterization of Aly08

The expression strain *E. coli* BL21-pET22b-Aly08 was grown in LB broth and Aly08 was purified by a Ni-NTA affinity column. The specific activity of purified Aly08 was 841.1 U/mg with high viscosity sodium alginate as substrate. Moreover, the purified enzyme Aly08 was analyzed by sodium dodecyl sulfate polyacrylamide gel electrophoresis (SDS-PAGE) and observed as a single band on the gel with an approximate Mw of 35 kDa (Figure 3), which was corresponding to the theoretical molecular mass of 32.89 kDa.

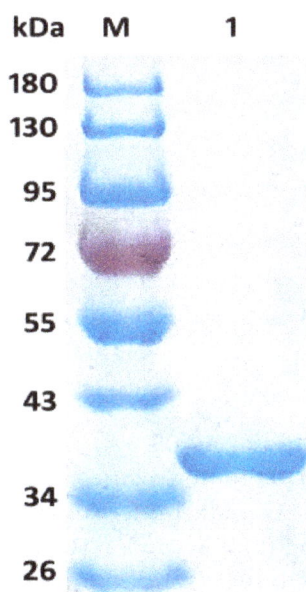

Figure 3. SDS-PAGE analysis of the recombinant enzyme Aly08. Lane M, protein marker; Lane 1, the purified Aly08.

Then, the characterization of purified Aly08 was analyzed as follows. The enzyme Aly08 showed maximum activity at 45 °C, and maintained activities of 82.8% and 48.7% when the enzyme was incubated at a low temperature for 1 h, 10 °C and 20 °C, respectively (Figure 4A,B). The optimal pH for Aly08 was found to be 8.35 (Figure 4C). In addition, Aly08 holds more than 60% of activity in a wide pH range from 7.0–11.0 after incubation in different buffers at 4 °C for 12 h, and was particularly stable under alkaline conditions. (Figure 4D) As previous study, most of the alginate lyases showed optimal pH and stability close to a neutral environment, such as AlyH1 from *Vibrio furnissii* H1 shows high activity at the optimal pH 7.5 and it was stable at pH 6.5–8.5. In addition, AlyH1 only retains about 20%

of residual activity when incubated at pH 9.5 for 12 h [35]. AlgNJU-03 from *Vibrio* sp. NJU-03 possessed a neutral optimal pH at 7.0 and its pH-stability range from 6.0 to 8.0 [20]. Those alginate lyases prefer neutral pH and they only show high activities within a narrow pH range after incubating for several hours and always exhibit instability under alkaline conditions. Another reported high-alkaline alginate lyase A1m from marine bacteria *Agarivorans* sp. JAM-A1m, exhibited high activity at pH 9.0 under glycine-NaOH buffer with 0.2 M NaCl added to the reaction mixture. However, A1m was only stable and maintain more than 60% of residual activity in a short period of 1 h over a narrow pH range of 7.0–9.0 [33]. Comparing with A1m, Aly08 was stable over 12 h with its 80% residual activity even at pH 9.0–11.0, while another enzyme derived from *Vibrio* sp. NJ-04 maintains good stability at pH 4.0–10.0, but the enzyme only maintains its maximum activity under neutral conditions (pH 7.0) [36]. Thus, Aly08 is an alkaline-stable lyase with industrial application potential as it has been proven to conduct catalysis reactions and maintain activity in a broader pH range.

In particular, the substrate preferred by Aly08 was determined by experimenting with three polymeric substrates (sodium alginate, polyM block and polyG block). Aly08 was found to prefer polyG blocks (1078.2 U/mg) rather than polyM blocks (297.1 U/mg) and native alginate (841.1 U/mg).

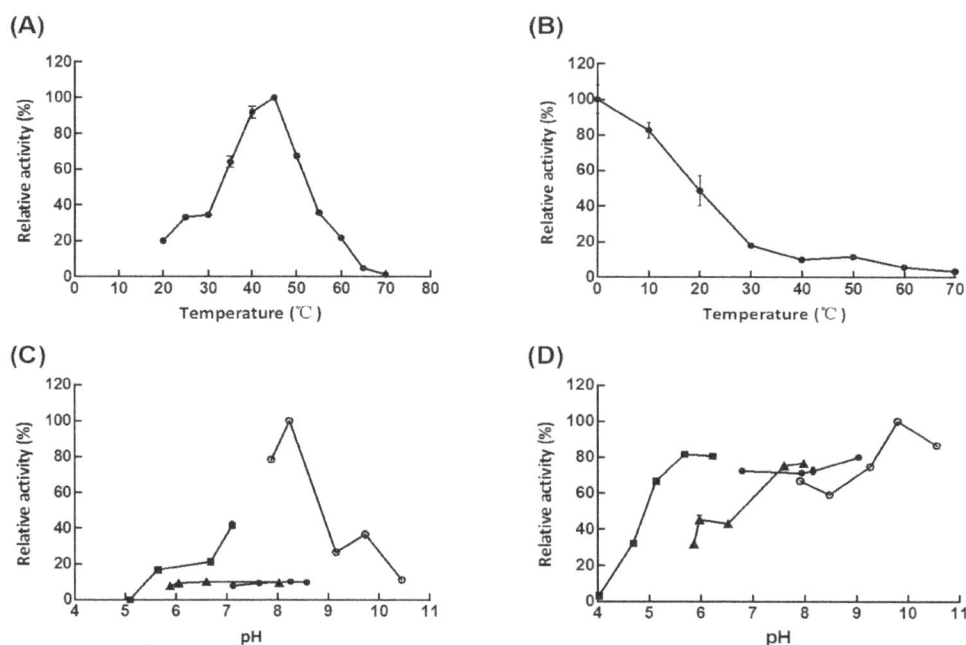

Figure 4. The biochemical characteristics of Aly08. (**A**) The optimal temperature of Aly08. (**B**) The thermal-stability of Aly08. (**C**) Optimal pH for the relative activity of Aly08 was determined in 20 mM Tris-HCl buffer (solid circle), 20 mM phosphate buffer (solid triangle), 20 mM citic-Na$_2$HPO$_4$ (solid square), or 20 mM glycine-NaOH buffer (hollow circle). (**D**) pH stability of Aly08 in 20 mM Tris-HCl buffer (solid circle), 20 mM phosphate buffer (solid triangle), 20 mM citic-Na$_2$HPO$_4$ (solid square), or 20 mM glycine-NaOH buffer (hollow circle).

Moreover, activities of Aly08 were enhanced by NaCl (different concentrations from 10 mM to 3 M), and the activity reached its maximum at 300 mM NaCl, at which point the activity was about eight times higher than the activity in the absence of NaCl (Table 2). Similarly, the activity of AlgM4 from *V. weizhoudaoensis* M0101 was increased about seven times at 1 M NaCl and activated by a concentration range of NaCl at 0–1 M [37]. For these alginates from marine bacteria, a certain level of NaCl concentration is essential for strain survival and enzyme activation. The effects of other metal ions on the activity of Aly08 were also shown in Table 2. Aly08 showed no obvious activated effect in the presence of NH$_4$$^+$, Li$^+$, Zn^{2+}, Ba^{2+}, and Co^{2+}, while Al^{3+} and Ni^{2+} showed no obvious inhibiting effects on relative enzymatic activity. Enzyme activity was activated by divalent ions, such as Ca^{2+} and Mn^{2+}. Different concentrations of KCl had little effect on the activities of Aly08 (Table S2). Interestingly, the enzyme activity of the reaction system containing divalent ions of Ca^{2+} was about twice as high as that of the reaction system without any ions. However, other reagents such as SDS and EDTA showed significant inactivation effects wherein relative activity was reduced to 50% and 55.9%, respectively.

Table 2. Effects of metal ions, EDTA and SDS on the activity of Aly08. Notes: Activity without addition of chemicals was defined as 100%. Data are shown as means ± SD ($n = 3$).

Reagent Added	Concentration (mM)	Relative Activity (%)
None	-	100.00 ± 0.24
	10	382.22 ± 2.64
	50	580.89 ± 4.36
NaCl	300	865.96 ± 26.46
	800	647.33 ± 11.25
	3000	361.97 ± 10.74
SDS	1	50.00 ± 3.32
EDTA	1	55.88 ± 8.60
$Al_2(SO_4)_3$	1	85.33 ± 10.63
KCl	1	99.67 ± 0.86
KCl	100	103.54 ± 0.16
$NiCl_2$	1	97.24 ± 2.20
$(NH_4)_2SO_4$	1	105.56 ± 1.37
$MnSO_4$	1	100.64 ± 1.78
Li_2SO_4	1	103.91 ± 0.93
$ZnCl_2$	1	114.27 ± 2.48
$BaCl_2$	1	120.74 ± 17.14
$CoCl_2$	1	130.86 ± 4.03
$MnCl_2$	1	166.39 ± 5.47
$CaCl_2$	1	281.18 ± 29.18

2.3. Thermo-Tolerance and Heat Recovery of Aly08

When we sought to determine the thermostability of Aly08, we found an interesting phenomenon. After ice-bath, the residual activity of heat-treatment enzymes was always higher than that of enzymes without an ice-bath (Figure 5A,B). After incubation at 30 °C and 40 °C for 1 h, Aly08 retained only 17.9% and 9.9% of its initial activity when directly assayed its activities. However, when the enzyme incubation at 0 °C for 30 min, the residual activity could recover to 43.8% and 39.4% in the same heat treatment condition (Figure 5A). Moreover, even the enzyme was boiled for 5 min, Aly08 was able to recover 78.3% of its initial activity after 30 min incubating in the ice-bath (Figure 5B).

To determine the optimal incubation temperature that contributed to the recovery of activity after boiling for 5 min, the enzyme was incubated for 30 min at various temperatures (0–80 °C). The recovered activity of the enzyme reached levels of 76.3%, 63.9%, and 30.1% after incubation at 0 °C, 10 °C, and 20 °C for 30 min, respectively. When the incubation conditions were above 30 °C, the recovery of activity was measured at less than 10% (Figure 5C).

To further determine the optimal incubation time that contributed to the recovery of activity, the enzyme was incubated at 0 °C for different times. The activity of Aly08 gradually increased with prolonged culture time at 0 °C. Aly08 was rapidly re-activated approximately 56.7% and 71.3% of its activities after incubation at 0 °C for 5 min and 10 min, respectively. After incubation for 20 min, the activity was restored to 77.9%, after which the activity recovery rate began to decrease (Figure 5D).

The thermo-stability experiment indicated that low temperature may contribute to the recovery of Aly08. The thermo-tolerance of Aly08 could promote effective storage and transportation as the inactivated enzyme with heat treatment is able to successfully restore most of its activity after incubation at 0 °C.

(A)

(B)

(C)

(D)

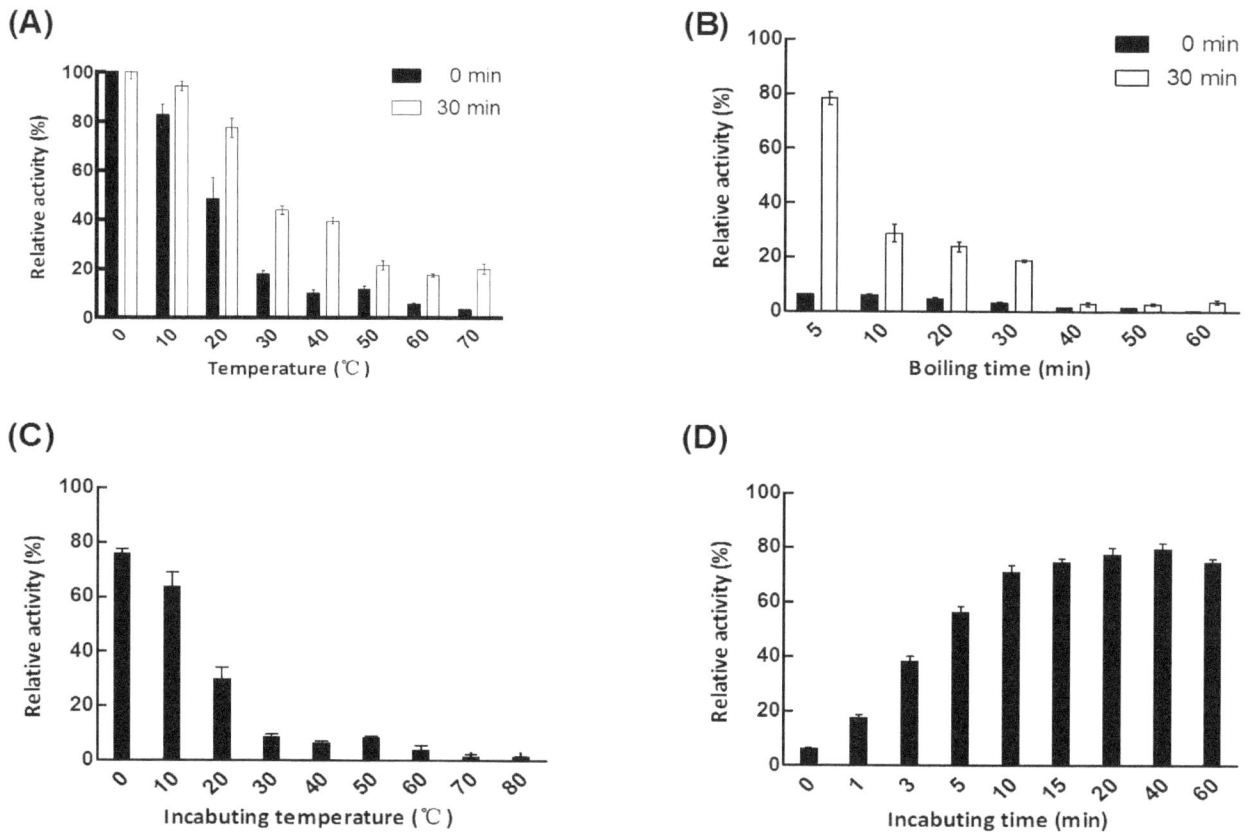

Figure 5. Thermo-tolerance and heat recovery of Aly08. (**A**) The difference of thermostability of enzymes incubation at ice-bath for 0 min (black columnar) and 30 min (white columnar). (**B**) Effects of boiling times on enzyme Aly08. Black and white columns indicate the activity of the heat-inactivated enzyme following ice-bath for 0 min and 30 min, respectively. (**C**) Effects of different incubation temperatures on the activity recovery of Aly08 under 5 min heat-inactivated conditions. (**D**) Effects of incubation time at 0 °C on the activity recovery of heat-inactivated Aly08. The enzyme activity without any treatment was 100%.

2.4. Action Pattern and Final Product Analysis

The action mode of Aly08 was determined by size-exclusion chromatography with a Superdex™ peptide 10/300 column (General Electric Company, Boston, MA, USA) using high-performance liquid chromatography (HPLC) platform (Figure 6). The hydrolysis pattern of Aly08 works as an endo-type because of the rapid depolymerization of substrates, the rise in polydispersity, and the production of intermediate oligosaccharides. Meanwhile, the action mode of Aly08 was further monitored by viscosity analysis (Figure S1). The viscosity of the alginate solution decreased rapidly during the first 5 min following the addition of Aly08 but changed little during subsequent time periods. During the whole observation period, the oligosaccharide content which was tested by A235 increased steadily. It can be further suggested that Aly08 is an endo-type enzyme in accordance with this finding.

The hydrolytic degradations were analyzed by thin-layer chromatography (TLC) method after the alginate was completely degraded (Figure 7A). In the hydrolysis proceeds, there was a gradual decrease of alginate polysaccharide and an accumulation of oligosaccharides with various DPs. And two clear spots of end product (2 h) appeared on the TLC plate, indicating that the migration rate was in good agreement with the alginate disaccharide (DP2) and trisaccharide (DP3) marker. The final degradation product was also determined by negative-ion electrospray ionization mass spectrometry (ESI-MS) (Figure 7B). Two main spectra were 351.1 m/z [ΔDP2-H]$^-$ and 527.2 m/z [ΔDP3-H]$^-$, corresponding to the molecular mass of the unsaturated alginate disaccharides and trisaccharides, respectively [38].

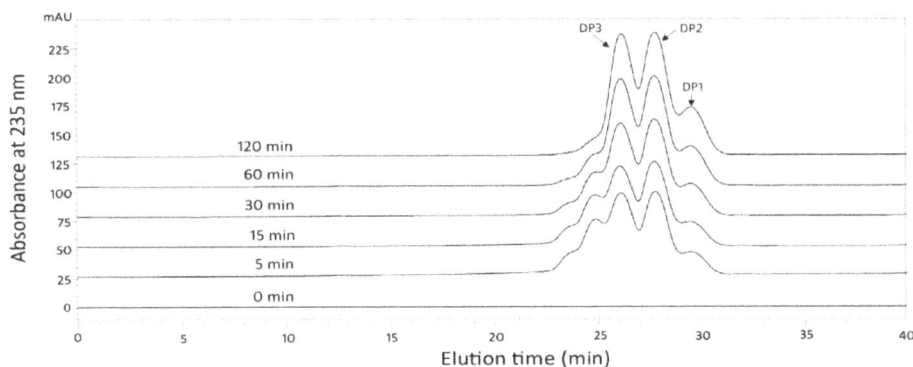

Figure 6. Degradation patterns of Aly08 toward sodium alginate. The elution positions of the unsaturated oligosaccharide product fractions with different degrees of polymerization are shown with arrows: DP1 represents unsaturated monosaccharide, DP2 represents unsaturated disaccharide, DP3 represents unsaturated trisaccharide.

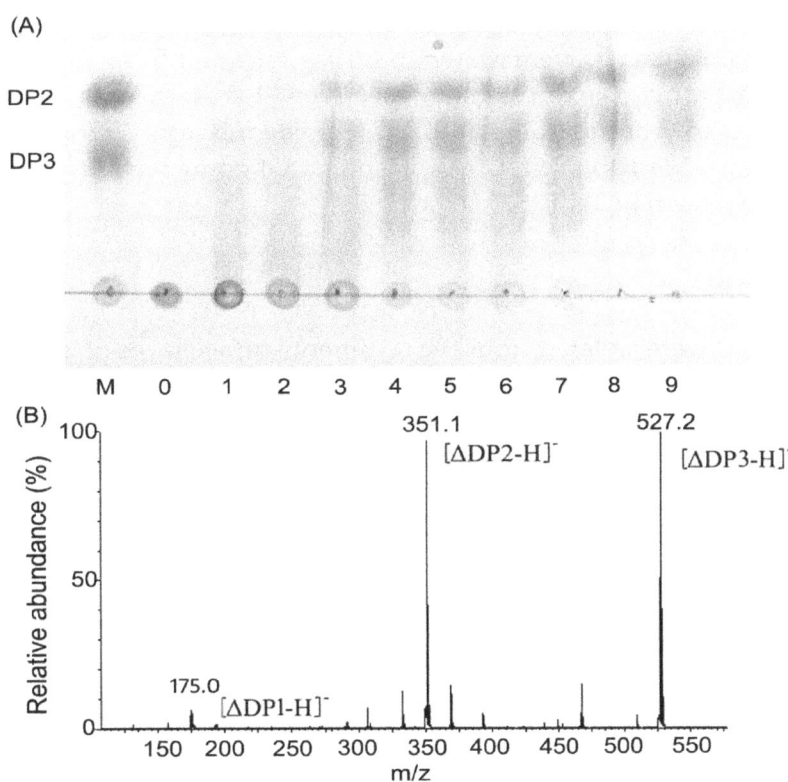

Figure 7. The hydrolytic products of Aly08. (**A**) TLC analysis of the hydrolytic products of Aly08. *Lane M*, standard alginate oligosaccharides (DP2-3); Line 0, alginate; Lane 1–9, hydrolytic products of Aly08 for different times (1, 2, 5, 10, 15, 30, 60, 90, and 120 min) toward 0.3% (*w/v*) high viscosity sodium alginate. DP2 and DP3 indicate alginate disaccharide and trisaccharide, respectively. (**B**) ESI-MS analysis of the end products of Aly08.

Through HPLC, viscosity, TLC and ESI-MS analysis, Aly08 was shown to degrade alginate polymer as an endo-type manner, eventually degrading alginate into disaccharides and trisaccharides. Previous studies have reported that enzymatic oligosaccharide products have a variety of specific biological activities and possess broad potential application prospects in many fields, such as antioxidant activities, regulation of plant root growth, anti-inflammatory activities, and reguation of lipoprotein metabolism [39–42]. The single homogeneous products in the progress of enzymatic production of oligosaccharides is conducive to the oligosaccharide purification and application. Thus far, most of the reported products of alginate lyase are mixtures of DP2–DP5, such as AlyA-OU02 from

V. spiendidus OU02, appear to take disaccharides, trisaccharides, and tetrasaccharides as the main hydrolytic products [28]. Additionally, the final degradation products of the alginate lyase FlAlyA from *Flavobacterium* sp. UMI-01 are DP2-DP5 [34]. Compared with those alginate lyase, the end products of Aly08 are a disaccharide and trisaccharide, which are advantageous for further separation and industrial high-efficiency production. Aly08 may have potential as a tool for the preparation of single homogeneous products of monosaccharides which have wide pharmaceutical applications.

3. Materials and Methods

3.1. Materials

High viscosity sodium alginate (20–50 kDa, 100–260 monosaccharide in polymer, M/G ratio: 1.66) and low viscosity sodium alginate (1–5 kDa, 5–26 monosaccharide in polymer the, M/G ratio: 1.66) was purchased from Bright Moon Seaweed Group (Qingdao, China), PolyM and PolyG blocks (purity: 95%) were purchased from Qingdao BZ Oligo Biotech Co., Ltd. (Qingdao, China). Standard alginate disaccharide and trisaccharide were also purchased from Qingdao BZ Oligo Biotech Co., Ltd. Standard monosaccharide (glucuronic acid) was purchased from Sigma. In addition, *E. coli* strains DH5α and BL21 (DE3) (Solarbio, Beijing, China) were grown in Luria–Bertani (LB) medium and used for plasmid construction and as a host for gene expression, respectively. LB broth supplemented with ampicillin (50 μg/mL) was used to grow both strains at 37 °C. The expression vectors used for gene cloning were pET-22b(+) (Novagen, Madison, WI, USA) plasmids. Oligonucleotides used for the cloning and expression of *aly08* were shown in Table S1.

3.2. Strains and Nucleotides

The sea mud samples were isolated from the sediment surface layer of Yellow Sea bottom (depth 36 m, E 120.13° N 35.76°, collected in May, 2017) and then immersed, diluted and spread on alginate sole-carbon selective medium plates [2 g $(NH_4)_2SO_4$, 30 g NaCl, 0.1 g $MgSO_4 \cdot 7H_2O$, 7 g K_2HPO_4, 3 g KH_2PO_4, 0.1 g $FeSO_4$, 5 g sodium alginate, dissolved in 1 L distilled water with 10 g agar, pH 7.0)]. At least 100 strains were isolated from the detectable colonies after incubation at 25 °C for 4 days, and then the strains were inoculated into agar-free selective medium for the purpose of identifying the activities of alginate lyases. A higher activity strain was screened out, 27F and 1492R primers were used to amplify the 16S rDNA gene of this strain. The 16S rDNA gene sequence of this strain was blasted to obtain the closely related sequence using the BLASTn algorithm program, the sequence was aligned with its closely related sequences using MEGA 6.0. According to the sequence alignment, this strain was classified as *Vibrio* sp. SY01. This strain has been preserved at the China Center for Type Culture Collection (CCTCC) under no. M2018769.

3.3. Sequence Analysis

In our previous study, genomic sequence analysis of *Vibrio*. sp SY01 showed a putative alginate lyase-encoding gene, *aly08*. The gene was cloned from the genome of strain SY01 and deposited in the Genbank database (accession number MH791447). The open reading frame (ORF) was identified with the program ORF finder (https://www.ncbi.nlm.nih.gov/orffinder/) and the signal peptide was analyzed using the SignalP 4.0 server (http://www.cbs.dtu.dk/services/SignalP/). The domain analysis of *aly08*, and its family analysis, is based on a comparison with the CDD (https://www.ncbi.nlm.nih.gov/cdd). The theoretically pI and Mw of *aly08* was determined by pI/Mw Tool (http://web.expasy.org/compute_ pi/). Afterwards, the BLAST algorithm program on NCBI was used to search for similar sequences of *aly08*. Multiple sequence alignment was constructed using ClustalX 2.1 (National Center for Biotechnology Information, Bethesda, MD, USA), and the phylogenetic tree was created using the bootstrapping neighbor-joining method of MEGA 6.0 (Center for Evolutionary Medicine and Informatics, The Biodesign Institute, Tempe, AZ, USA).

3.4. Heterologous Expression and Purification of Recombinant Aly08

In order to express Aly08, the primers (PyAly08-F and PyAly08-R) were used to amplify the genomic DNA of *Vibrio* sp. SY01 without a signal sequence or stop codon. The *aly08* gene was then ligated into the expression vector pET-22b(+) with recognition sites *Nde* I and *Xho* I. In addition, the recombinant plasmid pET22b-Aly08 with a C-terminal 6 × His-tag was transformed into the *E. coli* BL21 (DE3) and grown on media (50 µg/mL ampicillin). Single colonies of *E. coli* BL21-pET22b-Aly08 were picked and cultured in LB medium (50 µg/mL ampicillin) at 37 °C with shaking at 200 rpm until OD_{600} reached 0.6–0.8. In order to induce the expression of protein, and the incubation was continued for 20 h at 20 °C (containing 0.1 mM isopropyl β-D-thiogalactoside (IPTG)). The cultured supernatant was harvested by high-speed refrigerated centrifuge (Hitachi, Tokyo, Japan) system (12,000 rpm, 5 min, 4 °C) and loaded onto a Ni-NTA sepharose column (GE Healthcare, Little Chalfont, Buckinghamshire, UK) using the AKTA150 automatic intelligent protein purification system (GE Healthcare, Little Chalfont, Buckinghamshire, UK) which had been equilibrated with wash buffer (20 mM phosphate buffer (pH 7.6), 500 mM NaCl). The Ni-NTA sepharose column was then eluted with a linear gradient of imidazole (25-500 mM imidazole, 20 mM phosphate buffer, 500 mM NaCl, pH7.6) in order to collect the active fractions. The active fraction was further analyzed by 12% SDS-PAGE, and the PageRuler Prest Protein Ladder (Thermo Scientific, USA) was used as a protein standard marker. Afterwards, the protein concentration of purified Aly08 was determined by a BCA protein assay kit (Beyotime Biotechnology, Shanghai China), bovine serum albumin (BSA) was used as a standard.

3.5. Alginate Lyase Activity Assay

Absorption at 235 nm (A235) was used to measure the activity of Aly08 as previously described [43–45]. The appropriately diluted enzyme (100 µL) was mixed with 900 µL of 0.3% (*w/v*) sodium alginate solution (10 mM glycine-NaOH buffer and 100 mM NaCl, pH 8.35). Then, the reaction system was incubated at 45 °C for 10 min and terminated by boiling for 10 min, and its absorbance was measured on a NanoPhotometer Pearl-360 spectrophotometer (IMPLEN, Munich, Germany). Alginate lyase activity was determined by increasing A235 as the production of unsaturated double bonds occurred as the alginate lyase cleaved glycosidic bonds at the non-reducing end of the polymer chain. One unit (U) of enzyme activity was defined as the amount of enzyme required to increase A235 by 0.1 per minute, under the above conditions.

3.6. Biochemical Characterization of the Recombinant Enzyme

The enzyme and substrate was incubated under 10 mM glycine-NaOH buffer (pH 8.35) at various temperatures (10–70 °C) to obtain the optimal temperature for Aly08. The thermal stability of Aly08 was then assayed by measuring its activity after pre-incubation at various temperatures (10–70 °C) for 1 h. The influence of pH values on Aly08 was calculated by measuring the residual activities in different buffers. The following buffers were used: 50 mM Na_2HPO_4-citric acid (pH 4.6–7.0), Tris-HCl (pH 7.6–8.6), glycine-NaOH (pH 8.6–10.6), and Na_2HPO_4-NaH_2PO_4 (pH 6.6–7.6). The highest activities represent 100% enzyme activity. The pH stability of Aly08 was determined by measuring the residual activity after incubating the purified enzyme with various pH buffers (pH 4.6–10.6) at 4 °C for 12 h. Substrate solution (10 mM glycine-NaOH buffer and 100 mM NaCl, pH 8.35) with three different substrates [0.3% (*w/v*) sodium alginate, polyM block and polyG block] were prepared and used for assaying the activities of the purified enzyme Aly08 in order to determine the preferred substrate of Aly08. Afterwards, based on the protein concentration measured by the BCA protein kit, the specific activity of Aly08 with the different substrates was calculated. To measure the effects of chemical compounds and metal ions on enzymatic activity, different metal ions and chemical compounds were added to the reaction system with a final concentration of 1 mM. A reaction mixture containing no metal ion or chemical compounds was used as a control. All reactions were performed in triplicate and the reaction parameters were expressed as the mean ± standard deviation.

3.7. Thermo-Tolerance Properties of Aly08

Purified Aly08 was placed at different temperatures for 1 h and then divided into two parts. One part was directly measured for its activity, while the other part was measured for its activity following incubation in an ice-bath for 30 min. In order to further observe whether the difference in temperature stability is related to boiling time, different boiling times (5, 10, 20, 30, 40, 50, and 60 min) were selected to evaluate the residual activity of the purified enzyme. After that the heat-treatment group was divided into two parts, one part was directly measured for its activity, while the other part was measured for its activity following incubation in an ice-bath for 30 min. To determine the effects of different temperatures on the activity recovery of heat-inactivated Aly08, the enzyme was immediately incubated for 30 min between 0 °C and 80 °C after boiling 5 min. Moreover, the enzyme was boiled for 5 min and further incubated at 0 °C for different times (0–60 min) to evaluate the effect of the time of low temperature treatment on the recovery of enzyme activity after it was heat treated.

3.8. Analysis of Reaction Products and Hydrolytic Pattern

Low viscosity sodium alginate was used for analysis the reaction product of Aly08. In order to examine the hydrolysis pattern of Aly08, the products from different incubation times (0, 5, 15, 30, 60, and 120 min) were monitored at A235 using gel filtration chromatography with a Superdex^{TM} peptide 10/300 GL column with 0.2 M NH_4HCO_3 (flow rate: 0.6 mL/min) as an eluent on HPLC platform (LC-20A, Shimadzu, Japan) [46]. The mode of action was also analyzed using an Ostwald viscometer (No.1; Shibata Scientific Technology) with high viscosity sodium alginate as substrate. The equal component products (0.5 mL), which were degraded by Aly08 at 1, 5, 10, 15, and 30 min, were removed to characterize the viscosity and degradation products.

The hydrolytic degradation products were monitored using TLC method, wherein a reaction system containing 0.3% of high viscosity sodium alginate and Aly08 was constructed and the samples selected at different times (1, 2, 5, 10, 15, 30, 60, 90, and 120 min). The reaction products were analyzed using a TLC plate (TLC silica gel 60 F254, Merck KGaA, 64271 Darmstadt, Germany) with butanol/acetic-acid/water (2:1:1, by vol.) and color-developed with sulfuric acid/ethanol reagent (1:4, by vol.) after heating the TLC plate at 80 °C for 30 min. ESI-MS system (Thermo Fisher Scientific^{TM} Q Exactive^{TM} Hybrid Quadrupole-Orbitrap^{TM}, Waltham, MA, USA) was employed to further investigate the composition and degree of polymerization (DP) of the end products.

4. Conclusions

In conclusion, we purified and characterized a new alginate lyase, Aly08, from marine bacterium *Vibrio* sp. SY01. Its special characteristics (such as: thermo-tolerance and pH stability) make Aly08 a superior candidate for industrial applications. Further analysis will focus on analyzing the three-dimensional structure of Aly08 and exploring its molecular mechanisms.

Author Contributions: S.L. and Y.H. designed the experiments. Y.W., L.B., Y.R., Q.H. and Z.Y. conducted the experiments. S.L., X.C. and R.Y. analyzed the data. Y.W., S.L. and Y.H. wrote the main manuscript. All authors reviewed the manuscript.

References

1.   Scieszka, S.; Klewicka, E. Algae in food: A general review. *Crit. Rev. Food Sci. Nutr.* **2018**, 1–10. [CrossRef] [PubMed]

2.   Senturk Parreidt, T.; Muller, K.; Schmid, M. Alginate-based edible films and coatings for food packaging applications. *Foods* **2018**, *7*, 710. [CrossRef] [PubMed]

3. Wargacki, A.J.; Leonard, E.; Win, M.N.; Regitsky, D.D.; Santos, C.N.; Kim, P.B.; Cooper, S.R.; Raisner, R.M.; Herman, A.; Sivitz, A.B.; et al. An engineered microbial platform for direct biofuel production from brown macroalgae. *Science* **2012**, *335*, 308–313. [CrossRef] [PubMed]

4. Pawar, S.N.; Edgar, K.J. Alginate derivatization: A review of chemistry, properties and applications. *Biomaterials* **2012**, *33*, 3279–3305. [CrossRef]

5. Yamasaki, M.; Moriwaki, S.; Miyake, O.; Hashimoto, W.; Murata, K.; Mikami, B. Structure and function of a hypothetical *Pseudomonas aeruginosa* protein PA1167 classified into family PL-7: A novel alginate lyase with a beta-sandwich fold. *J. Biol. Chem.* **2004**, *279*, 31863–31872. [CrossRef] [PubMed]

6. Ertesvag, H. Alginate-modifying enzymes: Biological roles and biotechnological uses. *Front. Microbiol.* **2015**, *6*, 523.

7. Wong, T.Y.; Preston, L.A.; Schiller, N.L. Alginate lyase: Review of major sources and enzyme characteristics, structure-function analysis, biological roles, and applications. *Annu. Rev. Microbiol.* **2000**, *54*, 289–340. [CrossRef] [PubMed]

8. Li, S.Y.; Wang, Z.P.; Wang, L.N.; Peng, J.X.; Wang, Y.N.; Han, Y.T.; Zhao, S.F. Combined enzymatic hydrolysis and selective fermentation for green production of alginate oligosaccharides from Laminaria japonica. *Bioresour. Technol.* **2019**, *281*, 84–89. [CrossRef]

9. Sharma, S.; Horn, S.J. Enzymatic saccharification of brown seaweed for production of fermentable sugars. *Bioresour. Technol.* **2016**, *213*, 155–161. [CrossRef]

10. Kim, H.T.; Chung, J.H.; Wang, D.; Lee, J.; Woo, H.C.; Choi, I.G.; Kim, K.H. Depolymerization of alginate into a monomeric sugar acid using Alg17C, an exo-oligoalginate lyase cloned from *Saccharophagus degradans* 2-40. *Appl. Microbiol. Biotechnol.* **2012**, *93*, 2233–2239. [CrossRef] [PubMed]

11. Xu, F.; Dong, F.; Wang, P.; Cao, H.Y.; Li, C.Y.; Li, P.Y.; Pang, X.H.; Zhang, Y.Z.; Chen, X.L. Novel molecular insights into the catalytic mechanism of marine bacterial alginate lyase AlyGC from polysaccharide lyase Family 6. *J. Biol. Chem.* **2017**, *292*, 4457–4468. [CrossRef] [PubMed]

12. Garron, M.L.; Cygler, M. Uronic polysaccharide degrading enzymes. *Curr. Opin. Struct. Biol.* **2014**, *28*, 87–95. [CrossRef] [PubMed]

13. Kam, N.; Park, Y.J.; Lee, E.Y.; Kim, H.S. Molecular identification of a polyM-specific alginate lyase from *Pseudomonas* sp. strain KS-408 for degradation of glycosidic linkages between two mannuronates or mannuronate and guluronate in alginate. *Can. J. Microbiol.* **2011**, *57*, 1032–1041. [CrossRef] [PubMed]

14. Yang, M.; Yu, Y.; Yang, S.; Shi, X.; Mou, H.; Li, L. Expression and characterization of a new PolyG-specific alginate lyase from marine bacterium *Microbulbifer* sp. Q7. *Front. Microbiol.* **2018**, *9*, 2894. [CrossRef] [PubMed]

15. Chen, X.L.; Dong, S.; Xu, F.; Dong, F.; Li, P.Y.; Zhang, X.Y.; Zhou, B.C.; Zhang, Y.Z.; Xie, B.B. Characterization of a new cold-adapted and salt-activated polysaccharide lyase family 7 alginate lyase from *Pseudoalteromonas* sp. SM0524. *Front. Microbiol.* **2016**, *7*, 1120. [CrossRef]

16. Inoue, A.; Anraku, M.; Nakagawa, S.; Ojima, T. Discovery of a novel alginate lyase from *Nitratiruptor* sp. SB155-2 thriving at deep-sea hydrothermal vents and identification of the residues responsible for its heat stability. *J. Biol. Chem.* **2016**, *291*, 15551–15563. [CrossRef]

17. Jagtap, S.S.; Hehemann, J.H.; Polz, M.F.; Lee, J.K.; Zhao, H. Comparative biochemical characterization of three exolytic oligoalginate lyases from *Vibrio splendidus* reveals complementary substrate scope, temperature, and pH adaptations. *Appl. Environ. Microbiol.* **2014**, *80*, 4207–4214. [CrossRef]

18. Bonugli-Santos, R.C.; Dos Santos Vasconcelos, M.R.; Passarini, M.R.; Vieira, G.A.; Lopes, V.C.; Mainardi, P.H.; Dos Santos, J.A.; de Azevedo Duarte, L.; Otero, I.V.; da Silva Yoshida, A.M.; et al. Marine-derived fungi: Diversity of enzymes and biotechnological applications. *Front. Microbiol.* **2015**, *6*, 269. [CrossRef]

19. Li, S.Y.; Yang, X.M.; Bao, M.M.; Wu, Y.; Yu, W.G.; Han, F. Family 13 carbohydrate-binding module of alginate lyase from *Agarivorans* sp. L11 enhances its catalytic efficiency and thermostability, and alters its substrate preference and product distribution. *FEMS Microbiol. Lett.* **2015**, *362*. [CrossRef]

20. Zhu, B.W.; Sun, Y.; Ni, F.; Ning, L.M.; Yao, Z. Characterization of a new endo-type alginate lyase from *Vibrio* sp. NJU-03. *Int. J. Biol. Macromol.* **2018**, *108*, 1140–1147. [CrossRef]

21. Ogura, K.; Yamasaki, M.; Mikami, B.; Hashimoto, W.; Murata, K. Substrate recognition by family 7 alginate lyase from *Sphingomonas* sp. A1. *J. Mol. Biol.* **2008**, *380*, 373–385. [CrossRef]

22. Schiller, N.L.; Monday, S.R.; Boyd, C.M.; Keen, N.T.; Ohman, D.E. Characterization of the *Pseudomonas aeruginosa* alginate lyase gene (algL): Cloning, sequencing, and expression in *Escherichia coli*. *J. Bacteriol.* **1993**, *175*, 4780–4789. [CrossRef]

23. Thomas, F.; Lundqvist, L.C.; Jam, M.; Jeudy, A.; Barbeyron, T.; Sandstrom, C.; Michel, G.; Czjzek, M. Comparative characterization of two marine alginate lyases from *Zobellia galactanivorans* reveals distinct modes of action and exquisite adaptation to their natural substrate. *J. Biol. Chem.* **2013**, *288*, 23021–23037. [CrossRef]

24. Swift, S.M.; Hudgens, J.W.; Heselpoth, R.D.; Bales, P.M.; Nelson, D.C. Characterization of AlgMsp, an alginate lyase from *Microbulbifer* sp. 6532A. *PLoS ONE* **2014**, *9*, e112939. [CrossRef]

25. Yoon, H.J.; Hashimoto, W.; Miyake, O.; Okamoto, M.; Mikami, B.; Murata, K. Overexpression in *Escherichia coli*, purification, and characterization of *Sphingomonas* sp. A1 alginate lyases. *Protein Expr. Purif.* **2000**, *19*, 84–90. [CrossRef] [PubMed]

26. Huang, L.S.; Zhou, J.G.; Li, X.; Peng, Q.; Lu, H.; Du, Y.G. Characterization of a new alginate lyase from newly isolated *Flavobacterium* sp. S20. *J. Ind. Microbiol. Biotechnol.* **2013**, *40*, 113–122. [CrossRef] [PubMed]

27. Uchimura, K.; Miyazaki, M.; Nogi, Y.; Kobayashi, T.; Horikoshi, K. Cloning and sequencing of alginate lyase genes from deep-sea strains of *Vibrio* and *Agarivorans* and characterization of a new *Vibrio* enzyme. *Mar. Biotechnol.* **2010**, *12*, 526–533. [CrossRef]

28. Zhuang, J.J.; Zhang, K.K.; Liu, X.H.; Liu, W.Z.; Lyu, Q.Q.; Ji, A.G. Characterization of a novel polyM-preferred alginate lyase from marine *Vibrio splendidus* OU02. *Mar. Drugs* **2018**, *16*, 295. [CrossRef]

29. Miyake, O.; Ochiai, A.; Hashimoto, W.; Murata, K. Origin and diversity of alginate lyases of families PL-5 and -7 in *Sphingomonas* sp. strain A1. *J. Bacteriol.* **2004**, *186*, 2891–2896. [CrossRef] [PubMed]

30. Peng, C.N.; Wang, Q.B.; Lu, D.R.; Han, W.J.; Li, F.C. A novel bifunctional endolytic alginate lyase with variable alginate-degrading modes and versatile monosaccharide-producing properties. *Front. Microbiol.* **2018**, *9*, 167. [CrossRef] [PubMed]

31. Duan, G.F.; Han, F.; Yu, W.G. Cloning, sequence analysis, and expression of gene alyPI encoding an alginate lyase from marine bacterium *Pseudoalteromonas* sp. CY24. *Can. J. Microbiol.* **2009**, *55*, 1113–1118. [CrossRef] [PubMed]

32. Brown, B.J.; Preston, J.F.; Ingram, L.O. Cloning of alginate lyase gene (alxM) and expression in *Escherichia coli*. *Appl. Environ. Microbiol.* **1991**, *57*, 1870–1872. [PubMed]

33. Kobayashi, T.; Uchimura, K.; Miyazaki, M.; Nogi, Y.; Horikoshi, K. A new high-alkaline alginate lyase from a deep-sea bacterium *Agarivorans* sp. *Extremophiles* **2009**, *13*, 121–129. [CrossRef] [PubMed]

34. Inoue, A.; Takadono, K.; Nishiyama, R.; Tajima, K.; Kobayashi, T.; Ojima, T. Characterization of an alginate lyase, FlAlyA, from *Flavobacterium* sp. strain UMI-01 and its expression in *Escherichia coli*. *Mar. Drugs* **2014**, *12*, 4693–4712. [CrossRef]

35. Zhu, X.Y.; Li, X.Q.; Shi, H.; Zhou, J.; Tan, Z.B.; Yuan, M.D.; Yao, P.; Liu, X.Y. Characterization of a novel alginate lyase from marine bacterium *Vibrio furnissii* H1. *Mar. Drugs* **2018**, *16*, 30. [CrossRef]

36. Zhu, B.W.; Ni, F.; Ning, L.M.; Sun, Y.; Yao, Z. Cloning and characterization of a new pH-stable alginate lyase with high salt tolerance from marine *Vibrio* sp. NJ-04. *Int. J. Biol. Macromol.* **2018**, *115*, 1063–1070. [CrossRef] [PubMed]

37. Huang, G.Y.; Wang, Q.Z.; Lu, M.Q.; Xu, C.; Li, F.; Zhang, R.C.; Liao, W.; Huang, S.S. AlgM4: A new salt-activated alginate lyase of the PL7 family with endolytic activity. *Mar. Drugs* **2018**, *16*, 120. [CrossRef] [PubMed]

38. Zhu, B.W.; Ni, F.; Ning, L.M.; Yao, Z. Elucidation of degrading pattern and substrate recognition of a novel bifunctional alginate lyase from *Flammeovirga* sp. NJ-04 and its use for preparation alginate oligosaccharides. *Biotechnol. Biofuels* **2019**, *12*, 13. [CrossRef] [PubMed]

39. Yang, J.H.; Bang, M.A.; Jang, C.H.; Jo, G.H.; Jung, S.K.; Ki, S.H. Alginate oligosaccharide enhances LDL uptake via regulation of LDLR and PCSK9 expression. *J. Nutr. Biochem.* **2015**, *26*, 1393–1400. [CrossRef]

40. Falkeborg, M.; Cheong, L.Z.; Gianfico, C.; Sztukiel, K.M.; Kristensen, K.; Glasius, M.; Xu, X.; Guo, Z. Alginate oligosaccharides: Enzymatic preparation and antioxidant property evaluation. *Food Chem.* **2014**, *164*, 185–194. [CrossRef]

41. Zhang, Y.H.; Yin, H.; Zhao, X.M.; Wang, W.X.; Du, Y.G.; He, A.; Sun, K.G. The promoting effects of alginate oligosaccharides on root development in Oryza sativa L. mediated by auxin signaling. *Carbohydr. Polym.* **2014**, *113*, 446–454. [CrossRef] [PubMed]

42. Qu, Y.; Wang, Z.M.; Zhou, H.H.; Kang, M.Y.; Dong, R.P.; Zhao, J.W. Oligosaccharide nanomedicine of alginate sodium improves therapeutic results of posterior lumbar interbody fusion with cages for degenerative lumbar disease in osteoporosis patients by downregulating serum miR-155. *Int. J. Nanomed.* **2017**, *12*, 8459–8469. [CrossRef] [PubMed]

43. Qin, H.M.; Miyakawa, T.; Inoue, A.; Nishiyama, R.; Nakamura, A.; Asano, A.; Ojima, T.; Tanokura, M. Structural basis for controlling the enzymatic properties of polymannuronate preferred alginate lyase FlAlyA from the PL-7 family. *Chem Commun. (Camb)* **2018**, *54*, 555–558. [CrossRef]

44. Doi, H.; Tokura, Y.; Mori, Y.; Mori, K.; Asakura, Y.; Usuda, Y.; Fukuda, H.; Chinen, A. Identification of enzymes responsible for extracellular alginate depolymerization and alginate metabolism in *Vibrio algivorus*. *Appl. Microbial. Biotechnol.* **2017**, *101*, 1581–1592. [CrossRef] [PubMed]

45. Li, S.Y.; Wang, L.N.; Hao, J.H.; Xing, M.X.; Sun, J.J.; Sun, M. Purification and characterization of a new alginate lyase from marine bacterium *Vibrio* sp. SY08. *Mar. Drugs* **2016**, *15*, 1. [CrossRef] [PubMed]

46. Chen, P.; Zhu, Y.M.; Men, Y.; Zeng, Y.; Sun, Y.X. Purification and characterization of a novel alginate lyase from the marine bacterium *Bacillus* sp. Alg07. *Mar. Drugs* **2018**, *16*, 86. [CrossRef] [PubMed]

PERMISSIONS

LIST OF CONTRIBUTORS

Annarita Poli, Ilaria Finore, Gennaro Roberto Abbamondi, Barbara Nicolaus and Licia Lama
Institute of Biomolecular Chemistry, National Research Council of Italy, Via Campi Flegrei 34, 80078 Pozzuoli, Naples, Italy

Paola Di Donato
Institute of Biomolecular Chemistry, National Research Council of Italy, Via Campi Flegrei 34, 80078 Pozzuoli, Naples, Italy
Department of Science and Technology, University of Naples "Parthenope", Centro Direzionale Isola C4, 80143 Naples, Italy

Andrea Buono
Department of Engineering, University of Naples "Parthenope", Centro Direzionale Isola C4, 80143 Naples, Italy

Dwi Yuli Pujiastuti, Muhamad Nur Ghoyatul Amin and Mochammad Amin Alamsjah
Department of Marine, Faculty of Fisheries and Marine, Universitas Airlangga, Surabaya 60115, Indonesia

Jue-Liang Hsu
Department of Biological Science and Technology, National Pingtung University of Science and Technology, Pingtung 91201, Taiwan
Research Center for Austronesian Medicine and Agriculture, National Pingtung University of Science and Technology, Pingtung 91201, Taiwan

Annick Turbé-Doan, Emmanuel Bertrand, Craig B. Faulds, Anne Lomascolo, Giuliano Sciara and Eric Record
Biodiversité et Biotechnologie Fongiques, Aix-Marseille Université, INRAE, UMR1163 Marseille, France

Tahar Mechichi
Laboratoire de Biochimie et de Génie Enzymatique des Lipases, Ecole Nationale d'Ingénieurs de Sfax, Université de Sfax, Sfax 3029, Tunisia

Wissal Ben Ali and Amal Ben Ayed
Biodiversité et Biotechnologie Fongiques, Aix-Marseille Université, INRAE, UMR1163 Marseille, France
Laboratoire de Biochimie et de Génie Enzymatique des Lipases, Ecole Nationale d'Ingénieurs de Sfax, Université de Sfax, Sfax 3029, Tunisia

Yann Mathieu
Michael Smith Laboratories, University of British Columbia, Vancouver, BC V6T 1Z4, Canada

Guiyuan Huang, Qiaozhen Wang, Mingqian Lu, Chao Xu, Fei Li, Rongcan Zhang and Shushi Huang
Guangxi Key Laboratory of Marine Natural Products and Combinatorial Biosynethesis Chemistry, Guangxi Academy of Sciences, Nanning 530007, China

Wei Liao
Guangxi Key Laboratory of Marine Natural Products and Combinatorial Biosynethesis Chemistry, Guangxi Academy of Sciences, Nanning 530007, China
The Food and Biotechnology, Guangxi Vocational and Technical College, Nanning 530226, China

Ryuji Nishiyama, Akira Inoue and Takao Ojima
Laboratory of Marine Biotechnology and Microbiology, Faculty of Fisheries Sciences, Hokkaido University, Hakodate, Hokkaido 041-8611, Japan

Linna Wang and Yuejun Wang
Key Laboratory of Sustainable Development of Marine Fisheries, Ministry of Agriculture, Yellow Sea Fisheries Research Institute, Chinese Academy of Fishery Sciences, 106 Nanjing Road, Qingdao 266071, China

Ximing Xu
Institute of Bioinformatics and Medical Engineering, School of Electrical and Information Engineering, Jiangsu University of Technology, Changzhou 213000, China

Shengxiang Lin
Laboratory of Oncology and Molecular Endocrinology, CHUL Research Center (CHUQ) and Laval University, 2705 Boulevard Laurier, Ste-Foy, Ville de Québec, QC G1V 4G2, Canada

Mi Sun
Key Laboratory of Sustainable Development of Marine Fisheries, Ministry of Agriculture, Yellow Sea Fisheries Research Institute, Chinese Academy of Fishery Sciences, 106 Nanjing Road, Qingdao 266071, China
Laboratory for Marine Drugs and Bioproducts, Qingdao National Laboratory for Marine Science and Technology, Qingdao 266237, China

Yatong Wang, Yanhua Hou, Yifan Wang, Lu Zheng, Xianlei Xu, Kang Pan, Rongqi Li and Quanfu Wang
School of Marine Science and Technology, Harbin Institute of Technology, Weihai 264209, China

Giorgio Maria Vingiani, Adrianna Ianora and Chiara Lauritano
Marine Biotechnology Department, Stazione Zoologica Anton Dohrn, CAP80121 (NA) Villa Comunale, Italy

Pasquale De Luca
Research Infrastructure for Marine Biological Resources Department, Stazione Zoologica Anton Dohrn, CAP80121 (NA) Villa Comunale, Italy

Alan D.W. Dobson
School of Microbiology, University College Cork, College Road, T12 YN60 Cork, Ireland
Environmental Research Institute, University College Cork, Lee Road, T23XE10 Cork, Ireland

Jingjing Sun, Wei Wang, Junzhong Liu and Fangqun Dai
Key Laboratory of Sustainable Development of Polar Fishery, Ministry of Agriculture and Rural Affairs, Yellow Sea Fisheries Research Institute, Chinese Academy of Fishery Sciences, Qingdao 266071, China
Laboratory for Marine Drugs and Bioproducts, Laboratory for Marine Fisheries Science and Food Production Processes, Qingdao National Laboratory for Marine Science and Technology, Qingdao 266071, China

Congyu Yao
Key Laboratory of Sustainable Development of Polar Fishery, Ministry of Agriculture and Rural Affairs, Yellow Sea Fisheries Research Institute, Chinese Academy of Fishery Sciences, Qingdao 266071, China
College of Food Science and Technology, Shanghai Ocean University, Shanghai 201306, China

Zhiwei Zhuang
New Hope Liuhe Co. Ltd., Qingdao 266071, China

Jianhua Hao
Key Laboratory of Sustainable Development of Polar Fishery, Ministry of Agriculture and Rural Affairs, Yellow Sea Fisheries Research Institute, Chinese Academy of Fishery Sciences, Qingdao 266071, China
Laboratory for Marine Drugs and Bioproducts, Laboratory for Marine Fisheries Science and Food Production Processes, Qingdao National Laboratory for Marine Science and Technology, Qingdao 266071, China

Jiangsu Collaborative Innovation Center for Exploitation and Utilization of Marine Biological Resource, Lianyungang 222005, China

Haiyan Jin
Graduate School of Environmental and Human Science, Meijo University, Tempaku, Nagoya 468-8503, Japan

Yoshiko Hiraoka, Yurie Okuma, Elisabete Hiromi Hashimoto, Miki Kurita and Andrea Roxanne J. Anas
Faculty of Pharmacy, Meijo University, Tempaku, Nagoya 468-8503, Japan

Ken-Ichi Harada
Graduate School of Environmental and Human Science, Meijo University, Tempaku, Nagoya 468-8503, Japan
Faculty of Pharmacy, Meijo University, Tempaku, Nagoya 468-8503, Japan

Hitoshi Uemura and Kiyomi Tsuji
Kanagawa Prefectural Institute of Public Health, Shimomachiya, Chigasaki, Kanagawa 253-0087, Japan

Benwei Zhu, Fu Hu, Heng Yuan, Yun Sun and Zhong Yao
College of Food Science and Light Industry, Nanjing Tech University, Nanjing 211816, China

Haibin Zhang
Institute of Deep-Sea Science and Engineering, Chinese Academy of Sciences, Sanya 572000, China

Yanan Li and Xue Kong
Institute of Deep-Sea Science and Engineering, Chinese Academy of Sciences, Sanya 572000, China
College of Earth and Planetary Sciences, University of Chinese Academy of Sciences, Beijing 100039, China

Yueming Zhu, Yan Men, Yan Zeng and Yuanxia Sun
National Engineering Laboratory for Industrial Enzymes, Tianjin Institute of Industrial Biotechnology, Chinese Academy of Sciences, Tianjin 300308, China

Peng Chen
National Engineering Laboratory for Industrial Enzymes, Tianjin Institute of Industrial Biotechnology, Chinese Academy of Sciences, Tianjin 300308, China
University of Chinese Academy of Sciences, Beijing 100049, China

Yong Zhao, Wen-Xia Wang and Heng Yin
Liaoning Provincial Key Laboratory of Carbohydrates, Dalian Institute of Chemical Physics, Chinese Academy of Sciences, Dalian 116023, China

Hai-Hua Cong
College of Food Science and Engineering, Dalian Ocean University, Dalian 116023, China

Xian-Yu Zhu
Liaoning Provincial Key Laboratory of Carbohydrates, Dalian Institute of Chemical Physics, Chinese Academy of Sciences, Dalian 116023, China
College of Food Science and Engineering, Dalian Ocean University, Dalian 116023, China

Huai-Dong Zhang
Liaoning Provincial Key Laboratory of Carbohydrates, Dalian Institute of Chemical Physics, Chinese Academy of Sciences, Dalian 116023, China
Engineering Research Center of Industrial Microbiology, Ministry of Education; College of Life Sciences, Fujian Normal University, Fuzhou 350117, China

Yanan Wang, Xuehong Chen, Yining Ren, Qi Han, Yu Zhou and Yantao Han
Department of Pharmacology, College of Basic Medicine, Qingdao University, Qingdao 266071, China

Shangyong Li
Department of Pharmacology, College of Basic Medicine, Qingdao University, Qingdao 266071, China
Key Laboratory of Sustainable Development of Marine Fisheries, Ministry of Agriculture, Yellow Sea Fisheries Research Institute, Chinese Academy of Fishery Sciences, 106 Nanjing Road, Qingdao 266071, China

Xiaolin Bi
Department of Rehabilitation Medicine, Qingdao University, Qingdao 266071, China

Ruyong Yao
Central Laboratory of Medicine, Qingdao University, Qingdao 266071, China

Index